LARGE DIAMETER SHIELD TUNNELING TECHNOLOGY

LARGE DIAMETER SHIELD TUNNELING TECHNOLOGY

Wenbo Zhou
Huiming Wu
Shanghai Tunnel Engineering Co., Ltd., China

World Scientific
NEW JERSEY • LONDON • SINGAPORE • BEIJING • SHANGHAI • HONG KONG • TAIPEI • CHENNAI • TOKYO

Published by

World Scientific Publishing Co. Pte. Ltd.
5 Toh Tuck Link, Singapore 596224
USA office: 27 Warren Street, Suite 401-402, Hackensack, NJ 07601
UK office: 57 Shelton Street, Covent Garden, London WC2H 9HE

Library of Congress Control Number: 2023053649

British Library Cataloguing-in-Publication Data
A catalogue record for this book is available from the British Library.

大直径盾构法技术
Originally published in Chinese by China Communications Press Co. Ltd.

LARGE DIAMETER SHIELD TUNNELING TECHNOLOGY

ISBN 978-981-12-8595-0 (hardcover)
ISBN 978-981-12-8596-7 (ebook for institutions)
ISBN 978-981-12-8597-4 (ebook for individuals)

For any available supplementary material, please visit
https://www.worldscientific.com/worldscibooks/10.1142/13669#t=suppl

Desk Editors: Soundararajan Raghuraman/Amanda Yun

Typeset by Stallion Press
Email: enquiries@stallionpress.com

Foreword

Since the completion of the Tokyo Bay undersea tunnel in 1996, shield tunnels with a diameter of more than 14 m have gradually come to the attention of the public. Following Japan and Germany, the Netherlands, Russia, the United States, China and other countries have successively rolled out the construction of large-diameter shield tunnels, and the shield diameter has gradually increased to 17 m, with more new records expected in the near future.

Since the beginning of the 21st century, with the continuous development of multi-type, multi-level and integrated three-dimensional traffic networks in underground spaces, China has become a superpower in the construction and application of shield tunneling. In tunnel construction for metro sections, highways and railways, as well as for other purposes, China's application of shield tunneling has achieved outstanding results and remarkable achievements, and its design and construction of shield tunnels have been ranked among the most advanced in the world, greatly promoting the rapid development of civil engineering and transportation in China.

In 2004, when the construction of the Shangzhong Road Tunnel in Shanghai was started, a slurry balanced shield machine with a diameter of 14.87 m was introduced, marking a breakthrough in the construction of soft soil tunnels with a shield diameter of more than 14 m in China. Subsequently, this technology was vigorously promoted and applied in the Yangtze River Delta, the Yangtze River Economic Belt, the Guangdong–Hong Kong–Macao Greater Bay Area and other regions, and several representative projects, such as the Shanghai Yangtze River Tunnel, the Wuhan Sanyang Road Yangtze River Tunnel and the Zhuhai Hengqin Underwater Tunnel emerged. At present, China boasts the largest number of super-large-diameter shield tunnels in the world and owns a number of shield construction and equipment manufacturing technologies with independent intellectual property rights, breaking the technological monopoly of a few developed countries.

The development of large-diameter shield construction technology has solved some common obstacles encountered in shield construction. However, in special circumstances, such as complex engineering geological and hydrogeological conditions, there are still many new challenges in shield construction, such as shield launching and receiving technology, water and soil balance on the work during shield excavation with a super-large section, stability of shield tunneling in the tunnel section with an ultra-shallow overburden depth, shield and segment lining construction to advance in the vicinity of or across any operational metro, and risk control in cutter replacement under high water pressure. These have also become internationally recognized technical challenges in the industry, putting forward higher and more stringent requirements for the development of large-diameter shield tunneling technology.

Under the backdrop of the fast development of shield tunnel construction, this book was meticulously compiled by the author team, based on their decades of profound experience. Focusing on the construction technology of large-diameter shield tunneling, based on the latest results of technology research and development, and extensive engineering practices in recent years, this book comprehensively expounds and summarizes many research outcomes, including the latest design of large-diameter shield tunneling machines, the high-accuracy fabrication of large-size segments, and innovative technology of large-diameter shield tunneling. This book covers multiple aspects such as type selection of tunnel boring machines (TBMs) change to shield machine selection, the slurry screening process, the fabrication of shield segments, shield tunneling operation, the fabrication and erection of internal tunnel structures, the effective monitoring of construction information, and the refined and intelligent management of technical risks. The main members participating in the compilation of this book are all experts in the field of shield technology who have been long engaged in frontline work on large-diameter shield tunneling projects.

This book collects a massive amount of measured engineering data and engineering examples that have been carefully selected to explain professional knowledge in an easy-to-understand approach. This book will help front-line engineers and technicians better understand and master the construction technology of large-diameter shield tunneling and will serve as a valuable reference and learning material for design, construction and technical management personnel engaged in shield tunneling projects, thus promoting the improvement of China's large-diameter shield tunneling technology system.

As another brand-new business card in the field of engineering technology in China, large-diameter shield tunneling has been widely promoted and applied, marking that China's infrastructure construction capacity has leaped into the world's advanced ranks. With the wide application of digitalization, informatization, the Internet of Things, machine learning and big data technologies, intelligence and smartization represent the inevitable development trends of shield tunneling technology in the future. All tunnel practitioners will be motivated by the spirit of "start from scratch, and grow from a beginner to a master" to create a new development path for large-diameter shield tunneling, featuring highly scientific and technological content, good economic benefits, low resource consumption and minimized environmental pollution, thus fulfilling humanity's environmental goals!

Deeply impressed by the unremitting efforts and hard work of the author team, I would like to extend my warm congratulations on the publication of this book with the above words and would like to recommend it to my peers.

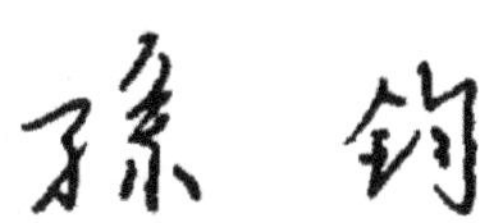

Mr. Sun Jun

Written in Tongji Garden, Hubin, in the Cold Dew period of the golden autumn in the year of 2019.

Mr. Sun Jun, First Class Honorary Professor and Honorary Dean of the Department of Geotechnical Engineering of Tongji University, Senior Academician of the Chinese Academy of Sciences (Division of Technological Sciences), and well-known scholar and technical expert in the field of geomechanics and geo-engineering, tunnel and underground engineering.

Preamble

Since the completion of the Shanghai Yangtze River Tunnel, shield tunneling with a diameter of more than 14 m has been rapidly promoted in the soft soil areas of the Yangtze River Delta, the composite strata zones in the Guangdong-Hong Kong-Macao Greater Bay Area, the Yangtze River Economic Belt, and others. At present, large-diameter shield tunneling is more and more widely applied in highways, railways, urban utility tunnels and other fields in China and is quickly becoming the most effective solution to building urban underground expressway networks. However, in the large-scale construction of this kind of tunnel, due to the large size of excavation sections, the technicians must cope with many tunneling challenges, such as ultra-burial depth, long distances and complex strata. In the process of continuously solving engineering puzzles, a series of new technologies, new processes, new materials and new equipment for large-diameter shield tunneling have been introduced, developed and applied, such as the launching and receiving of large-diameter shields, the balance of the excavation face with a super-large section, stability control of tunnels with an ultra-shallow overburden depth, risk control during construction when crossing any operational metro lines, and cutter replacement under high water pressure. The engineering practices in coping with numerous world-class challenges have contributed to China's great progress in large-diameter shield tunneling technology, which marks that China's large-diameter shield tunneling strength has reached the international advanced level.

This book aims to systematically sort out, summarize and refine knowledge relating to the large-diameter shield tunneling technology, on the basis of the abundant results from extensive large-diameter shield tunneling engineering practices, and develop a technical monograph with a certain scientific system to meet the new requirements of engineering design, construction, and management for urban shield tunnels in the new era. The book is divided into 10 chapters. Chapter 1 introduces the application development and technical innovation of large-diameter shield tunneling technology in China and abroad; Chapter 2 presents the latest design and main functions of large-diameter shield tunneling machines; Chapter 3 mainly introduces the integrated design of the slurry treatment system and its zero slurry discharge process; Chapter 4 describes the new technologies in the design and fabrication of high-precision steel forms and segment production; Chapter 5 discusses various related technologies such as launching and receiving large-diameter shields, excavation face stability, shield attitude management, segment erection, synchronous grouting and tunneling in special sections; Chapter 6 mainly introduces the construction of tunnel internal structures and connecting passages; Chapter 7 focuses on data management and control from the perspective of informationization and intelligence; Chapter 8 introduces the risk management and response strategies in large-diameter shield construction; Chapter 9 presents the typical large-diameter shield tunnels in the world and their key technologies; Chapter 10 is devoted to the outlook of the book and puts forward the development trends of tunnels and shields in the future.

I would like to extend my thanks to Academician Sun Jun for reviewing and writing a foreword for this book in his tight schedule and pay tribute to all the participants, experts and scholars who have worked hard and provided help. Urban large-diameter shield tunnels built by a large number of technicians through hardship showcase their unwavering spirit of craftsmanship and outstanding wisdom in the face of difficulties. This book is intended to pay tribute to all the designers, builders and researchers who have contributed their wisdom and sweat to the development of large-diameter shield tunneling technology!

This book is suitable for professionals and technicians in design institutes and construction companies, researchers in scientific research institutes, as well as others engaging in tunneling and related underground engineering fields. In view of the fact that the discipline of tunnel and underground engineering is in the process of development and transformation, this book introduces many of the latest theories and engineering practice results, provides examples and references for the construction of large-diameter shield tunnels in the future and broadens the application scope of the industry. There are inevitably some omissions and inaccuracies in the book. Any comments and corrections by both Chinese and foreign experts and readers will be appreciated.

Contents

Chapter 1

Introduction

With the large-scale construction of urban metro tunnels, river-crossing and sea-crossing traffic tunnels, urban rapid transit tunnels, water conservancy and hydropower tunnels and utility tunnels, large-diameter shield tunneling technology has been widely applied and developed by leaps and bounds. Faced with many challenges in the construction process of super-large-diameter shield tunnels because of their super-large tunnel sections, tunnel builders continue to struggle for solutions to the globally significant research topic related to the construction technology of such tunnels.

1.1 New demands in urban development for tunnels

1.1.1 Widespread application of shield tunneling

For nearly 200 years since the birth of shield tunneling, the development of shield technology abroad has experienced roughly four stages: the first-generation of shields, represented by manual shield machines; the second-generation of shields, represented by mechanical and air pressure shield machines; the third-generation of shields, represented by earth pressure balanced (EPB) shield machines in sealed structures; the fourth-generation of shield machines, featuring diversification, informatization and automation.

1.1.1.1 *First-generation shields and their application (1820s–1870s)*

In 1818, Marc Isambard Brunel, a French engineer, observed shipworms boring in a ship's wooden board and smearing the liquid they secreted around the hole they made in the wall. Inspired by this, he proposed the principle of shield tunneling and applied for a patent in Britain. In 1825, Brunel started to build the world's first shield tunnel under the Thames River in London with a rectangular cast iron shield machine. The section size of the tunnel was 11.4 m × 6.8 m. During its construction, an extraordinarily heavy water inrush occurred multiple times in the tunnel. The 458 m long tunnel was finally completed in 1843, after Brunel improved the machine. In 1869, a British engineer, Janes Henry Greathead, first used a circular open shield machine and cast-iron segments to build a 402 m long shield tunnel with a 2.18 m outer diameter under the Thames River for pedestrians (Fig. 1.1).

Fig. 1.1. Circular shield machine used by Greathead.

Source: Sourcing Network.

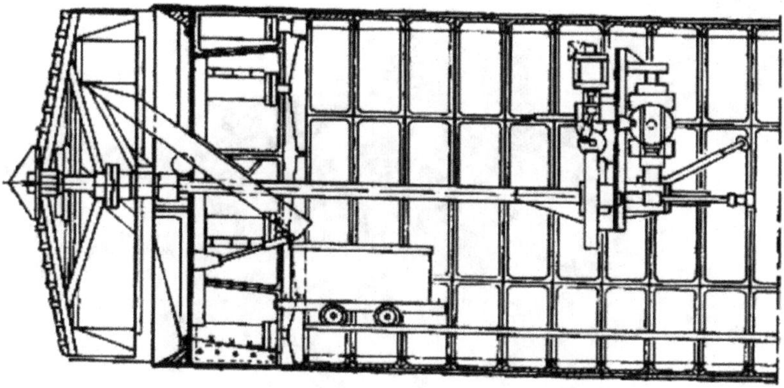

Fig. 1.2. Schematic diagram of the first mechanized shield machine.
Source: Sourcing Network.

1.1.1.2 *Second-generation shields and their application (1870s–1960s)*

The first patent for a mechanized shield machine appeared in 1876, which was jointly applied by two British engineers, John Dickinson Brunton and George Brunton. It was envisaged to cut soil with a hemispherical cutterhead and transfer muck onto the conveyor belt via a radially rotating bucket (Fig. 1.2). In 1886, when the London Metro was built in England, Greathead adopted the compressed air shield method for the first time to solve the puzzle of shield tunneling in a stratum with confined water and proposed the construction method of grouting at the outer side of the lining to fill structure voids at the shield tail. The use of compressed air extended the scope of application of the shield method to strata with confined water, effectively promoting the popularization and application of the shield method. In 1896, the British engineer Price improved the shield cutterhead by turning it into a spoke-type cutterhead (Fig. 1.3), connected to a drive motor with a long shaft. This mechanical shield machine had a good application effect in the clay stratum in London, laying the foundation for modern shield tunneling.

1.1.1.3 *Third-generation shields and their application (1960s–1980s)*

In 1964, John Bartlett of the British Mott, Hay and Anderson (MHA), a firm of consulting civil engineers, applied for the patent of the slurry-pressurized balanced shield machine. In 1967, Japan built a sewer tunnel

Fig. 1.3. Price's improved spoke-type cutterhead.
Source: Sourcing Network.

with the world's first slurry balanced shield machine made by Mitsubishi, which uses a cutterhead to cut soil and adopts hydraulic mucking. Based on this test shield machine of 3.1 m in diameter, Japan manufactured the slurry balanced shield machine with the largest diameter (7.29 m) at that time, which was successfully used in the construction of the Haneda Tunnel. Compared with the second-generation shield, the slurry balanced shield machine realizes balance in soil mass with the help of slurry mixed with bentonite, and muck is carried by slurry into the slurry treatment system. Although both the air pressure shield machine and slurry balanced shield machine can dig in water-bearing strata, the adverse effects of compressed air on the human body and the environmental pollution from slurry urged tunnel experts to search for new and better solutions, which led to the EPB shield machine coming into being. In 1974, the first screw EPB shield machine (Fig. 1.4) manufactured by Ishikawajima-Harima Heavy Industries (IHI), Japanese enterprise was put into use in Tokyo, with an outer diameter of 3.72 m and a tunneling length of 1900 m.

1.1.1.4 *Fourth-generation shields and their application (1980s–current times)*

After the 1980s, fourth-generation shield machines featuring diversification, informatization and automation gradually emerged. These shield

Fig. 1.4. World's first EPB shield machine.

machines have achieved a high degree of automation in the aspects of advance steering, posture adjustment, segment feeding and assembly, synchronous grouting, information management, fault diagnosis, etc. In the 1980s, the manufacturing of shield machines has been gradually dominated by Japanese and German enterprises. Japan focused on the diversification of shield machines, while Germany is committed to the development of hybrid shield machines. Due to its limited land and large population density, Japan already had a very high utilization rate of underground space in many large cities. Therefore, the Japanese engineers focused on diversifying shield machines for further development and utilization in the crowded underground spaces. The Japanese shield machines showed diversification in multiple aspects, such as section forms, functions and excavation methods, including rectangle, double-section overlap, three-section overlap (Fig. 1.5), and more sections; spherical shield, detachable two-section slurry shield, branch route shield, expanding shield and other varieties; and shaky, sway and more cutting modes. A single-function shield machine can hardly bore through a whole tunnel under the greatly varying stratum and hydrological conditions, under which a hybrid shield machine (Fig. 1.6) is the right choice. The functional components of different types of shield machines are integrated on one machine to realize switching between two or more different working modes, including the

Fig. 1.5. Japanese three-section shield mechanism.

Fig. 1.6. Hybrid shield machine of Herrenknecht AG.

open type, EPB type and slurry balanced type, by adjusting the support mode on the excavation face and the muck-conveying mechanism, i.e. a hybrid shield machine. The hybrid shield machines provide an effective solution for the construction of shield tunnels under different geological conditions in Europe.

1.1.1.5 *Application and development of shield tunneling technology in China*

The application and development of shield tunneling technology in China began in the early 1950s, when several small-diameter water intake tunnels were excavated with manual shield machines. In the 1960s, Shanghai Tunnel Engineering Co. Ltd. (hereinafter referred to as "Shanghai Tunnel") began to independently design, manufacture and apply the grid extruding shield machine for tunneling in soft soil. Since 1990, China's economic status has been significantly improved, and Beijing, Shanghai, Guangzhou, Shenzhen and other major cities have begun to vigorously develop urban rail transit to solve traffic congestion, stimulating wide application of the shield tunneling technology. From the introduction of shield machines at the very beginning, to continuous digestion and absorption in practices, and finally to independent design, manufacturing and export, the Chinese enterprises broke the monopoly of the shield technology by foreign companies, taking China's shield technology into the world's advanced ranks.

In 1953, Fuxin Coal Mine built a drainage tunnel in the northeast of China with a 2.6 m diameter manual shield machine and small precast concrete blocks. In 1957, Beijing built urban sewers with manual shield machines of 2.0 and 2.6 m in diameter. In 1963, Shanghai Tunnel successfully conducted a shield tunnel test in the quaternary soft water-bearing stratum at Tangqiao, Pudong (Fig. 1.7), with its self-developed 4.2 m diameter manual shield machine, finishing tunnel excavation for a length of 68 m. In 1966, the first underwater highway tunnel in China, the Dapu Road River Crossing Tunnel in Shanghai (Fig. 1.8), was excavated by Shanghai Tunnel for a tunneling length of 1,322 m with a 10.22 m diameter large grid extruding shield machine manufactured by Jiangnan Shipyard, supplemented by the pneumatically stabilized excavation face.

Fig. 1.7. Manual shield test at Tangqiao, Pudong, Shanghai.

Fig. 1.8. Dapu Road River Crossing Tunnel in Shanghai.

In 1990, in order to initiate the all-around construction of Shanghai Metro Line 1, seven EPB shield machines with a diameter of 6.34 m manufactured by the French FCB Company were introduced. In 1995, Shanghai Tunnel introduced a slurry pressurized shield machine with a

Fig. 1.9. "Advance" shield machine.

diameter of 11.22 m, manufactured by Mitsubishi Heavy Industries of Japan, for the excavation of a 1,300 m circular tunnel, forming the South Tunnel of East Yan'an Road. In October 2004, China's first EPB shield machine, "Pioneer", with completely independent intellectual property rights, was successfully developed and applied to the sectional tunnel in the West Extension of Shanghai Metro Line 2. In December 2008, the large slurry balanced shield machine "Advance" (Fig. 1.9) with a diameter of 11.22 m, independently designed and manufactured by China, was used for the construction of the second tube of the Dapu Road River Crossing Tunnel in Shanghai.

In recent years, in the context of the huge demand for shield machines in a new round of construction of railway tunnels, highway tunnels, metro tunnels, utility tunnels and others, the Chinese shield machine manufacturers, represented by Shanghai Tunnel Engineering Co. Ltd., China Railway Engineering Equipment Group Co. Ltd. and China Railway Construction Heavy Industry Co. Ltd., have successively developed a series of new shield equipment, including large cross-section rectangular pipe jacking machines, ground penetrating shield machines, quasi-rectangular shield machines and horseshoe-shaped shield machine (Fig. 1.10), contributing to a complete industrial chain of design, manufacturing and construction taking initial shape in the Chinese shield industry and gradually increasing the share of domestic shield machines in the domestic market. However, the

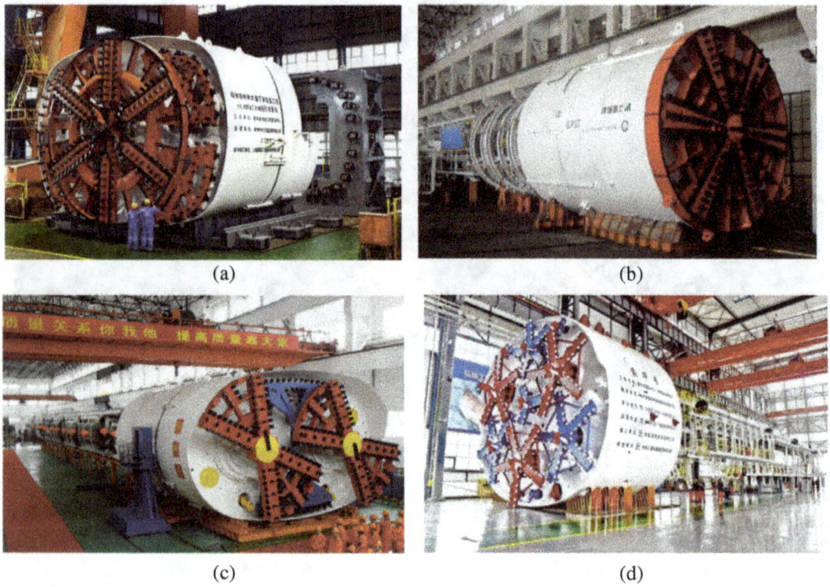

Fig. 1.10. New shield equipment developed by the Chinese. (a) Large-section rectangular pipe jacking machine, (b) ground penetrating shield machine, (c) quasi-rectangular shield machine and (d) horseshoe-shaped shield machine.

development of shield machine technology in China is facing many challenges, such as diversified geological conditions, normalization of river and sea crossings, a variety of structural sections and the complexity of the construction environment. There are still gaps with the shield machine manufacturers in Germany and Japan in terms of overall design, system integration technology and production technology of key components.

In the wake of the development of mechanization and electrification, shield machine technology has developed rapidly, experiencing the first generation of manual shields, the second generation of mechanized shields and the third and fourth generations of electrified and electronic shields. Integrating the optical, electrical, mechanical and hydraulic systems, the shield machines closely follow the innovations of industrial technology at each stage of development, as shown in Fig. 1.11. In the 21st century, reflecting a new era of information, digitalization and intelligence, with the development of the Internet of Things, big data, artificial intelligence and other technologies, the management and control of shield machine construction should also be intelligent technology-oriented and

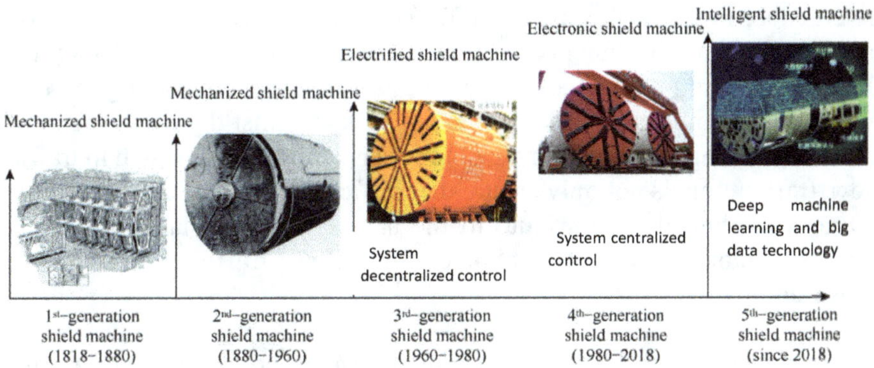

Fig. 1.11. Development history of shield equipment.

follow the information technology-based and intelligent management concept. To be specific, dynamic monitoring, early warning of risks and intelligent decision-making should be realized in shield tunneling by applying the technologies of big data, BIM and Internet of Things, as well as based on the massive and real-time data on quality, progress, safety, equipment status and others in tunnel construction. At the dawn of the Industrial Revolution 4.0, shield tunneling technology is bound to develop in the direction of industrial intelligence.

1.1.2 Significance of tunnels to urban traffic

Urban transportation is not only the infrastructure serving people's lives and production but also an important guarantee for urban economic development. Convenient, fast and developed transportation is an important symbol to measure the level of urban modernization. Although cities are different in geographical location, scale, development orientation, etc., a compatible and coordinated traffic system is essential for their modernization and development. A sound and efficient transport network, coupled with convenient and comfortable travel modes, plays an important role in the development of the urban economy and the improvement of urban life quality.

Urban traffic tunnels constitute an important part of the public transport network, mainly including metro tunnels and underground rapid transit tunnels. The urban traffic tunnels effectively realize urban traffic divergence, improve the urban traffic environment and shorten the distance between urban functional zones. Especially for the riverside cities, the river-crossing tunnels are important passageways connecting the two banks and key nodes over the traffic network. Against the background of a serious shortage of urban land resources and substantial growth in traffic flow, traffic tunnels not only ensure the traffic efficiency of the urban traffic network but also contribute to the saving of urban land resources, making them a preferred choice for urban traffic infrastructure construction.

(1) **Metro tunnel:** Metro tunnels constitute an important part of the public transport system in major cities and serve as the primary mode of transportation for the daily travel of urban residents. With the advantages of high transport capacity, small traffic interference, fast driving rates and environmental friendliness, they play an irreplaceable role in urban social and economic activities. With the worsening urban road traffic congestion, major cities around the world have gradually reached a consensus that the fundamental way out of urban traffic congestion is to give priority to developing urban public transport systems with rail transit as the backbone.

London, England, built the first metro line in the world — the Metropolitan Railway — in 1863. Paris, Moscow, New York, Tokyo and other major cities in the world also began to vigorously develop rail transit. Paris, France, opened its first metro line in 1900. As of 2010, Paris had 14 trunk lines and 2 branch lines, with a total length of 215 km and an annual passenger ridership of 1.506 billion. In 1904, New York, the United States, opened its first metro line with a total length of 14.6 km and 28 stations. At present, the total length of operational metro lines in New York City exceeds 369 km. As the first city operating metro transport in Asia, Tokyo opened its first line in December 1927. At present, Tokyo Metro operates 13 lines with a total length of 312.6 km, contributing a proportion of rail transit of up to 90% to its public transportation network.

China initiated its metro construction as one of the moves against the war. The first metro line was completed and put into operation in Beijing in September 1969, but it was not open to the public for a long time. Major cities in China are currently undertaking rapid development of their rail transit construction. As of December 31, 2018, 36 cities, including Shanghai, Beijing, Shenzhen and Guangzhou, had opened 174 rail transit lines with a total length of 5,582 km, an increase of 18.6% over 4,706 km at the end of 2017. Among them, Shanghai (705 km) ranks first in the length of urban rail transit in service, followed by Beijing (626 km), Guangzhou (456 km), Nanjing (378 km), Chongqing (313 km), Wuhan (304 km) and Shenzhen (286 km). The total rail transit length of these seven cities exceeds 50% of the total figure in China. The operational urban rail transit lines in China are still dominated by metros, with a proportion of operating length increasing from 80.85% in 2010 to 88.50% in 2018.

For future urban rail transit planning, the proportion of public transport trips in the central urban area of Shanghai during the "13th Five-Year Plan" period is planned to reach 55%, of which rail transit ridership accounts for 60% of the public transport network. Shanghai plans to continue to promote the construction of the rail transit network, reaching a total rail transit length of 800 km by 2020. Wuhan put forward the "60/60" passenger transport target in its long-range planning of urban rail transit, that is, the proportion of public transport trips reaching 60% and the rail transit trips accounting for 60% of the public transport trips. As an important carrier of urban rail transit, metro tunnels will continue to play a cornerstone role in the development of urban transport network (Fig. 1.12).

(2) **Underground rapid transit tunnel:** Urban underground rapid transit first emerged in the form of mountain tunnels and underwater tunnels, built to maintain the integrity of the urban traffic network when encountering rivers, lakes, seas and mountains in the construction of the urban rapid transit network. With the continuous breakthroughs of tunnel construction technology and the growing demand for traffic space in modern cities since the end of the 20th century, the development and utilization of underground space have received extensive attention, and urban underground rapid transit has become the best choice to ensure driving rates without occupying urban ground space. Moreover, such tunnels are equipped with ventilation stations to filter and decompose particulates and other harmful substances from automobile exhaust and discharge the treated and almost pollution-free gas (Fig. 1.13).

The construction of modern urban underground rapid transit began in 1927 with the Holland Tunnel under the Hudson River in New York City, United States. It was bored using the shield method in two tubes for one-way traffic, with a length of 2,680 m and

(a)

(b)

Fig. 1.12. Urban metro tunnels and stations in China. (a) Metro tunnel under construction and (b) metro station in service.

2,551 m, respectively, and first applied with the mechanical full transverse ventilation technology. Afterward, a transit tunnel with a total length of 3.4 km was built in the downtown business district of Seattle in the United States in 1990, becoming the first fully enclosed underground bus rapid transit system put into operation in the world. Since viaducts ran through the downtown area of Boston with a great impact on the urban environment, the government of the City of Boston began to dismantle the urban viaducts built in the 1950s in 1995 and develop underground rapid transit instead. An underground rapid transit route with 8–10 lanes and a length of 2.4 km was built under Central Avenue. After the removal of the viaducts, the ground

(a)

(b)

Fig. 1.13. Urban underground rapid transit tunnels. (a) Underground rapid transit passageway and (b) traffic status.

was used as green spaces and moderately developed urban plots to promote the ecological and sustainable development of the city (Fig. 1.14). Although the construction process encountered problems such as water leakage, traffic congestion and soaring project costs, the completed tunnel has greatly improved the local traffic and urban environment.

Since Japan put forward the concept of underground space development in the 1980s, its major cities, such as Tokyo, began to vigorously build underground rapid transit tunnels. The Central Circular Route of Tokyo's Shuto Expressway (Fig. 1.15) is 47 km long in total, with the 11 km Shinjuku Route in the west and the 9.4 km

(a)

(b)

Fig. 1.14. Comparison of Central Avenue in Boston before and after viaduct demolition. (a) Before demolition of viaduct and (b) after demolition of viaduct.

Shinagawa Route in the south as its important components. The Central Circular Shinjuku Route, built at about 40 m underground, is a two-way, four-lane rapid transit line running from Kumano cho, Itabashi ku, Tokyo, to Aobadai, Meguro ku, via three important commercial centers: Ikebukuro, Shinjuku and Shibuya. The construction contractor has built most of the tunnel using shield machines with a diameter of 11.2–13 m, which has solved the construction difficulties such as underground expanding excavation, shield tunneling on sharp curves and shield machine making U-turns underground. The construction of the Central Ring Shinagawa Route started in 2006. Two EPB shield machines with diameters of 12.55 m and 12.53 m were used to bore the south and north tunnels, respectively (Fig. 1.16), and a tunneling length of up to 8 km in a single operation cycle

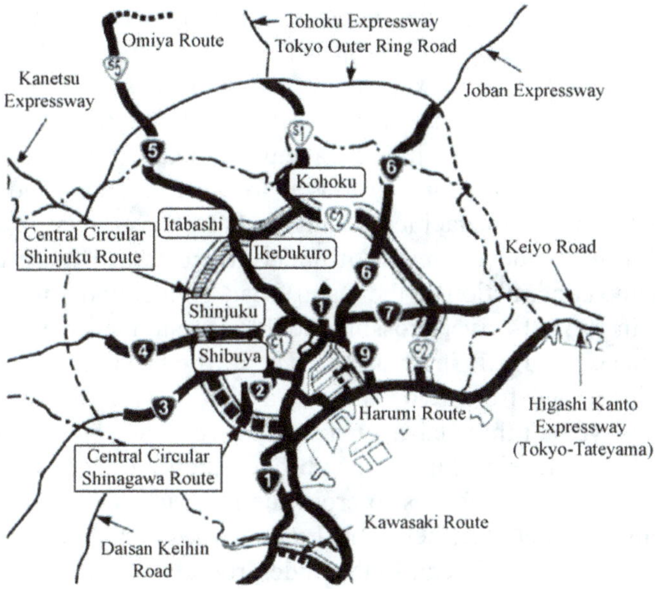

Fig. 1.15. Schematic diagram of "Central Circular Route of Shuto Expressway" in Tokyo, Japan.

Fig. 1.16. Shield construction of Central Circular Shinagawa Route.

solved a variety of environmental constraints faced by shield construction in densely populated urban areas. In addition, the construction of rapid transit tunnels such as the Yokohama North Route, the Yamatogawa Route of the Hanshin Expressway and the Kanetsu–Tomei Section of the Tokyo Outer Ring Road has also been launched.

The vigorous development of urban underwater tunnels and underpasses in China began at the beginning of the 21st century, mainly driven by the extensive application of shield tunneling. In view of the worsening traffic congestion and insufficient road capacity in Chinese first-tier cities, Academician Qian Qihu of the Chinese Academy of Engineering proposed to build underground expressways in China's megacities. Shanghai has always been in a leading position in the development of underground traffic space in China. Since the construction of the Dapu Road Tunnel under the Huangpu River in 1965, 14 river-crossing, double-lane tunnels have been built and opened to traffic in Shanghai. The traffic capacity of tunnels has also developed from two lanes on a single layer to four lanes on double layers or three lanes on a single layer. The construction of the urban river-crossing tunnels has broken the bottleneck that citizens could only rely on ferries to cross the river in the past, making the exchange and contact between the two sides of the Huangpu River closer and more efficient. In the underground space of the urban core area, Shanghai has also built the Bund Passageway, Yingbin 3rd Road Tunnel and Zhuguang Road Tunnel and is digging the Beiheng Passageway, which together constitute an urban underground rapid transit network. Other cities in China, such as Shenzhen, Nanjing, Wuhan, Hangzhou, Changsha and Yangzhou, have also successively built a large number of underground rapid transit tunnels through shield tunneling. Urban underground rapid transit tunnels can strengthen the direct connection between key urban areas and ensure the traffic efficiency of urban roads, thus making them an important means of leveraging urban underground space.

1.1.3 New trends of modern traffic tunnels

With the process of urbanization occurring at a faster pace in China, the number of cities with a population of 10 million has gradually increased. The contradiction between the growing travel demands of the people and inadequately developed urban transport has become considerably more prominent. The rising level of urban modernization requires the continuous improvement and upgrading of urban transport infrastructure. However, restricted by the urban spatial layout and construction land, the expansion of surface roads and the construction of urban viaducts and bridges have no longer met the functional requirements of modern cities. The development of underground space and the construction of a multi-type, multi-level and integrated urban three-dimensional traffic network system have become important directions for urban traffic development in the future. Rigid constraints such as the shortage of urban ground resources, the improvement of urban motorization and the sharp increase in citizens' trips have prompted modern traffic tunnels to gradually embark on the path of sustainable development with the new demands of "large-section, long-distance, deep-burial and multi-function".

(1) **Large-section tunnels:** According to statistics, as of July 2018, the ownership of motor vehicles in China had reached 319 million, of which more than 70% were automobiles. At present, automobile ownership in 58 cities nationwide has exceeded one million, and in seven top-ranked cities, including Beijing, Chengdu, Chongqing and Shanghai, it has exceeded three million. In the past 10 years, the continuous and rapid growth of urban motor vehicle ownership has been the direct cause of urban traffic congestion. In order to alleviate urban traffic congestion and meet the increasing travel demands of urban residents, modern traffic tunnels have begun to develop in the direction of large sections.

    Urban metro tunnels in China are basically "single lane/track in single tube" shield tunnels with an outer diameter of 6.2 m and an inner diameter of 5.5 m. The river-crossing tunnel on Wuhan Metro Line 8, bored through in August 2017, is the largest "double lanes/track/ in single tube" metro shield tunnel in China at present, with an outer diameter of 12.1 m and an inner diameter of 11.1 m, built in a double-layer lining structure of "prefabricated segment + cast-in-situ lining" (Fig. 1.17). The urban river-crossing highway tunnels built in the early stages are mainly in the form of four lanes of two-way traffic in double tubes, such as the Shanghai Dapu Road Tunnel and its second tube, the Shanghai East Yan'an Road Tunnel, and the Wuhan Yangtze River Tunnel. A single tube generally accommodates two lanes, with a width of 7 m and a design speed of 40–60 km/h. With the continuous progress in the manufacturing of shield main bearings, processing of precision components and welding technology, the shield machines have gradually been upgraded from the medium-sized, with a diameter of 11 m, to the large-sized, with a diameter of 14 m, promoting the urban underground rapid transit tunnels into the era of large sections. For example, the Shanghai Yangtze River Tunnel is a double-tube, single-layer, two-way, six-lane tunnel with a design speed of 80 km/h and with three lanes of 3.75 m in width,

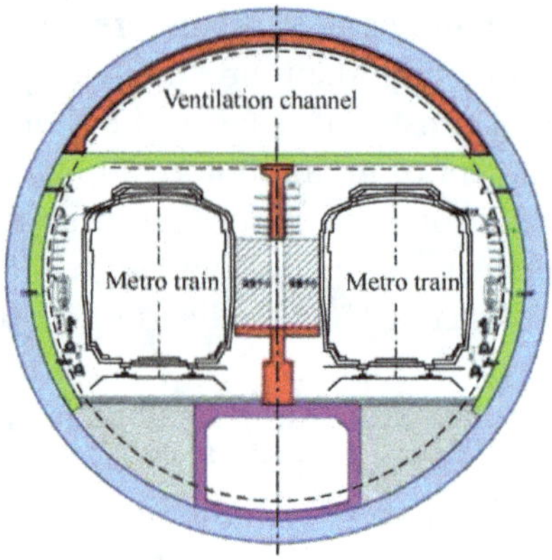

Fig. 1.17. Cross-section of Wuhan Metro Line 8.

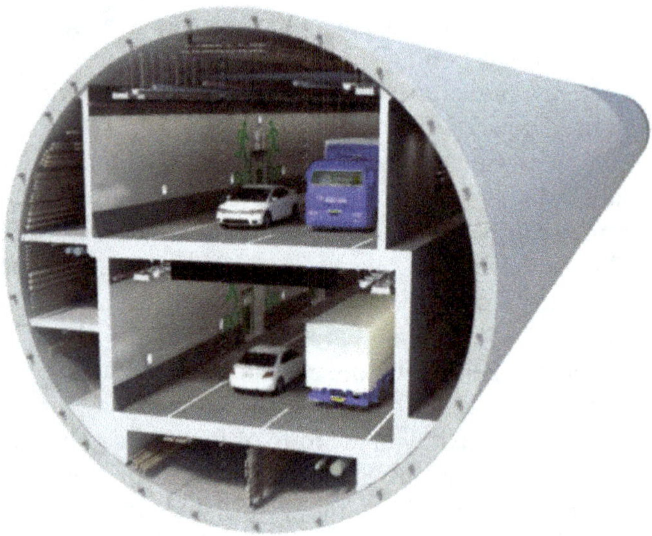

Fig. 1.18. Cross-section rendering of Seattle SR99 Highway Tunnel.

respectively, in a single tube; the Shanghai Jungong Road Tunnel is a double-tube, double-layer, two-way, eight-lane tunnel with a design speed of 80 km/h; the SR99 highway tunnel in Seattle, USA, built with an EPB shield machine has a diameter of 17.45 m and is designed as a single-tube, double-layer, four-lane tunnel (Fig. 1.18).

Large-section tunnels can greatly improve the traffic efficiency of urban traffic and save underground space resources, so they have become the first choice for modern urban shield tunnel construction.

(2) **Long-distance tunnels:** It has become a typical model of development for riverside cities to expand the space for urban development by balancing resources on both sides of rivers. Therefore, lake- and river-crossing tunnels have become important links between the two sides. Numerous lake- and river-crossing tunnels have been built under the Yangtze River, the Huangpu River in Shanghai, the Qiantang River in Hangzhou, the Xiangjiang River in Changsha, the Slender West Lake in Yangzhou, the Xuanwu Lake in Nanjing, the East Lake in Wuhan and other lakes in China. Underground rapid transit tunnels have also been built under the core areas of Shanghai, Shenzhen and other cities. Many countries in the world have built many famous river- and sea-crossing tunnels, such as the English Channel Tunnel between Britain and France, the Qinghan Submarine Tunnel in China and the Tokyo Bay Submarine Highway Tunnel in Japan.

With the expansion of urban traffic infrastructure and the implementation of the national "urban-rural integration development" strategy, the construction of long-distance traffic tunnels has started, concentrating on the peripheral areas of cities. The Shanghai Yangtze River Tunnel and the Shanghai Riverside Passageway, two representative projects, have single-tube lengths of 7.47 km and 5.09 km, respectively, for their circular tunnel sections. The famous Tokyo Bay Submarine Highway Tunnel (Fig. 1.19) in Japan, including a 9.5 km long submarine tunnel and a 4.4 km long bridge, is connected by two artificial islands. The double-tube tunnel is 9.5 km long, with an average burial depth of 16 m and an average water depth of 27.5 m.

(3) **Deeply buried tunnels:** The large-scale development and utilization of urban shallow underground space and the overburden requirement for river- and sea-crossing tunnels are the main reasons for the development of modern traffic tunnels toward large burial depths. Amid the development of new urban tunnel projects in deeper underground spaces, the Chongqing Hongtu Station — Liyuchi Station section has become the metro section tunnel with the largest burial depth in China, reaching up to 84 m. The Shanghai Hongmei South Road Tunnel (Fig. 1.20) starts from the north of the intersection between Yongde Road and Hongmei Road and ends at the Xizha Highway

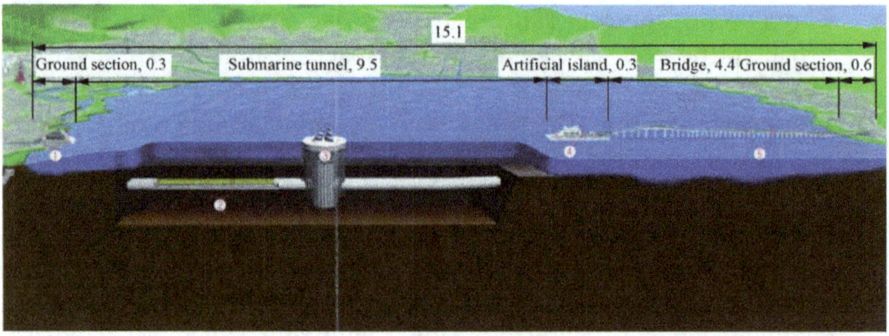

Fig. 1.19. Tokyo Bay Submarine Highway Tunnel (dimensions in km).

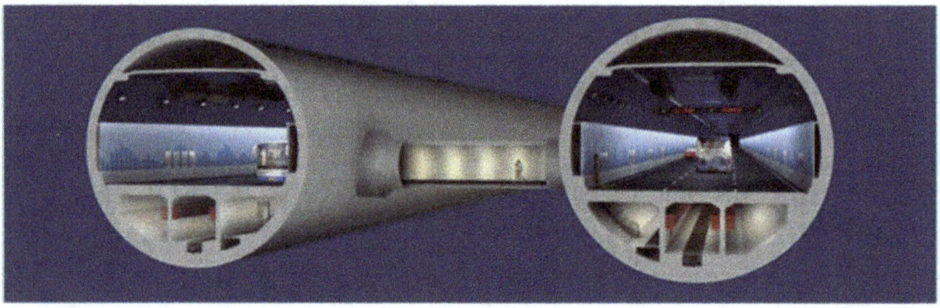

Fig. 1.20. Shanghai Hongmei South Road Tunnel.

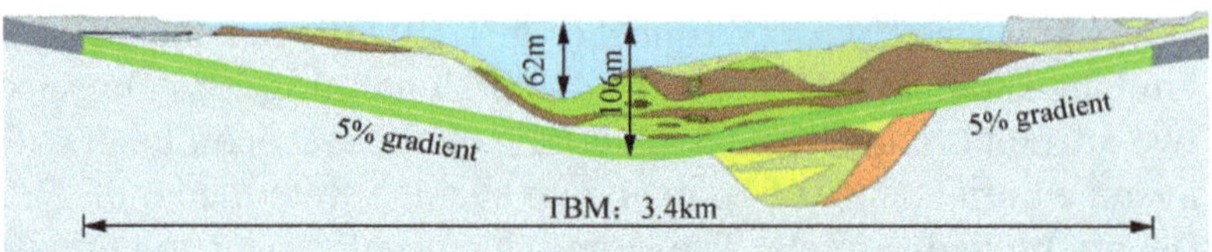

Fig. 1.21. Longitudinal section of Bosporus Highway Tunnel.

Interchange, running for a total length of 5,260 m. The main line was bored with a shield of 14.93 m in diameter to cross the river. This two-way, six-lane tunnel is the longest and features the largest burial depth up to 59 m under the Huangpu River.

The road tunnel underneath the Bosphorus Strait in Turkey (Fig. 1.21) is also called the "Eurasia Tunnel" because it crosses Europe and Asia. With a total length of 13.5 km and a maximum depth of water over the tunnel of 62 m, it is located deepest among those bored with slurry balanced shield machines, reaching the largest burial depth of 106 m.

(4) **Multi-function tunnels:** Under the background of the "negative growth" of construction land proposed by many cities in China, the land resources available for the construction of urban transportation facilities are on the decrease. How to utilize tunnel space comprehensively and efficiently, reduce engineering construction costs and waste of underground space resources and meet the requirements of diversified transportation modes and utility pipeline laying are new challenges to the development of modern traffic tunnels. According to the principles of "modern, safe, efficient, green and economic", multi-function tunnels have become a solution to meet the new demand for modern traffic tunnels, calling for the technology of comprehensive and intensive utilization of tunnel space.

The circular cross-section of the Shanghai Yangtze River Tunnel is divided into three layers: the upper layer dedicated to ventilation ducting, the middle layer as the space for a three-lane carriageway, and the lower layer for 220 kV river-crossing cables. The conditions are reserved for future rail transit lines, realizing the integration of high-voltage cables, highway transport and rail transit. The Sanyang Road River Crossing Tunnel on Wuhan Metro Line 7, which was completed in June 2018, is the world's first super-large-diameter shield tunnel built for both highway and railway purposes while also meeting the traffic needs of rail transit and highway. The Zhuhai Hengqin Tunnel has achieved a layout that integrates a carriageway, 220 kV high-voltage cables, water supply pipeline and communication optical cables, as well as providing space for trams in a super-large-diameter tunnel with an outer diameter of 14.5 m. The tunnel was put into traffic in November 2018.

A series of multi-function tunnels have also been built around the world, among which the most typical one is the Stormwater Management and Road Tunnel (SMART) in Malaysia (Fig. 1.22). It is an integrated system of flood dredging and road management, built as a major national project in Kuala Lumpur, the capital of Malaysia. Its main part is a super-large-section tunnel with a length of 9.7 km and an inner diameter of 11.8 m. A 3 km section in the middle of the tunnel is used concurrently for flood discharge and traffic, serving as a two-way four-lane traffic tunnel when flood discharge is suspended to ease traffic congestion in the urban area.

Fig. 1.22. SMART Tunnel in Malaysia.

The SMART Tunnel has a three-layer structure: the bottom layer as a permanent drainage channel; the second floor, usually open to traffic. In the case of a five-year flood, the second floor may serve as a drainage channel; in the case of an extreme rainstorm, all the traffic lanes are closed, turning the whole tunnel into a flood discharge channel, basically solving the problem of waterlogging drainage in the urban center (Fig. 1.23).

1.2 Construction and development of large-diameter tunnels in the world

1.2.1 Large-diameter tunnels around the world

The construction of 14 m class large-diameter shield tunnels began in 1989 with the Tokyo Bay Submarine Highway Tunnel in Japan. Eight years later, Hamburg, Germany, began to build the first 14 m class large-diameter shield tunnel in Europe, the fourth tunnel of the Elbe River. Since the beginning of the 21st century, the Netherlands, Russia, the United States and China have successively joined the construction of 14 m class large-diameter shield tunnels, and the diameter of the shield machines has gradually increased to 17 m. The shield machine manufacturers, represented by Herrenknecht AG of Germany, Mitsubishi Heavy Industries, Kawasaki Heavy Industries, Hitachi Zosen Corporation and

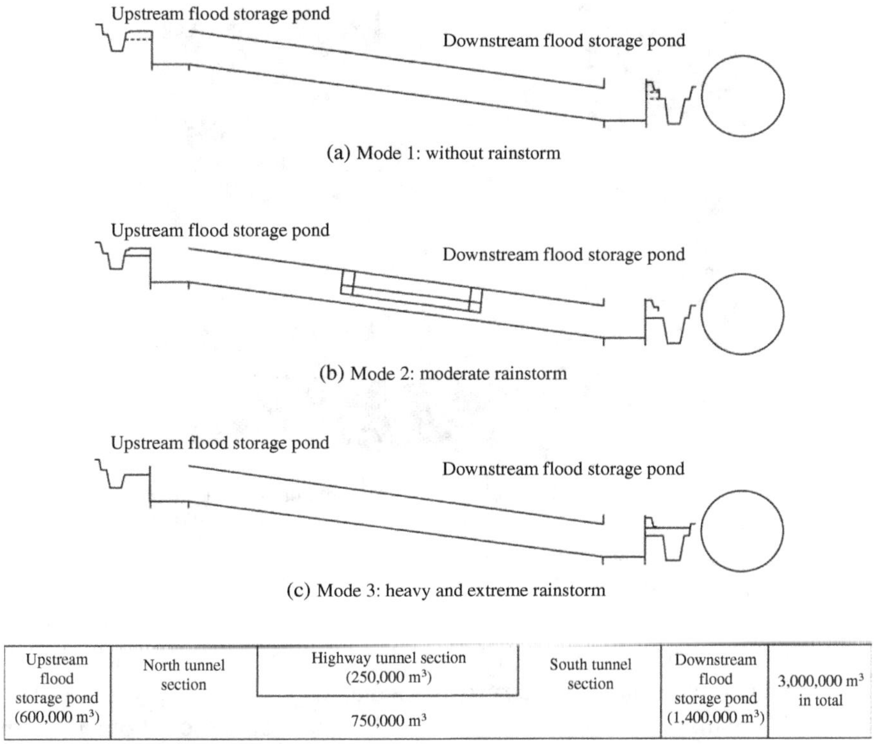

Fig. 1.23. Schematic diagram of functional modes of SMART Tunnel in Malaysia.

Ishikawajima-Harima Heavy Industries of Japan, have manufactured an array of large-diameter shield machines, including slurry balanced shield machines, EPB shield machines and hybrid shield machines. As of December 2018, there have been sixteen 14 m class large-diameter shield tunnel projects abroad. The typical ones are as follows:

(1) **Tokyo Bay Submarine Highway Tunnel, Japan:** The first 14 m class large-diameter shield tunnel project in the world, the Tokyo Bay Submarine Highway Tunnel in Japan, was started in 1989. Eight 14.14 m diameter slurry balanced shield machines ran for long-distance tunneling in the soft alluvial and proluvial cohesive soil layers under a high water pressure. Two tunnels, about 9.5 km long, were connected thanks to the subsea docking technology (Fig. 1.24). The tunnel is in a double-layer lining structure, and the outer lining

Fig. 1.24. Slurry balanced shield machine with a diameter of 14.14 m for Tokyo Bay Submarine highway tunnel.

is assembled by segments. The outer diameter of the tunnel is 13.9 m, and the inner diameter is 12.6 m. The eight shield machines, were produced by Japanese enterprises such as Kawasaki Heavy Industries and Mitsubishi Heavy Industries. With a main shield of about 13.5 m in length and 3,200 t in weight, the shield machines were equipped with an advanced automatic tunneling management system, an automatic measurement management system and an automatic segment assembly system.

(2) **Fourth Tunnel of the Elbe River, Germany:** In 1997, the fourth tunnel of the Elbe River was built in Hamburg, Germany, with a 14.2 m diameter hybrid slurry balanced shield machine manufactured by Herrenknecht. The stratum crossed is dominated by a clay layer with rich water content and a gravel layer with a high permeability coefficient, and the minimum overburden thickness of the tunnel is only 7 m at the river bed. The shield machine was equipped with a leading small cutterhead with a diameter of 3 m in the center of the main cutterhead and a crusher capable of crushing boulders with a diameter of 1.2 m under the slurry tank. In addition, the cutterhead facilities can be replaced under atmospheric pressure. Another new technology was introduced, the sonic softground probing system (SSP), which can generate three-dimensional reflection images

Fig. 1.25. Hybrid slurry balanced shield machine with a diameter of 14.2 m for the fourth tunnel of the Elbe River.

within a range of 20–30 m in front of the cutterhead during shield tunneling. After the completion of the fourth tunnel under the Yibei River in 2003, the shield machine was reused for tunneling 2.5 km in the Lefortovo Tunnel, a single-tube, single-layer, three-lane tunnel in Moscow, Russia, after maintenance, and an additional two 2.2 km tunnels in the Silverpine Forest in the west of Moscow (Fig. 1.25).

(3) **"Green Heart" Tunnel, the Netherlands:** In 2000, the "Green Heart" Tunnel in the Netherlands was started (Fig. 1.26), as a part of a high-speed railway from Amsterdam to Brussels. It is a single-tube double-track railway tunnel with a total length of 7,156 m, three shafts in the middle and a maximum burial depth of 30 m. A 14.87 m diameter air pressure-regulating slurry balanced shield machine manufactured by the French company NFM was used for the construction, with a daily tunneling length of 10 m in the stratum dominated by saturated sand. Later, Shanghai introduced this shield machine to dig the Shangzhong Road Tunnel.

(4) **M30 Loop Tunnel in Madrid, Spain:** Before the construction commencement of the Madrid M30 Loop Tunnel in 2004, the 14 m class large-diameter EPB shield machine had not been developed. The stratum crossed by this tunnel is dominated by hard, fissured gray or

Fig. 1.26. Slurry balanced shield machine with a diameter of 14.87 m used in the "Green Heart" tunnel in the Netherlands.

green marl rock clay and a gypsum layer. The depth of the groundwater level line is about 40 m from the ground surface, while the average overburden thickness of the tunnel is about 30 m. This tunnel is a double-tube, single-layer, two-way, six-lane one of 3.67 km in length, 13.45 m in inner diameter, 0.6 m in thickness and 2 m in ring width. The EPB shield machines of 15.2 m and 15.0 m in diameter, respectively, used for the north and south lines of the tunnel (Fig. 1.27), were manufactured by Herrenknecht AG of Germany and Mitsubishi Heavy Industries of Japan. In the design of the shield machine, Herrenknecht designed the internal and external cutterheads with opposite rotation directions, while Mitsubishi Heavy Industries chose the design concept of the overall cutterhead and installed a central mixer in the earth chamber of the shield machine to prevent the accumulation of muck.

(5) **SR99 Highway Tunnel in Seattle, the United States:** In order to build a highway tunnel as a replacement of the Alaskan Way Viaduct in the downtown of Seattle, the SR99 Highway Tunnel was built with the largest-diameter EPB shield machine in the world at that time, Bertha, with a diameter of 17.45 m (Fig. 1.28). In July 2013, the Bertha shield machine manufactured by Hitachi Zosen Corporation of Japan was launched. Only half a year later, Bertha broke down due

(a)

(b)

Fig. 1.27. EPB shield machine for M30 Loop Tunnel. (a) 15.2 m diameter shield machine manufactured by Herrenknecht, Germany and (b) 15.0 m diameter shield machine manufactured by Mitsubishi Heavy Industries.

to damage to its main bearing and sealing system. The construction contractor spent the next two years excavating an emergency repair shaft and lifting the Bertha head out for maintenance. After replacing the main bearing and installing new seals for the outer and inner rings

(a)

(b)

Fig. 1.28. 17.45 m diameter EPB shield machine for SR99 Highway Tunnel in Seattle. (a) EPB shield machine and (b) cutterhead of shield machine.

of the main bearing, the shield machine resumed advance in December 2015. In April 2017, Bertha began to be fully disassembled after completing the excavation, with a total length of 2,825 m. Hitachi Zosen Corporation presented some cutters and control panels of Bertha to the Museum of History & Industry in Seattle.

See Table 1.1 for a list of 14 m large-diameter shield tunnels abroad (as of December 2019).

Table 1.1. 14 m class large-diameter shield tunnels abroad (as of December 2019).

Tunnel	Location	Diameter of shield machine (m)	Type of shield machine	Manufacturer	Tunneling length (m)	Remarks
Tokyo Bay Submarine Highway Tunnel	Tokyo, Japan	14.14	Slurry balanced shield machine	Kawasaki, Mitsubishi, IHI, Hitachi	9,500 × 2	In service
Azabu Tunnel on Tokyo Metro Line 7	Tokyo, Japan	14.18	Slurry balanced shield machine	IHI	364 + 777	In service
Fourth Tunnel of the Elbe River	Hamburg, Germany	14.2	Slurry balanced shield machine	Herrenknecht	2,561	In service
"Green Heart" Tunnel	Amsterdam, the Netherlands	14.87	Slurry balanced shield machine	NFM	7,156	In service
Lefortovo Tunnel	Moscow, Russia	14.2	Slurry balanced shield machine	Herrenknecht	2500	In service
Tunnels in Silverpine Forest	Moscow, Russia	14.2	Slurry balanced shield machine	Herrenknecht	2,200 × 2	In service
M30 Loop Tunnel	Madrid, Spain	15.2/15.0	EPB shield machine	Herrenknecht, Mitsubishi	3670 × 2	In service
Water Diversion Tunnel of the Niagara River	Ontario, Canada	14.4	Hard rock TBM	Robbins	10,400	In service
Seville SE-40 Highway Tunnel	Sevilla, Spain	14.0	EPB shield machine	NFM	2,180 + 1,900	In service

(*Continued*)

Table 1.1. (*Continued*)

Tunnel	Location	Diameter of shield machine (m)	Type of shield machine	Manufacturer	Tunneling length (m)	Remarks
Al Sparvo Tunnel	Appennini, Italy	15.55	EPB shield machine	Herrenknecht	$2,600 + 2,564$	In service
SR99 Tunnel	Seattle, the United States	17.45	EPB shield machine	Hitachi	2,825	In service
Waterview Highway Tunnel	Auckland, New Zealand	14.41	EPB shield machine	Herrenknecht	4,800	In service
Caltanissetta Highway Tunnel	**Sicily, Italy**	**15.08**	**EPB shield machine**	**NFM**	**$3,900 \times 2$**	**Under construction**
Santa Lucia Tunnel	Tuscany, Italy	15.87	EPB shield machine	Herrenknecht	7,500	Under construction
Tokyo Outer Ring Road	Tokyo, Japan	16.1	EPB shield machine	Mitsubishi, Kawasaki	16,200	Under construction
Westgate Highway Tunnel	Melbourne, Australia	15.6	EPB shield machine	Herrenknecht	$2,800 + 4,000$	Under construction

1.2.2 Large-diameter tunnels in China

In 2004, the construction of the Shangzhong Road Tunnel was started in Shanghai, which embarked on a construction wave of 14 m class large-diameter shield tunnels in China. The 14 m class large-diameter shield tunnels represented by the Shanghai Yangtze River Tunnel, the Shanghai Jungong Road Tunnel and the Shanghai Bund Passageway were all built to cross the soft soil layer in Shanghai. In the past decade, Nanjing, Hangzhou, Yangzhou, Zhuhai, Wuhan and other cities have also started to build 14 m class large-diameter tunnels to meet the increasing traffic demand, contributing to the flourishing development of large-diameter shield tunnels in China. As of December 2019, China has twenty-eight 14 m class large-diameter shield tunnels completed or under construction, and the diameter of the shield machines has also been expanded to 17.6 m. The typical tunnels are as follows:

(1) **Shangzhong Road Tunnel, Shanghai:** The Shangzhong Road River Crossing Tunnel (Fig. 1.29) was bored by the 14.87 m diameter slurry pressure balanced shield machine that had been used for the "Green Heart" tunnel in the Netherlands. The shield machine crossed the saturated muddy clay and muddy silty clay layers at the bottom of the Huangpu River. It is a double-tube, double-layer, two-way, eight-lane road tunnel reaching the maximum and minimum burial depths of 45 m and 8.6 m, respectively. With a length of about 1,250 m, the shield section is 14.5 m and 13.3 m, respectively, in the outer and inner diameters. It is built with 2 m wide assembled rings with staggered joints.

Fig. 1.29. Shanghai Shangzhong Road Tunnel.

As the first 14 m class large-diameter shield tunnel in China, the construction contractor, applying the accurate control method for slurry pressure balance, successfully solved the subsequent problems of longitudinal deformation and ovality deformation of large-diameter shield tunnels, developed a kind of single-liquid slurry with high absolute density, shear resistance, high early strength and anti-liquefaction, ensured the timeliness of grouting through automatic synchronous grouting control, and reduced the impact of shield machine construction on the surrounding environment. The continuous promotion and progress of these technologies have effectively guaranteed the rapid popularity and smooth construction of 14 m class large-diameter shield tunnels in China. In 2008, this shield machine was used again for the construction of two river-crossing tubes of the Jungong Road Tunnel on Shanghai Middle Ring Road, with a single-tube length of 1,490 m, which opened to traffic in 2010.

(2) **Shanghai Yangtze River Tunnel:** At the beginning of 2005, the construction of the Shanghai Yangtze River Tunnel (Fig. 1.30) was started by two slurry pressure balanced shield machines with the world's largest diameter of 15.43 m at that time and a tunneling length of up to 7.47 km in a single operation cycle, which created another world's record. It is a double-tube, single-layer, two-way, six-lane tunnel with a design speed of 80 km/h, and the rail transit space is provided in the hollow rectangle component. The tunnel has an outer diameter of 15 m, an inner diameter of 13.7 m, a segment ring width of 2 m and a thickness of 0.65 m. It is built with universal wedge-shaped segments made of C60 concrete and assembled with staggered joints to reach an impermeability grade of P12. The shield machine was driven into a formation composed of soft muddy clay, muddy-silty clay, clayey silt and sandy silt to a maximum burial depth of 55 m.

Targeting the engineering characteristics of "large-section, long-distance and deep-burial", the construction contractor has developed many new technologies and materials in the aspects of tunnel anti-floating, excavation face stability control, tunnel structure design, tunnel axis-control, etc. The shield machine for the east tunnel route was launched in September 2006 and finished tunneling through in 20 months; the shield machine for the west tunnel route was launched in January 2007 and finished tunneling through in 19.5 months.

(a)

(b)

Fig. 1.30. Shanghai Yangtze River Tunnel. (a) Tunnel completed and put into operation and (b) tunnel under construction.

The average advance rate of the shield machine was 12.5 m/d, and the maximum was recorded as 13 rings per day, 72 rings per week, or 241 rings per month. At the end of October 2009, the whole line of the Shanghai Yangtze River Tunnel opened to traffic.

(3) **Nanjing Yangtze River Tunnel:** The Nanjing Yangtze River Tunnel (renamed as the Nanjing Yingtian Avenue Yangtze River Tunnel) (Fig. 1.31) was built following the solution of "tunnel in left branch + bridge in right branch". The 3,020 m long shield tunnel is designed with a double-tube, single-layer, two-way, six-lane arrangement with an outer diameter of 14.5 m and an inner diameter of 13.3 m. Two slurry pressure balanced shield machines with a diameter of 14.93 m manufactured by German Herrenknecht were launched from the north bank to cross the composite stratum composed of silty fine

Fig. 1.31. Weiqi Road Yangtze River Tunnel, Nanjing.

sand, gravelly sand and round gravel with a high permeability coefficient, bringing huge difficulty in stability control of the excavation surface and demanding high wear resistance of the cutterhead and cutter. In addition, there were also other technical challenges, such as cutter replacement in pressurized chamber under high water pressure and highly permeable strata and the construction of a large-diameter shield under a super-shallow overburden.

(4) **Bund Passageway, Shanghai:** In 2007, the construction of the Bund Passageway was started in Shanghai, known as the "heart bypass operation", to solve the traffic jams in the city center of Shanghai. The Bund Passageway is the first tunnel constructed using a 14 m class large-diameter EPB shield machine in China (Fig. 1.32). The shield tunnel has a total length of 1,098 m and is designed as a single-tube, double-layer, two-way, six-lane arrangement with an outer diameter of 13.95 m, an inner diameter of 12.75 m and a segment thickness of 0.6 m.

The engineering characteristics of the Bund Passageway lie in the construction of a large-diameter shield machine under an ultra-shallow overburden, with the average overburden thickness in about one-third of the tunnel section being less than 9 m; the construction crosses "one bridge, two tunnels and 33 buildings (structures)" in the central area of the city. Through a series of measures, such as improvement of soil cut by the shield machine, parameter optimization for tunneling under the ultra-shallow overburden, synchronous slurry injection adaptive to strata and the stability control of tunnel

Fig. 1.32. Shanghai "Bund Tongtai" shield machine received.

rings, an effective construction technology system has been established for large-diameter shield tunnels in urban core areas. In 2009, this shield machine was also used in the construction of the Yingbin 3rd Road Tunnel of the Shanghai Hongqiao Integrated Transportation Hub, successfully crossing the main runway of Hongqiao Airport, which is more stringent on settlement control.

(5) **Tuen Mun — Chek Lap Kok Subsea Tunnel, Hong Kong:** The Tuen Mun — Chek Lap Kok Subsea Tunnel in Hong Kong is the largest and longest subsea highway tunnel with the greatest burial depth in Hong Kong. A 17.6 m Herrenknecht hybrid shield machine digging for about 0.8 km southward from the northern manual reclamation section was replaced by two 14 m diameter shield machines digging for about 4.2 km southward until the tunnel was bored through (Fig. 1.33). In June 2015, a 17.6 m diameter shield machine was launched and arrived at the ventilation shaft at the south end of the reclamation section after five months of tunneling. This 17.6 m diameter shield machine was the largest in the world, equipped with many new systems and equipment, including the built-in "Mobydic" sensing system of the cutterhead, enabling real-time monitoring of cutter wear and geological mapping in front of the excavation face; the mechanical snake-shaped arm to carry a camera; and lighting and cutting equipment, a high-pressure water gun.

See Table 1.2 for a list of 14 m class large-diameter shield tunnels in China (as of December 2019).

Fig. 1.33. 17.6 m-diameter hybrid shield machine for Tuen Mun — Chek Lap Kok Subsea Tunnel in Hong Kong.

1.3 Technological innovations of large-diameter shield tunneling

1.3.1 Latest design of shield tunneling machines

1.3.1.1 *General layout of large-diameter shield machine*

A shield machine is generally composed of a main shield and a frame. Although the main shield is structured in limited dimensions, the shield machine frame generally exceeds 100 m to meet the construction needs. The core components and systems, such as the cutterhead, main drive, manual brake, thrust cylinder, erector, shield and shield tail, constitute the main shield, and the other systems are located on the frame. In order to make scientific and reasonable use of the limited space created by shield machines, the main shield and frame are designed in a layered layout. The top layer is relatively open to accommodate a ventilation channel and measuring system. The bottom layer is arranged with a frame-traveling mechanism, segment lifting appliance, etc. The middle layer is centrally arranged with a central control chamber, power supply and distribution system, thrust system, segment erecting system, synchronous grouting system, shield tail grease system and centralized lubrication system. All the floors are connected with stairs as the footway, and cable trays are laid for wiring.

Table 1.2. 140 Class large-diameter shield tunnels in China as of December 20196.

Tunnel	Diameter of shield machine (m)	Type of shield machine	Manufacturer	Tunneling length (m)	Remarks
Shangzhong Road Tunnel, Shanghai	14.87	Slurry balanced shield machine	NFM	1,250 × 2	In service
Shanghai Yangtze River Tunnel	15.43	Slurry balanced shield machine	Herrenknecht	7,470 × 2	In service
Shanghai Jungong Road Tunnel	14.87	Slurry balanced shield machine	NFM	1,490 × 2	In service
Weiqi Road Yangtze River Tunnel, Nanjing	14.93	Slurry balanced shield machine	Herrenknecht	3,020 × 2	In service
Bund Passageway, Shanghai	14.27	EPB shield machine	Mitsubishi	1,098	In service
Yingbin 3rd Road Tunnel, Shanghai	14.27	EPB shield machine	Mitsubishi	1,860	In service
Hangzhou Qianjiang Tunnel	15.43	Slurry balanced shield machine	Herrenknecht	3,245 × 2	In service
Weisan Road Yangtze River Tunnel, Nanjing	14.93	Slurry balanced shield machine	CCCC Tianhe	4,135 + 3,557	In service
Slender West Lake Tunnel, Yangzhou	14.93	Slurry balanced shield machine	Herrenknecht	1,280	In service
Changjiang West Road Tunnel, Shanghai	15.43	Slurry balanced shield machine	Herrenknecht	1,430 × 2	In service

(Continued)

Table 1.2. (*Continued*)

Tunnel	**Diameter of shield machine (m)**	**Type of shield machine**	**Manufacturer**	**Tunneling length (m)**	**Remarks**
Hongmei South Road Tunnel, Shanghai	14.93	Slurry balanced shield machine	Herrenknecht	3,390 × 2	In service
Lung Shan Tunnel, Hong Kong	14.10	EPB shield machine	NHI (NFM)	2,400 × 2	Completed
Zhuhai Hengqin Tunnel	14.93	Slurry balanced shield machine	Herrenknecht	1,082 × 2	In service
Sanyang Road River Crossing Tunnel, Wuhan	15.76	Slurry balanced shield machine	Herrenknecht	2,590 × 2	In service
Shanghai Riverside Passageway	15.43	Slurry balanced shield machine	Herrenknecht	5,090 × 2	In service
Contract Section II of Beiheng Passageway, Shanghai	15.56	Slurry balanced shield machine	Herrenknecht	2,761 + 3665	Under construction
Tuen Mun — Chek Lap Kok Subsea Tunnel, Hong Kong	17.60/14.00	Slurry balanced shield machine	Herrenknecht	800 + 4,200 × 2	Under construction
Zhuguang Road Tunnel, Shanghai	14.45	EPB shield machine	Herrenknecht	1,390	In service
Shantou Bay Tunnel	15.01/15.03	Slurry balanced shield machine	Herrenknecht/ CREG	3,048 × 2	Under construction

Chunfeng Road Tunnel, Shenzhen	15.8	Slurry balanced shield machine	CREG	3,583	Under construction
Zhoujiazui Road Tunnel, Shanghai	14.93	Slurry balanced shield machine	Herrenknecht	2,572	In service
Wuhu South River Crossing Tunnel	14.93	Slurry balanced shield machine	Herrenknecht	3,850 × 2	Under construction
Oujiang River North Tunnel, Wenzhou	14.93	Slurry balanced shield machine	Herrenknecht	2,664	Under construction
Meizizhou Tunnel, Nanjing	15.46	Slurry balanced shield machine	Herrenknecht	1,800 × 2	Under construction
Heping Avenue Tunnel, Wuhan	15.93	Slurry balanced shield machine	Herrenknecht	1,310	Under construction
Jinan Yellow River Tunnel	15.76	Slurry balanced shield machine	Herrenknecht	2,519 × 2	Under construction
Heyan Road River-crossing Tunnel, Nanjing	15.03	Slurry balanced shield machine	Herrenknecht/ CCCC Tianhe	2,970 × 2	Under construction
Mawan Sea-crossing Tunnel, Shenzhen	15.43	Slurry balanced shield machine	Herrenknecht	2,060 × 2	Under construction

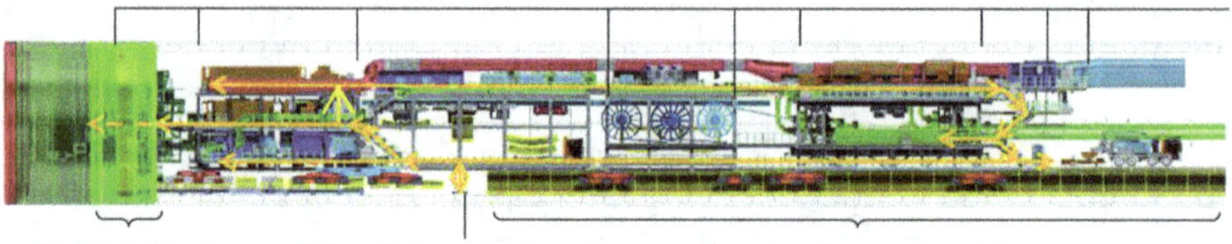

Fig. 1.34. Integrated layout of shield tunneling and construction of road structures (dimension unit: mm).

Due to the layout requirements of slurry pipelines, water inlet and outlet pipelines and cables, as well as the requirements of material conveying in the shield tunneling process, the internal space of tunnels is narrow, and it is extremely difficult to carry out synchronous construction of the internal structures while shield tunneling. The erection of internal structures, such as flue slabs, intermediate diaphragm walls, side walls and lane/track slabs, increases the difficulty of construction organization. In the past, only tunneling was considered in the layout design of shield tunneling machines, and the internal structures were constructed after the completion of shield tunneling, which increased the construction period and cost for the sake of a sufficient construction space. The currently prevailing layout design of shield machines should fully consider the construction integration with synchronous erection of tunnel internal structures so as to improve construction efficiency. Figure 1.34 shows an example of the integrated construction of shield machines: the front of the shield machine for tunneling and hoisting of precast road structural members of hollow rectangles and the back for the installation of pavement slabs, the cast-in-situ operation of the side wall and other construction procedures.

Scientific and reasonable planning and coordination between the construction of tunnel internal structures and shield tunneling, the rational layout of the shield machine frame and the addition of a turnaround platform can greatly improve the efficiency of conveying construction materials. In order to ensure normal transport inside tunnels during the construction of the side wall and lane/track slab, the "gate-type" formwork trolley (Fig. 1.35), serves concurrently shield tunneling and transport organization for three-dimensional cross-operation as the technical guarantee for the integrated shield tunneling.

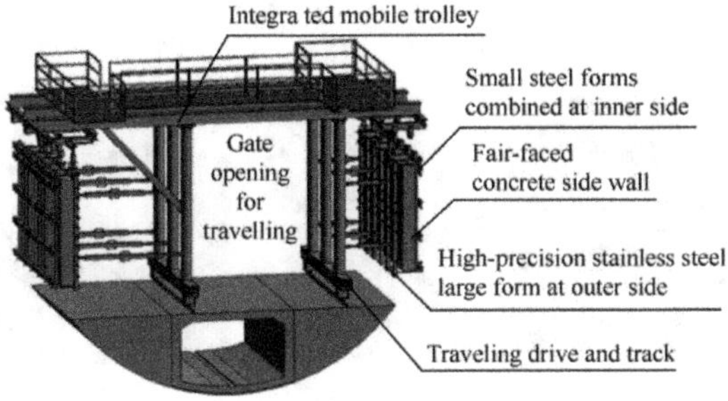

Fig. 1.35. "Gate-type" form-work trolley.

1.3.1.2 *Cutter replacement under atmospheric pressure*

Cutters for the shield machine used in the composite stratum are prone to wear and need to be frequently replaced, so the large-diameter hybrid shield machine is usually equipped with the cutterhead supporting cutter replacement under atmospheric pressure to improve the efficiency and safety of cutter replacement. The cutterhead is changed from a single-layer structure to a box structure. The staff can enter the box through the internal access for cutter replacement under atmospheric pressure.

The 14 m class shield machines used in China are generally products manufactured by Herrenknecht of Germany. Since the first-generation technology was applied to engineering practice in 1997, the replaceable cutter has developed into the fourth-generation technology, as shown in Table 1.3.

1.3.1.3 *Double-mode shield*

Generally, according to different working modes, tunnel boring machines (TBMs) are categorized into open hard rock TBMs, slurry balanced shield machines and EPB shield machines. A dual-mode shield machine is a multi-functional tunnel boring machine that can be switched between different modes according to different geological conditions to better adapt to complex and varied geological conditions. At present, the dual-mode shield machines that have been put into use mainly include the "hard rock open mode/soft soil shielded mode" and "earth pressure mode/slurry mode", as shown in Figs. 1.36 and 1.37.

The Port of Miami Tunnel in the United States is a case of successful application of large-diameter dual-mode shield machines (Fig. 1.38).

Table 1.3. Technologies of cutters designed replaceable under atmospheric pressure.

Item	Characteristics	Typical tunnel	Photo of shield machine
1st generation	Partial cutters (except the center one) to be replaced under atmospheric pressure	Fourth Tunnel of the Elbe River in 1997: 14.2 m diameter shield machine	
2nd generation	Full-section cutters to be replaced under atmospheric pressure	Shanghai Yangtze River Tunnel in 2006: 15.43 m diameter shield machine **(scraping cutter)**	
		Nanjing Metro Line 10 River-crossing Tunnel in 2011: 11.64 m diameter shield machine (gear cutter)	
3rd generation	Full-section double-edged hobs to be replaced under atmospheric pressure (interchangeable disc cutter and gear cutter)	Bosphorus Strait Road tunnel in 2013: 13.71 m diameter shield machine	
4th generation	Full-section single-edged hobs to be replaced under atmospheric pressure (interchangeable disc cutter and gear cutter)	Sanyang Road River Crossing Tunnel in 2015: 15.76 m diameter TBM	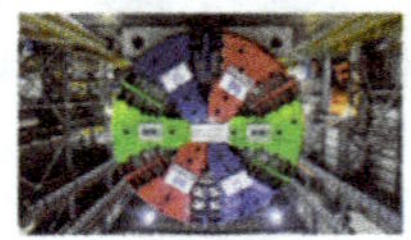

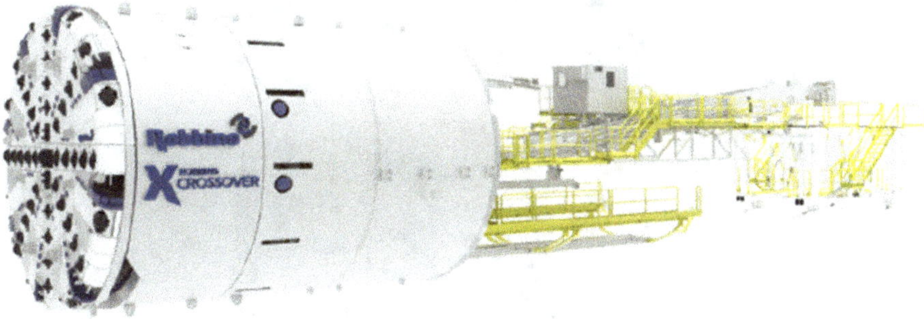

Fig. 1.36. TBM in hard rock open mode/shield machine in soft soil closed mode.

Fig. 1.37. Schematic diagram of shield machine in earth pressure mode/slurry mode.

Despite the original plan to use an EPB shield machine, the results of the geological survey before construction revealed that, under the maximum water head of 37 m, the whole tunnel would pass through eight different kinds of strata, from loose sandstone to strongly cemented limestone, and the Neogene sedimentary carbonate rock with a low integrity and high variability in strength, porosity and cohesion, of which the seventh layer, composed of coral limestone, 24–34 m in thickness and porous, is extremely unstable.

The complex geological conditions called for a dual-mode shield machine with both earth pressure and slurry to adapt to varied strata. Finally, the French contractor, Bouygues, selected a 12.86 m diameter earth pressure/slurry dual-mode shield machine to roll out a new excavation mode, named the water control process (WCP), which can realize the transport of excavated materials by means of a spiral transport integrating the internal slurry tank and rock crusher while ensuring the water pressure is under control.

The shield machine started tunneling in the middle of November 2011 and completed the west route on July 31, 2012. Subsequently, the shield machine made a U-shaped turnaround and began tunneling on the east

(a)

(b)

Fig. 1.38. Shield machine for Port of Miami Tunnel in the United States. (a) Cutterhead of dual-mode shield machine and (b) construction by double-mode shield machine.

route on October 29, 2012. The tunnel was bored through on May 6, 2013, and officially put into operation in August 2014.

1.3.2 High-precision fabrication of large-size segments

1.3.2.1 *Fabrication of high-precision steel formwork for segments*

The steel form for large-diameter segments is generally of the double-wedge universal design, and the inclined straight mandrel and embedded inner nut boss are used to construct the segment hand hole. Due to their super-large diameter, the arc length of blocks is large, and the whole side plate of the steel form is a long and narrow structure, so attention should be paid to deformation control during the fabrication process.

In terms of steel formwork design, compared with ordinary steel formwork, the steel form for large-diameter segments has a larger arc length and increases in weight, leading to higher requirements for its design. A larger arc length of the segment is the cause of the distortion-proneness of the side plate structure of the steel form and brings difficulty in ensuring the arc surface accuracy of the base, as well as an effect on the reliability of sealing between the base and the side plate. With the increase in segment weight, the stress on the steel form base changes, thus putting forward higher requirements for the strength and stiffness of the steel form base.

In terms of the fabrication of steel-form structural parts, the non-annealing stress relief metal processing process is adopted to effectively avoid the problem of large internal stress occurring on the workpiece due to excessive welding affecting the processing accuracy of the end and side plates and even leading to deformation or cracking of the workpiece. Before fabrication, the solid modeling of the end and side plates should be done with computer-aided digital 3D modeling software, and several processing steps should be carried out to guarantee the flatness and surface roughness of each structural part, including the processing of the datum bottom plane, the processing of the panel cavity surface, the rough processing, the finish processing and ultra-fine processing.

After the final assembly of the steel forms, the geometric measurement should be conducted by three-dimensional laser scanning technology to realize data acquisition and data processing, and the key scanning parameters should be optimized through point cloud modeling, calculation and analysis to improve the final assembly accuracy of the steel forms (Fig. 1.39).

1.3.2.2 *Fabrication process of large-sized segments*

The large-diameter concrete segments are generally made of high-performance concrete (HPC) to meet the comprehensive technical indicator requirements such as high strength, high permeability, high crack resistance and high durability. Through extensive experimental research and engineering application of large-diameter tunnel segment fabrication,

Fig. 1.39. Fabrication of high-precision segment steel forms.

considerable experience has been gradually accumulated in the concrete preparation technology for high-performance segments. The progressive production process of large-diameter tunnel segments, the in-depth research on raw materials and the development of internet technology drive constant improvement in the preparation technology of high-performance concrete.

With the wide application of automatic flow-line production in the fabrication of small-diameter segments in recent years, it can be expected that such an automatic process would be increasingly applied to large-diameter segments in the future. Based on the intelligent control system, the flow-line technology of segment fabrication is utilized to control the forms in circulation on the production line, along which locomotives or translation trolleys are operated on the track to pass the preset positions, thus realizing the whole process of segment fabrication. As the most important key link in the whole fabrication process, the control of segment concrete vibration is related to the internal and external quality of segments, such as compactness, smoothness, flatness and air voiding conditions. The main difference between large-diameter segments and small-diameter segments lies in the different vibration technology under large volume scenarios. It is necessary to comprehensively analyze the characteristics of various vibration modes to make corresponding choices. Large-diameter segments generate a large amount of hydration heat due to their large volumes, so more accurate temperature control is required in the steam curing of segments. The intelligent temperature control system

(a)

(b)

Fig. 1.40. Fabrication of large-size segments. (a) Steel forms of segments and (b) curing of segments.

for steam curing can effectively improve the curing quality. Another new technology involves the application of fiber concrete in large-diameter tunnel segments, which can improve the quality of the segments, enhance safety performance indicators and improve the tunnel quality. With the improvement of production management technology, information management technology has also been gradually introduced in segment production. The fabrication technology, in combination with the information technology, automation technology and modern management technology, can improve or even change all links, including operation, management, product development and production, of enterprises engaging in the fabrication of members, making progress toward construction industrialization (Fig. 1.40).

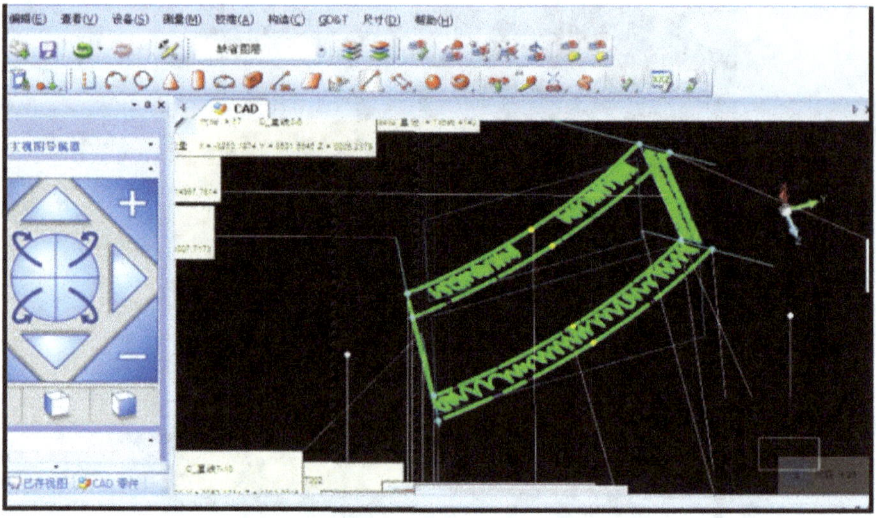

Fig. 1.41. Measurement technology for large-size segments.

1.3.2.3 *Measurement technology of large-size segments*

Since traditional measurement tools such as rulers can hardly meet the requirements due to the large size and high accuracy of large-diameter segments, 3D laser scanning technology has been used to measure the dimensions of steel forms and segments. With the help of high-precision instruments to obtain a large number of measurement points, the dimensional data of steel forms and segments are finally generated through a series of calculations. The multi-dimensional geometric information of steel forms and segments is helpful to evaluate the segment quality more objectively and from all aspects and effectively control the links that should be paid attention to during production (Fig. 1.41).

1.3.3 Innovations of large-diameter shield tunneling

1.3.3.1 *Stability control of excavation face*

During shield tunneling, excavation-induced stratum deformation is prevented by applying a support stress equal to the original stratum stress to the soil layer ahead of the excavation face. However, due to the complexity of stratum conditions and the variability of the construction situation, any improper setting and control of the support pressure on the excavation face may lead to different engineering accidents. A too small support pressure on the excavation face may cause soil mass collapse, while a too large support pressure may cause soil mass heave. Therefore, it is particularly important to control the stability of the excavation face during the shield construction. For the slurry pressure balanced and EPB shield machines, there are different methods to control the stability of the excavation face.

For the large-diameter EPB shield machine, the stability of the excavation face is mainly affected by factors such as earth chamber pressure and advance rate. In order to ensure the stability of the excavation face, a certain amount of muck improver (foam or bentonite) needs to be injected into the excavation face to reduce the water permeability of the soil mass ahead, improve its plastic fluidity and better control the earth chamber pressure. During construction, the foam concentration should be controlled at 2–3%, the foaming rate should be controlled at 20–30%, and the injection rate should be controlled at 30%.

For the large-diameter slurry pressure balanced shield machine, the speed and quality of slurry membrane formation depend on the density, viscosity, stability and filtration characteristics of the slurry. In order to ensure that the slurry membrane formed meets the requirements, the accuracy of slurry pressure fluctuation during construction should be controlled within ±0.05 bar; the slurry density should be generally controlled at 1.15–1.25 g/cm^3, reaching 1.30 g/cm^3 under high environmental protection requirements; and the viscosity of slurry should not be less than 20 Pa · s. When it is necessary to open the chamber for operation, the requirements for the slurry indicators on the excavation surface are more stringent, and high-quality fresh slurry should be fed for full replacement on the excavation surface.

With an excavation diameter of up to 15.43 m, the Qianjiang Tunnel in Hangzhou was built in diversified strata, subject to a large water and soil pressure difference on the excavation face. Affected by the high tidal range of 10 m (Fig. 1.42), improper control is likely to lead to serious consequences of instability or even collapse of the excavation face. Precise and dynamic control of slurry support pressure to maintain the stability of the excavation face has become a major technical challenge. Based on the proactive tunneling method, a high-precision control technology for the stability of the excavation face of the super-large-section shield machine has been created for the construction of the

Fig. 1.42. Tidal bore of the Qianjiang River in Hangzhou.

Qianjiang Tunnel, with a control accuracy within 8 kPa, which has successfully tackled the technological barrier of the stability of the excavation face for super-large-section shield tunnels.

1.3.3.2 *Synchronous grouting technology*

Compared with the shield machines of conventional diameters, large-diameter shield tunneling brings a wider range of impacts and produces a severer disturbance to the surrounding environment during construction. Therefore, it is necessary to control the shield grouting process more effectively, so that the slurry materials, grouting equipment and grouting parameters can meet the requirements of full and uniform filling of synchronous grouting for large-diameter shield machines, and it is also necessary to strictly control grouting parameters and construction quality. Slow-setting mortar is generally used for synchronous grouting of large-diameter shields, which is mainly composed of sand, fly ash, bentonite, water and additives, and the early and late strengths of the slurry should meet the corresponding requirements. Slump is an important indicator of workability for synchronous grout construction, and its size depends on such factors as the properties of slurry materials, stratum environmental conditions and grouting pressure. Slump control during construction is generally required to be 120 ± 30 mm. Figure 1.43 shows the testing process for the physical and mechanical properties of synchronous grouting materials.

In terms of construction method, at present, the synchronous grouting of large-diameter shields is generally conducted in the "double control" mode of grouting pressure and grouting amount. The synchronous

(a) Slump test (b) Density test (c) Fluidity test

(d) Bleeding rate test (e) Unconfined compressive strength test (f) Shear yield strength test

Fig. 1.43. Tests of synchronous grouting properties.

grouting amount is controlled by the number of strokes of the piston pump, and the grouting pressure is controlled within a preset range and is timely adjusted according to the monitoring value indicated on the pressure sensor near the slurry outlet so as to ensure that the slurry is injected into the structure voids formed when segments come out of the shield tail.

A too small synchronous grouting pressure is likely to cause slow filling and, subsequently, insufficient filling of the voids at the shield tail, resulting in excessive ground settlement; a too large grouting pressure may cause slurry invasion into the surrounding soil layer, resulting in ground heave and cracking in concrete segments. Six or eight synchronous grouting points are symmetrically arranged at the shield tail. Slurry feeding follows the principle of "decrease of feeding amount from the top to the bottom" and is adjusted in combination with the displacement of the formed tunnel and ground settlement. The grouting rate is 110–130%, and the grouting pressure is generally 0.3 MPa higher than the formation pressure under normal advance. When the shield passes through sensitive buildings (structures), the grouting rate is increased to 130–150%, and the grouting amount needs to be reduced in the reinforcement area. The reasonable setting of grouting pressure and grouting amount can contribute to constraining the tunnel structure and ensuring the longitudinal and transverse stability of the tunnel structure.

1.3.3.3 *Advance at small-radius curves*

Due to the restricted development of underground space in the urban core area, the alignment of new large-diameter shield tunnels is inevitably designed with small-radius curves. There are many challenges in large-diameter shield tunneling on small radius curves, such as axis control risk, segment ring quality risk, shield tail leakage risk, material transport risk and high difficulty in safety control when crossing buildings (structures) in curve sections.

In view of these challenges, in addition to the intensive design along the shield machine length, segment type selection and construction control measures are also critical for tunneling on small-radius curves. In the segment type selection, the segment ring width may be reduced or the wedge amount may be adjusted to reduce the lateral component force and the segment stress concentration. Moreover, the longitudinal stiffness of the tunnel is increased by longitudinal tensioning measures, providing sufficient reaction force for the shield machine to advance and reducing the tunnel offset from the arc. The construction control measures mainly include the control of offset correction, synchronous grouting pressure and grouting amount. Additionally, a certain pre-offset should be set, and the bolt re-tightening measure should be taken to control the voids between segments and the shield tail.

The minimum plane radiuses of circular curves in the second tube of the Shanghai Dapu Road Tunnel and the Beiheng Passageway by shield tunneling are 380 m and 500 m, respectively, and the ratios of the plane curve radius to the shield machine diameter are 33.9 and 32.1, respectively. Under the backdrop of increasingly dense urban underground resources, small-radius construction technology provides more flexibility and freedom for tunnel design and planning and alignment control, and it also lays a solid foundation for the development of urban underground expressways in the future.

1.3.3.4 *Protection for crossing sensitive surroundings*

It is inevitable for the large-diameter shield machine tunneling in the central urban area to pass through a large number of important buildings (structures). The Bund Passageway in Shanghai was built through a narrow route between the Bund Historic and Cultural Conservation Zone and the Huangpu River. The shield machine was tunneled at a short distance from the Astor House Hotel, the Garden Bridge of Shanghai, the Exotic Building Clusters in the Bund, the operational Metro Line 2, pedestrian underpasses and other public facilities. The shield section of the Beiheng Passageway under construction in Shanghai is designed to cross 136 buildings, 3 existing rail transit lines, 4 flood control walls of the Suzhou River, and several large-diameter utility pipelines.

Protection solutions should be developed in categories and zones, according to the tunnel burial depth, stratum conditions, the conditions of existing buildings (structures) and their plane clear distance from the tunnel for tunneling through sensitive surroundings by large-diameter shield machines. Specifically, the core technology for side-crossing of the shield machine through buildings (structures) at a short distance is referred to as micro-disturbance construction control. By virtue of soil mass improvement technology, precise control of tunneling parameters, optimized connection of various processes and other methods, disturbance by the shield machines to the surrounding soil mass can be reduced, and the settlement and deformation of the buildings (structures) can be controlled. In the case of side-crossing of the shield machine through buildings (structures) at a very short distance, the impact of shield tunneling on buildings can be effectively reduced with shelter piles separating the tunnel from buildings or by grouting to break the influence line. In cases where the shield machine crosses beneath buildings (structures), it is necessary to strictly control the shield machine axis, support pressure on the excavation face, synchronous grouting pressure, grouting amount and other parameters in cooperation with the reinforcement and protection of buildings and strengthened monitoring.

1.3.3.5 *Shield launching and receiving*

Shield launching and receiving have always been high-risk procedures in shield construction. Large-diameter shield machines suffer from doubled construction risk due to their large sizes and deep working shafts. In order to ensure the safety of shield launching and receiving, it is necessary to reinforce the soil mass at tunnel portals with cement-based methods, such as high-pressure jet grouting and tri-axial mixing piles. Other reinforcement methods, such as freezing and cut-off walls, were applied in some tunnels due to their special geological conditions. Both the Shanghai Yangtze River Tunnel and the Zhuhai Hengqin Tunnel have been applied with the tri-axial mixing pile ground reinforcement method for shield launching. The Nanjing Weiqi Road Yangtze River Tunnel and the Shanghai Beiheng Passageway have been applied with the freezing method between the end reinforcement area and the cut-off wall to ensure the safety of tunnel portal breaking and shield launching.

In order to simplify the construction process of shield receiving, the Shanghai Yangtze River Tunnel has been applied with the method of shield receiving underwater to eliminate the step of manually chiseling the tunnel portal and reduce the risk of water and soil pressure imbalance on both sides of the working shaft during shield receiving. As supporting actions for this process, the soil mass at the tunnel portal was reinforced before receiving, and the diaphragm wall at the provided portal ring position was built with GFRP rebar instead of ordinary rebar, so that the shield machine can directly cut the retaining wall and enter the working shaft. After the shield machine enters the receiving shaft, with the loss of friction around the shield machine and the reduced front water pressure, the segments are loosened from their originally compressed state under the expansion action of the water-stop rubber strip and the pulling action of the shield tail. Therefore, the segments close to the receiving shaft are equipped with special structures such as shear pins and pre-stressed bolts.

1.3.3.6 *Underground docking of shield machines*

Underground docking of shield machines can effectively solve the problems of low reliability and low construction efficiency of equipment encountered in long-distance tunneling by shield machines, and such an approach has been applied to the English Channel Tunnel, the Tokyo Bay Submarine Highway Tunnel and the Storebaelt Tunnel in Denmark. The shield machine used in the Shiziyang Tunnel of the Guangzhou — Shenzhen — Hong Kong Express Rail Link is China's first application of the underground docking technology of shield machines, which is characterized by a special construction organization mode of "opposite tunneling, underground docking and disassembly in tunnel".

Underground docking of shield machines is divided into two types: direct docking (i.e. mechanical connection) and auxiliary docking (i.e. civil engineering connection). The selection of docking methods should fully consider factors such as stratum conditions, confined water and tunnel alignment. High-precision docking measurement technology and sealing technology at docking parts are key to underground docking. Taking the construction of the Shiziyang Tunnel as an example, the relatively favorable stratum section selected in the center of the Shuidao River was reinforced and treated for water stop by the shield advance reinforcement equipment in the docking area. The method of front civil engineering connection was directly used, which successfully realized the underground docking of large-diameter slurry balanced shield machines.

1.4 Introduction to chapters of the book

It can be seen from the relevant literature and engineering experience in China and abroad that there are many difficulties and obstacles in the construction of super-large-diameter shield tunnels. However, there is a lack of systematic research on the related technologies in China, and the construction and advancement of super-large-diameter shield machines are facing great technological challenges in China and around the world. This book systematically unravels and summarizes the research results and emerging key construction technologies of large-diameter shield tunnels in recent years. The book is organized into the following chapters.

Chapter 1 mainly introduces the construction and development of large-diameter shield tunnels around the world in view of the new demand of modern traffic for tunnels and expounds on the development and innovation of large-diameter shield tunneling technology in the new era.

Chapter 2 describes the selection and classification of shield machines and the characteristics of EPB and slurry balanced shield machines, clarifies the basis and principles of shield machine selection and summarizes the coordination between the overall layout of shield machines and other links.

Chapter 3 discusses the design basis of the slurry treatment system, the design of the classification subsystem, the calculation of the technical parameters, the classification of the slurry treatment equipment, the composition of the slurry feeding system and the specific indicators of the slurry management monitoring based on the actual engineering works.

Chapter 4 introduces the design, manufacturing and inspection processes of high-precision steel forms, describes in detail the raw material of segments — high-performance concrete — and summarizes the current state of the new technology of segment fabrication.

Chapter 5 mainly includes the launching and receiving of large-diameter shield machines, excavation face stability, shield machine attitude management, segment erection technology, synchronous grouting technology, tunneling in special sections, etc., in an attempt to effectively ensure the safe tunneling of large-diameter shield machines.

Chapter 6 mainly describes the key technologies related to tunnel internal structures and connecting passages, focusing on their construction.

Chapter 7 focuses on the data processing technology, site information management, remote intelligent management and control in shield construction and life cycle management of shield machine equipment in an attempt to realize the scientific and safe excavation of large-diameter shield tunnels from the perspective of informatization and intelligentization.

Chapter 8 mainly introduces the main risk management and response solutions of large-diameter shield tunneling, including advanced geological prediction, cutter management and shield tail sealing management, and summarizes the risk management and control measures of large-diameter shield tunneling.

Chapter 9 analyzes and discusses the characteristics, difficulties and key technologies of five typical large-diameter shield tunnels in the world, including the Shanghai Yangtze River Tunnel and the Seattle SR99 Highway Tunnel.

Chapter 10 mainly focuses on the growing demand for the future development of underground space, introduces the innovation trends of shield tunneling and puts forward the prospect of intelligent shield machine construction, showing the development directions of large-diameter shield tunnels from multiple dimensions.

Chapter 2
Shield Machine Selection

Type selection of shield machines is one of the key tasks for shield tunneling to fulfill the requirements of safety, environmental protection, high quality, high economic efficiency and rapid construction, so selection should be made in holistic consideration of safety, reliability, technological advancement and economy. At present, the earth pressure balanced (EPB) shield machine and slurry balanced shield machine are commonly used in large-diameter shield tunneling in China and abroad. A reasonable selection of large-diameter shield machines based on the specific geological conditions can effectively reduce construction risks. This chapter focuses on the analysis of the classification of shield machines, the characteristics of the EPB shield machine and the slurry balanced shield machine, puts forward the scientific basis and principles for type selection of super-large-diameter shield machines and gives suggestions on their overall layout and supporting function settings.

2.1 Types and characteristics of shield machines

2.1.1 Classification of shield machines

Shield machines can be classified according to the working strata, balance modes, tunnel section shapes and sizes, etc.:

(1) They are categorized into soft soil shield machines, hybrid shield machines and hard rock shield machines, according to the conditions of the working strata.

(2) They are categorized into air pressure shield machines, EPB shield machines and slurry balanced shield machines, according to the pressure balance modes on the working face.
(3) They are categorized into semi-face shield machines, single-face shield machines, multi-face shield machines, rectangle-face shield machines and quasi-rectangle shield machines, according to the tunnel section shapes.
(4) They are categorized into mini shield machines (1 m $\leq \varphi$ (diameter of shield machine) < 3.5 m), small shield machines (3.5 m $\leq \varphi <$ 6 m), medium shield machines (6 m $\leq \varphi <$ 11 m), large shield machines (11 m $\leq \varphi <$ 14 m) and super-large shield machines ($\varphi \geq$ 14 m), etc., according to the diameters of shield machines.

According to the engineering practices of super-large-diameter shield tunneling, most of the shield machines working in soft soil and composite strata are made in Germany and Japan, and most of the hard rock TBMs in hard rock strata are made by Robbins Company of the United States. At present, the commonly used super-large-diameter shield machines can be divided into EPB shield machines, slurry balanced shield machines and hard rock TBMs, as shown in Table 2.1.

With the increasing application of shield machines in tunnel excavation, in order to adapt to more complex strata, shield machines have "evolved" to varying degrees. In recent years, major shield machine manufacturers around the world have launched different types of machines, including hybrid shield machines and multi-mode shield machines by German Herrenknecht and the crossover shield machines designed by Robbins Company of the United States.

The hybrid shield machines developed by Herrenknecht represent a major improvement on the traditional slurry balanced shield machines. This kind of machine is dedicated to highly efficient tunneling in heterogeneous strata. The Tuen Mun Tunnel in Hong Kong was bored using a 17.6 m diameter hybrid shield machine, which is not only equipped with an air tank for pressure adjustment on the excavation face but also empowered with a number of innovative technologies to cope with high water pressure up to 5 bar.

The multi-mode shield machines support free switching of two support modes (Fig. 2.1) on the working face, i.e. earth pressure balance and slurry balance, to adapt to varying stratum conditions with a short

Table 2.1. Types of super-large-diameter shield machines, conditions of working strata and application cases.

Type	Appearance	Working strata and applications
Earth pressure balance shield machine		(1) It is suitable for low-permeability strata dominated by clay, sludge and silt; it may be used in heterogeneous or unstable strata dominated by sand-gravel, sand bed and water, subject to improvement of soil mass; and it may be used in composite strata partially containing soft soil and rock layer in the tunnel section, subject to a rational number of hobbing cutters. (2) Application: the SR99 highway tunnel in Seattle, USA, with a shield diameter of 17.45 m.
Slurry balanced shield machine		(1) It is suitable for the high-permeability and high water pressure heterogeneous strata dominated by sand and sand-gravel, and it may be used in composite strata partially containing soft soil and rock layer in the tunnel section, subject to a rational number of hobbing cutters. (2) Application: Tuen Mun — Chek Lap Kok Subsea Tunnel, Hong Kong, with a shield diameter of 17.6 m.
Hard rock tunnel boring machine (TBM)	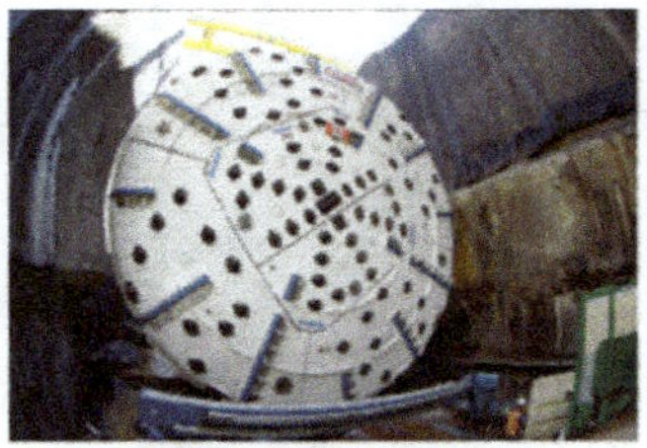	(1) It is suitable for full-section hard rock with uniaxial compressive strength of over 100 MPa (hobbing cutters work with the help of huge contact pressure). (2) Application: Water Diversion Tunnel of the Niagara River, Canada, with a shield diameter of 14.4 m.

Note: The earth pressure balanced shield machine and slurry balanced shield machine with the specially configured cutters can be widely used in the composite stratum with soft soil and rock layers in the tunnel section, under which the machine is referred to as a hybrid earth pressure/slurry balanced shield machine.

Fig. 2.1. Multi-mode shield machine.

switching time and low cost. The typical Klang Valley Mass Rapid Transit Project (KVMRT) tunnels in Kuala Lumpur, Malaysia, were bored using double-mode shield machines, successfully solving the puzzles of tunneling in karst limestone strata.

The crossover shield machines rolled out by Robbins Company of the USA in 2015 (Fig. 2.2) enable quick switching between "open" and "shielded" operation modes to adapt to different stratum types. This mode of shield machines mainly includes XRE (hard rock/earth pressure), XSE (slurry/earth pressure) and XRS (hard rock/slurry), supporting flexible switching depending on changes in stratum conditions to cope with different stratum environments.

2.1.2 Characteristics of EPB shield machine

2.1.2.1 *Fundamentals*

The pressure-bearing medium of an EPB shield machine is the soil in the sealed earth chamber. By setting a pressure chamber plate in the front of the mechanical shield machine, the earth chamber and the screw conveyor are filled with the excavated soil, and the advance and mucking rates of the screw machine are controlled to maintain a certain pressure on the soil in the earth chamber and maintain a dynamic balance between the water

(a) Hard rock/earth pressure shield machine (b) Slurry/earth pressure shield machine (c) Hard rock/slurry shield machine

Fig. 2.2. Crossover shield machines.

and soil pressures of the excavation face. The working principle is that the cutterhead rotates to cut soil on the excavation face and conveys the excavated soil into the earth chamber through the cutterhead opening, and the soil falling to the bottom of the earth chamber is transferred to the belt conveyor through the screw conveyor and then delivered to the muck car parked on the rail. While the shield machine advances under the action of the thrust cylinder, the shield shell plays a temporary supporting role for the excavated tunnel that has not been lined, bearing the earth pressure from the surrounding soil layer and the groundwater pressure and retaining groundwater outside the shield shell. The excavation, mucking, lining and other operations are carried out under the shield. The earth pressure balanced shield machine is composed of a shield shell, cutterhead, cutterhead drive, screw conveyor, belt conveyor, segment erector, man lock, thrust system, etc., as shown in Fig. 2.3.

2.1.2.2 *Cutterhead and cutter configuration*

The long cutterhead cantilevers used to cover the large excavation diameter of the super-large shield tunneling suffer from excessive uneven stress, which is the common cause of the wear of the cutterhead and rotary main bearing. In the cutterhead design, multiple factors should be considered, including the structure form, support mode, opening ratio and layout of the cutterhead.

The cutterhead structure of EPB shield machines is classified into panel, spoke and spoke-panel types, as shown in Fig. 2.4. The panel-type cutterhead is applicable to the complex and varying stratum with large underground water pressure, prone to collapse, dominated by soil, sandstone and weathered rock in an uneven soft and hard texture, and cutters

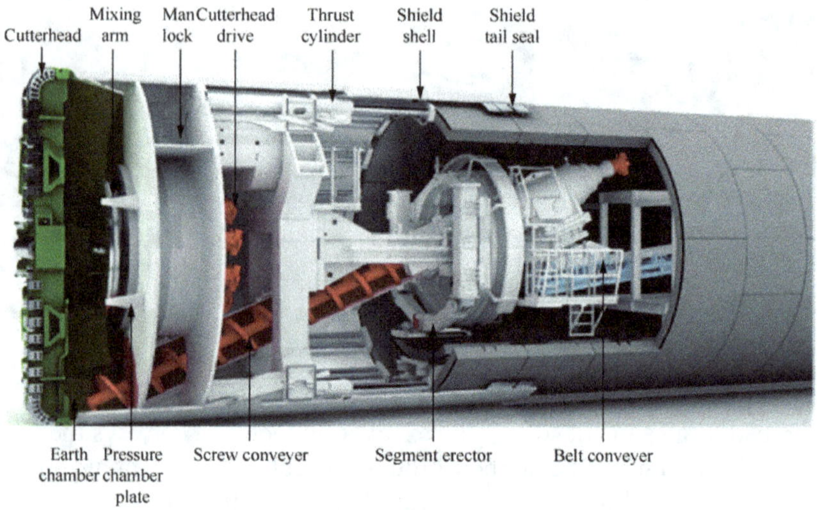

Fig. 2.3. Structure of EPB shield machine.

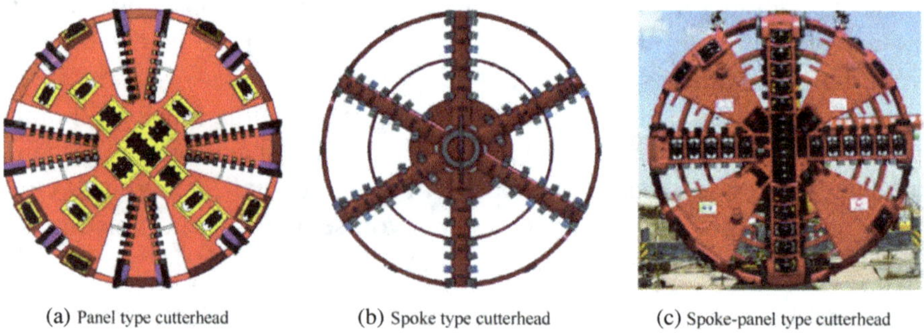

Fig. 2.4. Structures of cutterhead.

can be replaced safely and reliably. Its shortcomings lie in the large difference between the measured earth pressure in the earth chamber and the earth pressure on the excavation face, the difficulty in managing earth pressure, the poor mobility of soil mass, the proneness to mud caking, leading to wear of the cutterhead, the difficulty in advance, the damage to the shield machine and other problems.

The spoke-type cutterhead is applicable to a simple soft soil layer dominated by sand and soil under an earth pressure close to that on the excavation face, which is convenient for earth pressure management.

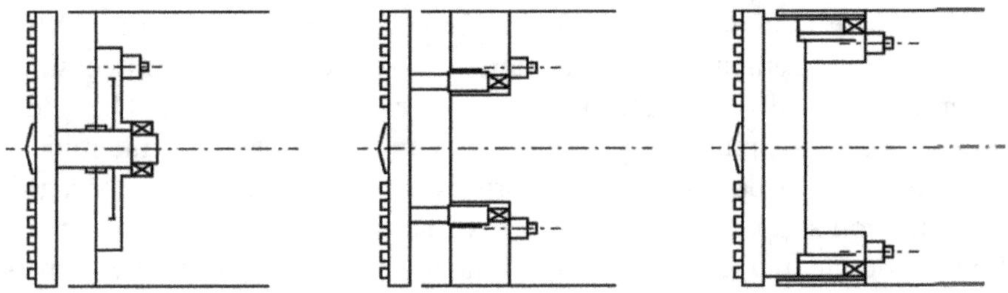

Fig. 2.5. Support modes of cutterhead.

With smooth mobility of soil mass, clay adhesion, bonding and blockage can be effectively prevented, causing a small load on cutters, so the service life is extended. Its shortcomings mainly lie in the low safety of cutter replacement and the demanding reinforcement of soil mass, resulting in high costs. In addition, the overall stiffness of the cutterhead is low, and the hobbing cutter is installed to adapt to the spoke width, which restricts its application in weathered rock formations, formations in uneven soft and hard textures or hard rock formations.

The spoke-panel type cutterhead has the characteristics of both the panel and spoke types. The selection of cutterhead type should depend on soil conditions and construction conditions. Compared with small-diameter earth pressure balanced shield machines, the super-diameter shield machines bring risks due to the more complex working strata, the lower wear resistance of the cutterhead and the more difficult stability control on the excavation face. In consideration of the disadvantages of the spoke type cutterhead, i.e. difficulty in replacing cutters and small cutterhead rigidity, which increases the risk of settlement, the panel type or spoke-panel type cutterheads are preferable for super-diameter shield machines.

Cutterheads are supported in three ways: center support, intermediate support and peripheral support, as shown in Fig. 2.5. The center support is placed on the axis. Its advantages include a small bearing structure, a small sealing area and easy quality assurance. Its disadvantages lie in its small rigidity around the cutterhead, large deformation under heavy load and eccentric load, so it is not suitable for large-diameter shield machines. The peripheral support is placed on the peripheral shell structure of the shield machine. Its advantages include large support stiffness, but its disadvantages lie in the positively related bearing diameter and cutterhead diameter with a small difference, the complex sealing structure and the stringent requirements for mechanical production and processing. Generally, it is only used for small-diameter shield machines. The intermediate support is placed on the intermediate beam to support the middle part of the cutterhead. This support form is between the center support and the peripheral support in terms of rigidity, strength and sealing and brings about moderate manufacturing requirements for the cutterhead drive bearing and machinery, so it is mainly used for super-large-, large- and medium-diameter shield machines.

The cutterhead opening ratio represents the proportion of the opening area of the cutterhead panel in the cutterhead area. Muck from cutterhead cutting is conveyed to the earth chamber through these cutterhead opening grooves. The opening ratio is an important indicator for the cutterhead design of the EPB shield machine. A too small opening ratio will cause problems such as mud caking, muck clogging, an increase in total thrust and torque, etc. On the contrary, a too large opening ratio will easily cause problems such as large formation losses. Therefore, the opening shape, size and configuration must be determined according to the geological conditions, stability of the excavation face and excavation efficiency.

Whether the cutters are suitable for the engineering geological conditions directly affects the tunneling effect of the shield machines. The cutters of the shield machines are generally classified according to the cutting principle, as shown in Fig. 2.6.

Hobs break rock by squeezing, shearing, grinding and other methods generated by rolling and sliding. They can be classified into tooth hobs

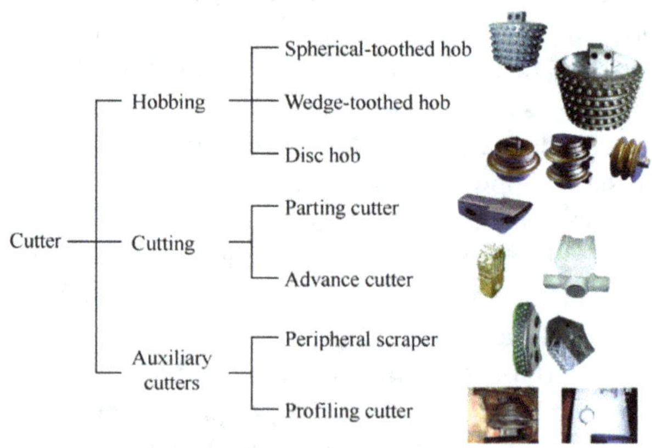

Fig. 2.6. Classification of shield machine cutters.

and disc hobs according to their shapes. At present, the most commonly used are disc hobs. Disc hobs are further classified into single-edge hobs, double-edge hobs and multi-edge hobs according to the number of blades and their different rock breaking capacities, as shown in Table 2.2.

Cutters break rock by shearing action generated by scraping and tearing. There are many kinds of such cutters, the main ones being scrapers, parting cutters, toothed blades and advance cutters, as shown in Table 2.3.

The cutter configuration is different for different stratum conditions. For example, soft soil cutters are allocated in soft soil areas, such as in Shanghai and Tianjin; advance tools with high strength (such as shell cutters) are allocated in sand gravel composite strata, such as in Beijing and Shenyang; hobs and cutting tools are allocated in the uneven composite strata of hard rock and soft rock in Guangzhou and Shenzhen. The cutter

Table 2.2. Applications and mounting positions of hobs.

Type	Working stratum	Rock breaking capacity (MPa)	Mounting position
Single-edge hob	Hard rock, medium hard rock	>80	Main part at the front side of the cutterhead
Double-edge hob	Soft rock	30–80	Cutterhead center or narrow space at the front
Triple-edge hob	Soft rock	30–80	Cutterhead center or for profiling

Table 2.3. Applications and mounting positions of cutters.

Type	Working stratum	Mounting position
Parting cutter	Soft soil	Both sides of cutterhead opening groove
Scraper	Soft soil and hard rock	Periphery of cutterhead arc
Center toothed blade	Soft rock	Cutterhead center (exchangeable with hobs)
Front toothed blade	Soft rock	Cutterhead front (exchangeable with hobs)
Advance cutter	Unconsolidated stratum, such as sandy gravel	Cutterhead front
Fishtail cutter	Soft soil	Cutterhead center
Profiling cutter	Offset correction, turning, etc. in soft soil	Cutterhead edge

layout should meet the stratum conditions and also consider the influence of such factors as cutter spacing, cutter height, balanced wear and balanced stress on the cutterhead.

2.1.2.3 *Dosing system and mixing device for soil mass improvement*

Soil mass improvement refers to the construction method of injecting water, foam, bentonite, polymers and other additives into the cutterhead surface, earth chamber or screw conveyor through the configured bentonite dosing system or foam system. The additives are mixed with muck by the action of rotary mixing of the cutterhead or screw conveyor to achieve good flow plasticity, appropriate consistency, low water permeability and small friction of muck.

The super-large-diameter shield machines are accompanied by instability of the excavation face due to the large area of the excavation face, the large difference in earth chamber pressure distribution and the multiple and complex cutting soil layers. In order to mitigate the adverse effects of the nonlinear distribution of earth pressure and the difference in physical and mechanical properties among soil layers, the earth chamber is provided with a mixing device.

2.1.2.4 *Rapid and continuous mucking system*

An earth pressure balanced shield machine is equipped with a screw conveyor for mucking, which is composed of a telescopic cylinder, mucking cylinder, hydraulic motor, screw shaft, mucking gate, etc., and functions mainly to continuously discharge soil in the earth chamber; prevent loss of water in the soil mass as a seal plug and keep stable earth pressure in the earth chamber; adjust the mucking rate, control the dynamic balance of earth pressure in the earth chamber and ensure the continuous and normal advance of the shield machine. The muck at the screw conveyor outlet is conveyed by a belt conveyor to a muck car parked on the rail.

A rapid and continuous mucking system was developed by engineers, as shown in Fig. 2.7, to help with the mucking of the super-large-diameter earth pressure balanced shield machines working in densely populated urban areas. The working principle of this system is as follows: erect a

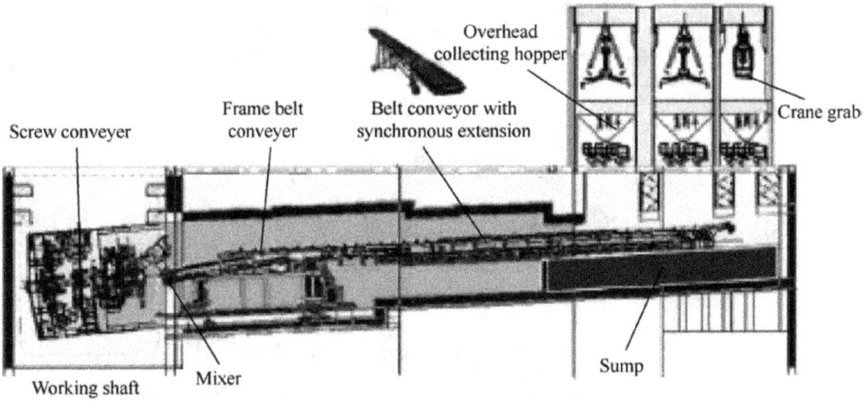

Fig. 2.7. Rapid and continuous mucking system.

horizontal belt conveyor with synchronous extension in the tunnel, with the connecting bolts of the circular tunnel segments, to horizontally transfer muck cut by the shield machine to the sump of the working shaft; use a vertical grab bucket to vertically transfer muck in the sump to the overhead collecting hopper near the ground in the working shaft; evenly mix muck in the overhead collecting hopper using a horizontal screw mixer and discharge the mixed muck into a muck car under the gravity by opening the valve at the bottom discharge port and controlling the mucking rate; after the muck car is fully loaded, close the valve at the discharge port, and move the fully loaded muck car before the next muck car is placed ready for loading.

2.1.2.5 *Shield tail seal*

The shield tail seal is installed on the inner side of the shield shell tail and between the shield tail and segments. A wire brush sealing device is adopted as the shield tail water stop, integrating spring steel, wire brush and stainless steel metal mesh into one structure. The shield tail grease pump feeds grease between wire brush seals to improve the water stop performance. The shield tail sealing system in the form of three wire brushes plus one steel plate brush is generally adopted for the super-diameter shield machines, as shown in Fig. 2.8.

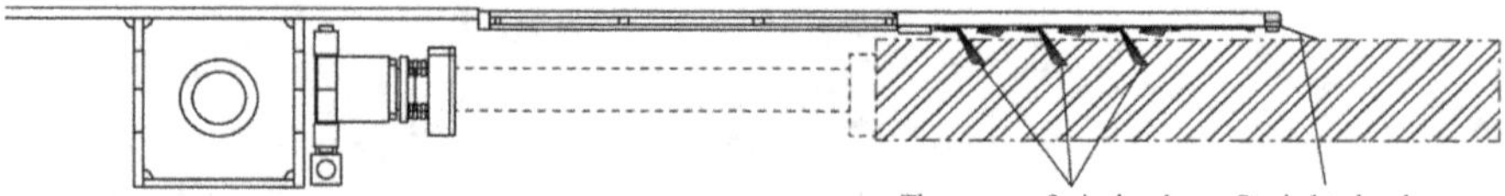

Fig. 2.8. Shield tail sealing system.

2.1.3 Characteristics of slurry balanced shield machine

2.1.3.1 *Fundamentals*

A slurry balanced shield machine is a mechanical shield machine with a partition behind the cutterhead to form a slurry pressure chamber between the cutterhead and the partition and to deliver pressurized slurry to the slurry tank. When the slurry tank is filled with slurry at a certain pressure and a layer of slurry membrane is formed on the excavation face, the stability of the excavation face is ensured through pressurization and a pressure maintaining device. While the slurry membrane is formed, the cutterhead continuously rotates to cut the soil containing the slurry membrane, which is mixed with the slurry in the slurry tank using a mixing device into high-concentration slurry, and then the mixture is transported to the ground through a slurry pump and slurry pipeline. This type of shield machine is mainly composed of a shield shell, cutterhead, sealed slurry tank, shield machine jack, electrical and operation control system, grouting system, segment erector, shield tail sealing device, slurry circuit, etc.

At present, the slurry balanced shield machines adopt one of two systems, namely the German system and the Japanese system. The main difference between these two systems is that the slurry tank of the German system is equipped with a pressure chamber, while the slurry tank of the Japanese system only contains slurry. The super-large-diameter slurry balanced shield machines used in China are mostly the German system, whose shield machine structure shown in Fig. 2.9.

2.1.3.2 *Cutterhead and cutter configuration*

Slurry balanced shield machines are generally designed with panel-type cutterheads, and the principles and considerations of cutterhead support, cutter type selection and layout are similar to those of earth pressure balanced shield machines. River crossings and crossing tunnels are generally

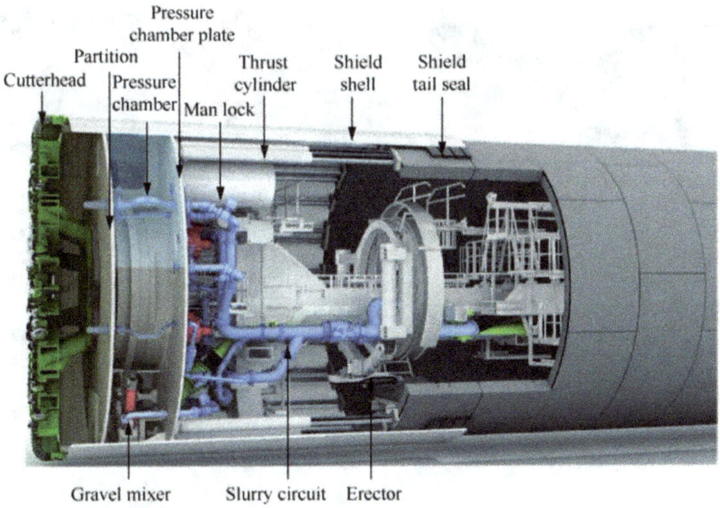

Fig. 2.9. Structure of slurry balanced shield machine.

built using slurry balanced shield machines in China. Because it is difficult to open the chamber for cutter replacement after stratum reinforcement underwater, the risk of traditional cutter replacement in a pressurized chamber is huge. For safety reasons, it is preferred to replace cutters in a closed state without contacting slurry and under atmospheric pressure. Therefore, super-large-diameter slurry balanced shield machines are usually equipped with cutters that are replaceable under atmospheric pressure.

The cutter replacement process is illustrated in Fig. 2.10. The staff enters the cutterhead spider space from the center of the main drive for cutter replacement. The procedure is as follows: connect the lifting frame (with bolts) to the cutter's fixing plate, unfasten the bolts on the fixing plate, and then operate the lifting frame (with bolts) to lower the cutter and close the pressure isolation door; remove the worn cutter, and lift the new cutter behind the pressure isolation door to the installation position; and finally, open the door to install the cutter. After the cutter replacement is finished, the frame is moved to the position of the next cutter.

2.1.3.3 *Slurry treatment*

The slurry treatment system is an indispensable supporting system for tunneling by slurry balanced shield machines. The process is to pump

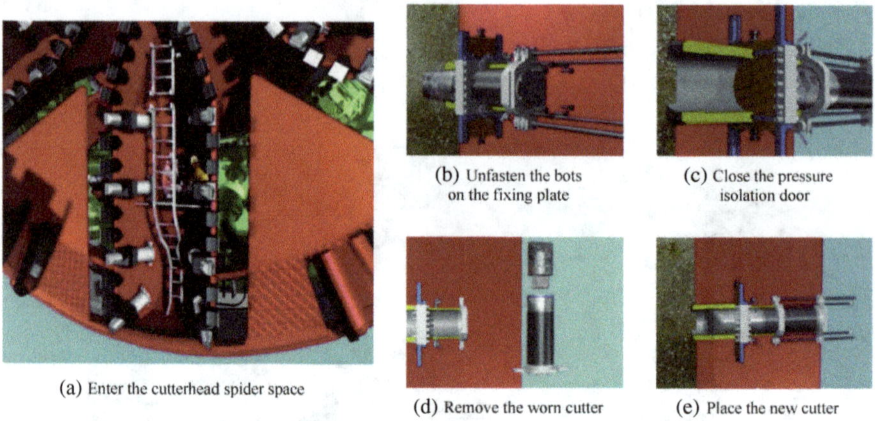

Fig. 2.10. Schematic diagram of cutter replacement process.

slurry up to the requirements of the indicators to the excavation face, mix the slurry with the soil cut by the shield machine and then transport it to the ground for separation and adjustment. The treated slurry is recycled and then pumped into the excavation face of the shield machine to repeat the cycle. With the rising requirements for environmental protection and the increasingly strict management of slurry discharge in cities, the traditional practice of direct slurry discharge is prohibited, so it is necessary to introduce centrifuges or filter presses to achieve dry mucking.

When the slurry pressure is greater than the groundwater pressure, the slurry on the excavation face theoretically penetrates into the soil according to Darcy's law, forming suspended particles in a certain proportion to soil voids. Under the effects of "blocking" and "bridging", suspended particles are captured and accumulate on the contact surface between soil and slurry to form a slurry membrane. In the theory of slurry balance, the formation of a slurry membrane is very important.

The slurry system is usually composed of two parts: a slurry conveying system and a slurry treatment system.

Among them, the slurry conveying system is usually composed of a slurry feeding pump, a sludge discharge pump, pipeline, slurry control valves, etc. The slurry treated by the slurry treatment system up to the requirements in varied indicators is temporarily stored in the adjustment tank and is pumped to the shield machine excavation face by the slurry delivery pump set on the ground in the transport system. The slurry balanced shield machine discharges the slurry mixed with muck, which is transported to the ground treatment system by the slurry discharge pump installed on the shield frame and the relay pump in the tunnel.

The function of the slurry treatment system is to remove the large-size particles and retain the fine clay particles from the slurry discharged by the slurry balanced shield machine through separation treatment. After meeting the requirements of the construction indicators, the slurry is delivered to the shield machine working face by the slurry feeding pump on the ground so as to achieve the purpose of slurry circulation. If the separated slurry does not meet the pumping requirements, it is treated until it meets the construction indicators to realize slurry circulation.

2.2 Basis and principles of type selection

2.2.1 Basis for type selection of shield machines

The type selection of shield machines is mainly based on the engineering geological conditions and hydrogeological conditions, taking into account the tunnel section size, tunneling length, burial depth, curve radius, gradient, surrounding buildings (structures) and surrounding environmental conditions. The main basis for type selection is composed of the following:

(1) Hydrogeological conditions: rock and soil stratification, characteristic indicator parameters and physical and mechanical properties of each layer, groundwater level, underground river and creek, etc.
(2) Environmental conditions along the route: site conditions, landform, ground and underground pipelines, buildings (structures) and their structural characteristics.
(3) Tunnel design parameters: tunnel length, burial depth, section shape and size, tunnel internal structure type, design route, alignment, gradient, etc.
(4) Construction conditions: site scale, construction water and power supplies, mucking, construction period, etc.
(5) Suitable auxiliary construction methods: precipitation method, air pressure method, freezing method, drilling and blasting method, grouting method, etc.
(6) Technical and economic analysis.

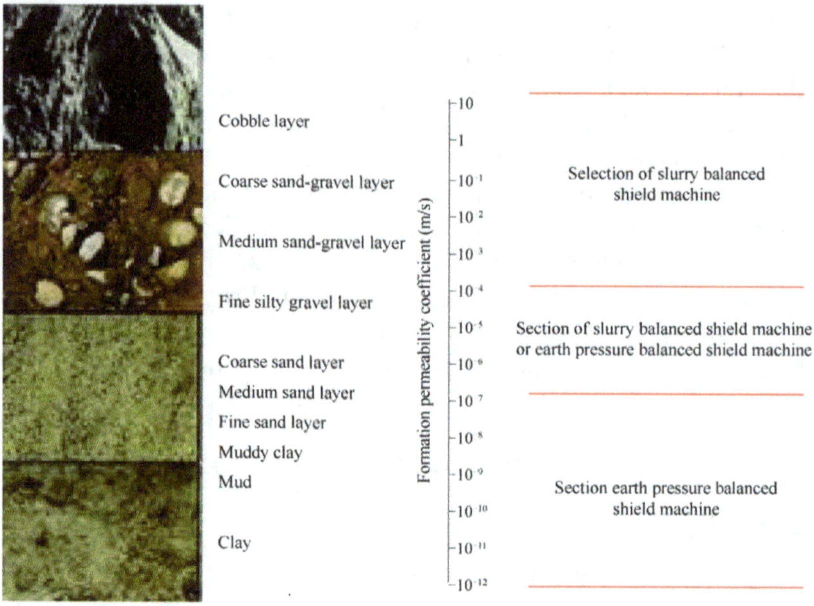

Fig. 2.11. Relationship between formation permeability coefficient and type selection of shield machines.

Permeability coefficient, rock and soil particle size distribution, and underground water pressure in engineering geological and hydrogeological conditions are all important factors in the type selection of shield machines. The relationship between the formation permeability coefficient and the type selection is shown in Fig. 2.11. According to the construction experience gathered around the world, when the formation permeability coefficient is less than 10^{-7} m/s, the earth pressure balanced shield machine should be selected; when the formation permeability coefficient is between 10^{-7}–10^{-4} m/s, both the earth pressure balanced shield machine and the slurry balanced shield machine are suitable; when the formation permeability coefficient is greater than 10^{-4} m/s, the slurry balanced shield machine should be selected. See Fig. 2.12 for the relationship between the shield machine types and the stratum particle size grading. The earth pressure balanced shield machine is applicable to clay, muddy soil and improved coarse sand and fine sand; the slurry balanced shield machine is applicable to gravel-coarse sand, coarse sand and fine sand. Generally, with a large amount of fine particles, muck is prone to impermeability in the flow-plastic state, enabling full filling in the earth

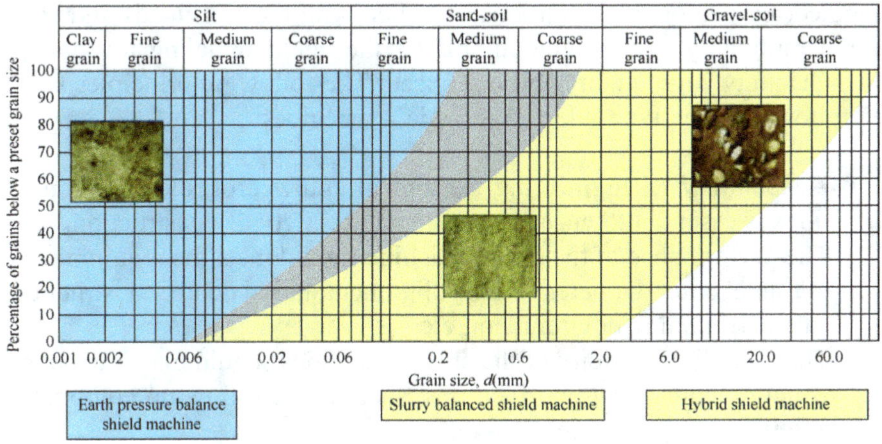

Fig. 2.12. Relationship between shield machine type and stratum particle size grading.

chamber and pressure building up in the earth chamber to balance the soil pressure on the excavation face. When the total amount of silt and clay particles in the rock and soil reaches more than 40%, the earth pressure balanced shield machine is usually selected; otherwise, the slurry balanced shield machine is more reasonable. The absolute size of silt particles is usually limited to 0.075 mm. When the stratum is rich in water, according to the construction experience, the earth pressure balanced shield machine has poor adaptability to the stratum with high water pressure (above 0.5 MPa) since the screw conveyor can hardly form an effective soil plug effect, and the muck gushing occurs at the muck discharge gate at a high probability, causing a pressure drop in the earth chamber and subsequent collapse of the excavation face. Generally, when water pressure is higher than 0.5 MPa, the slurry balanced shield machine should be used. If the soil pressure balanced shield machine is to be used, a special study should be carried out.

There are many constraints on the type selection of shield machines, and the corresponding supporting construction methods should be considered when necessary.

2.2.2 Principles for type selection of shield machines

The correct type selection is key to the success of tunneling using shield machines. The type selection must be closely combined with the engineering geological conditions, matched with the workloads and economic rationality, so as to give full play to the shield tunneling advantages of high safety, high quality and fast speed. The following principles shall be followed in the type selection of shield machines:

(1) Adaptability to engineering geological and hydrogeological conditions is essential to ensuring the absolute safety of construction.
(2) The performance of the shield machine should meet the requirements of tunnel outer diameter, length, alignment, burial depth, construction site, surrounding environment, etc.
(3) The tunneling capacity of the shield machines should match the construction equipment and facilities, such as the follow-up equipment and the launching base.
(4) Performance, reputation and technical services of shield machine manufacturers should be considered.

The most important principle is to ensure the stability of the cutting face and construction safety. Based on a study and analysis of the machine types and their main technical parameters, and on the premise of safety and reliability, technical advancement and economic rationality are considered to ensure the selection of the best shield construction method and the most suitable shield machine. In order to ensure the smooth completion of tunnels, the type selection of shield machines should be conducted carefully. Generally, the main process of type selection is as follows:

(1) Conduct preliminary type selection according to design section parameters.
(2) Select the supporting systems according to the lining type.
(3) Check the stability of the excavation face.
(4) The following conditions should be analyzed comprehensively: (i) outer diameter of shield machine; (ii) overburden thickness and groundwater level; (iii) soil conditions, including grain size, stratification conditions and soil parameters; (iv) construction period, cost, environmental factors, design route, water supply and drainage, ventilation, power source and other conditions.
(5) Check the types and applicability of auxiliary construction methods, including the air pressure method, precipitation method, chemical grouting method and freezing method.

(6) Conduct a comparative analysis of the combinations of machine type and auxiliary methods.
(7) Make the final selection of the tunnel boring machine.

2.3 General layout and coordination

2.3.1 Shield machine frame layout under efficient construction conditions

2.3.1.1 *Typical shield machine frame layout*

The frame is an important part of large shield machines. It carries the power supply and distribution system, hydraulic power system, grease lubrication system, grouting system, prefabricated members, synchronous grout and other materials transport, pipeline extension, etc. Moreover, it also accommodates space for construction measurements and ventilation. The reasonable layout of the frame in the limited space is an effective guarantee for efficient construction.

At present, the 15 m diameter class shield machines are mostly circular. During normal advance, the frame is in the circular segment linings. In order to ensure construction safety, a certain space allowance is required between the outer contour of the frame and the segment linings, usually not less than 30 cm. The circular contour formed by the frame layout due to such a restriction is generally called the equipment circle. The equipment circle is concentric with the segment lining circle. The frame and all the installed components must be within the equipment circle (except for the hull in direct contact with segments), as shown in Fig. 2.13. The shaded part shows the safety space.

The frame-traveling mechanism has many forms, such as the hull type and wheel type.

In order to make rational use of the limited space and improve construction efficiency, the frame layout must comprehensively consider all systems and adopt a layered layout. In order to facilitate installation and disassembly, the frame is usually divided into a moderate number of sections, say, 3–5, since too many sections result in complex installation, uncompact structure and redundant systems; a too large length of single sections due to a reduced number of sections brings inconvenience to construction on small-radius curves.

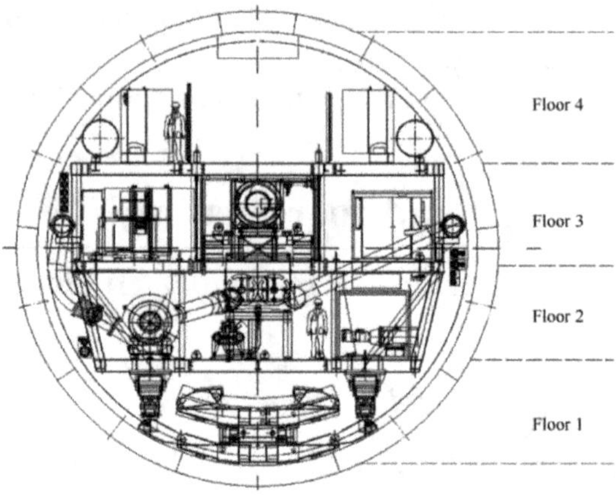

Fig. 2.13. Shield machine equipment circle.

Taking as example the 15.43 m diameter slurry balanced shield machine manufactured by Herrenknecht Company for the Shanghai Riverside Passageway, it is divided into three frame sections, of which the No. 1 frame is located behind the main shield and is mainly divided into four layers. The bottom floor is the frame walking wheel, the hull and the segment feeder; the second floor is equipped with a slurry pump, grout bucket, grouting pump, main hydraulic oil tank and power station; the third floor is equipped with a grease system, transformer, power distribution cabinet and control room; the fourth floor accommodates the cutterhead drive control cabinet, cooling system, air duct and measuring channel.

The No. 2 frame is a connecting beam between the No. 1 frame and the No. 3 frame (Fig. 2.14) and is equipped with a mounting crane for hollow rectangle components, a segment transport crane and synchronous construction.

The No. 3 frame moves on the auxiliary hulls on both sides of the hollow rectangle component. With the advance of the shield machine, the rear hull moves to the front one by one for cycling. The No. 3 frame is mainly equipped with hose reels, cable reels, etc., for slurry pipeline connectors, connecting cranes, industrial water supply, industrial gas supply, waste water pipeline, etc., and emergency generators, mainly enabling functions of extending various pipelines, placing various cables, etc.

In the Nos. 2 and 3 frames, ordinary trucks, double-headed vehicles and other engineering vehicles can drive on the hollow rectangle

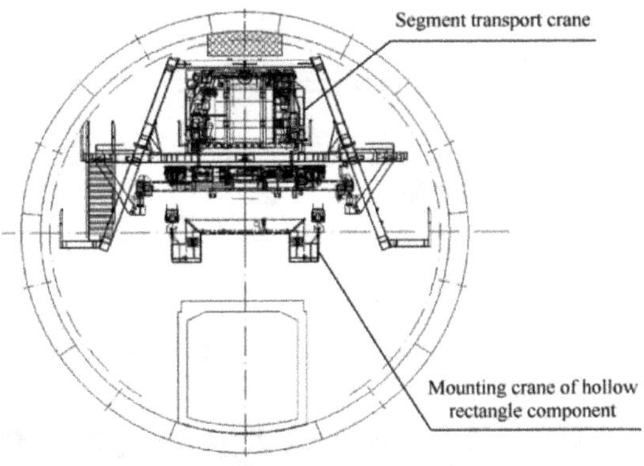

Fig. 2.14. Section of No. 2.

components to transport segments and other construction materials to the segment hoisting position ahead. The segment transport crane can transport the segments to the rotary platform behind the segment feeder or transport the mobile grout bucket to the front of the No. 1 frame.

As shield tunneling is a cyclic work process and all parts of the process are required to be closely connected, the transport vehicles must be dispatched and operated precisely according to the preset space and order, and unique special transport vehicles are designed to continuously transport construction materials, such as segments, hollow rectangle components and grease. All these tasks are completed by various systems on the frame, so a reasonable frame layout is key to guaranteeing efficient construction.

2.3.1.2 *Material transport of frame section*

In the above typical frame system, material transport efficiency is one of the prerequisites for efficient construction. Super-large shield machines generally support synchronous construction. General transport vehicles can directly drive into the frame through hollow rectangle components to achieve rapid transport (Fig. 2.15). Except for special double-headed vehicles, general engineering vehicles face the problem of turning around in the frame. In order to achieve the goal of rapid and safe construction, a turnaround platform is added at the tail of the No. 3 frame, as shown in Fig. 2.16.

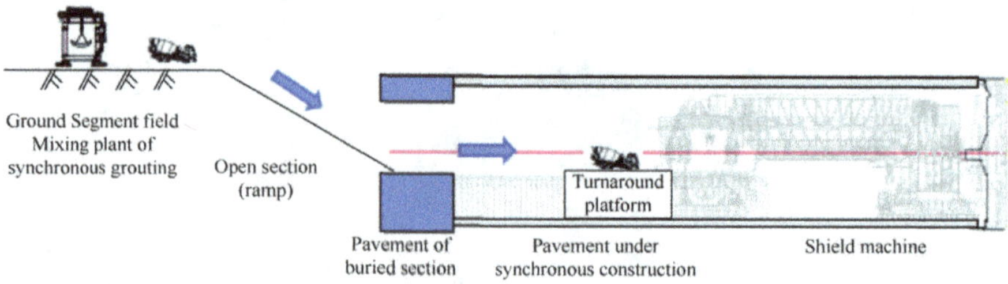

Fig. 2.15. Schematic diagram of longitudinal transport route of construction materials.

Fig. 2.16. Turnaround platform.

The turnaround platform is towed at the rear of the No. 3 frame and is equipped with a hull traveling mechanism, hull hoist, segment traveling block wheel, etc., to realize the turnaround operation of segment transport trucks, synchronous grout transport vehicles and other engineering vehicles, greatly facilitating the operation of transport vehicles and improving construction efficiency.

In order to reduce the number of vehicles entering the frame, a slurry relay pump (Fig. 2.17) is installed in the front of one side of the turnaround platform. Mixer trucks can directly drive to the turnaround platform to deliver synchronous slurry to the grout relay pump, which directly conveys grout into the fixed grout bucket of No. 1 frame through the pipeline. This not only reduces the working pressure of the segment crane but also enables the segment and synchronous grout to be transported separately at the same time, thus greatly improving construction efficiency.

Fig. 2.17. Synchronous grout relay pump.

2.3.2 Coordination between the shield launching state and the structure of the buried section

Before the shield launching, the main systems of the main shield and frame should be installed and debugged. The main shield is directly hoisted in the launching working shaft, while the frame is mainly installed in the buried structure section adjacent to the working shaft. Before the installation of the frame, the structure of the buried section should be completed. In order to ensure the smooth installation of the frame, the internal space size of the buried section must be compatible with the size of the frame; that is, the internal width and height of the buried section must be greater than the diameter of the frame equipment circle, with a certain margin. Moreover, the horizontal alignment of the tunnel exit section should be considered. If the tunnel exit section is curved, its curve radius, length of frame sections and the relative relationship with the internal width of the buried section should be comprehensively considered to ensure the smooth construction of the shield machine. In addition, the bottom slab of the buried section is sloped to be consistent with the design gradient of the tunnel axis.

Super-large shield machines are generally equipped with large-sized frames. The roof of the buried section is usually designed with two lifting holes with sizes consistent with the lifting size of the frame. The front and rear ends lifting holes correspond to the frames at the front and rear ends to facilitate the installation of the frame, as shown in Fig. 2.18. Taking the shield machine hoisting construction of three carriages as an example, the

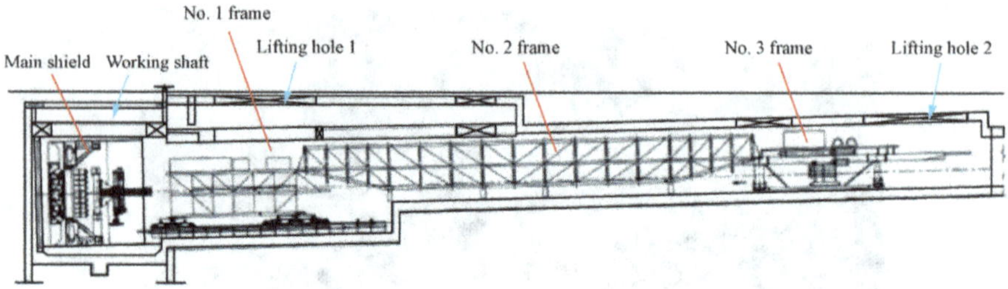

Fig. 2.18. Schematic diagram of launching state of 3-section frame.

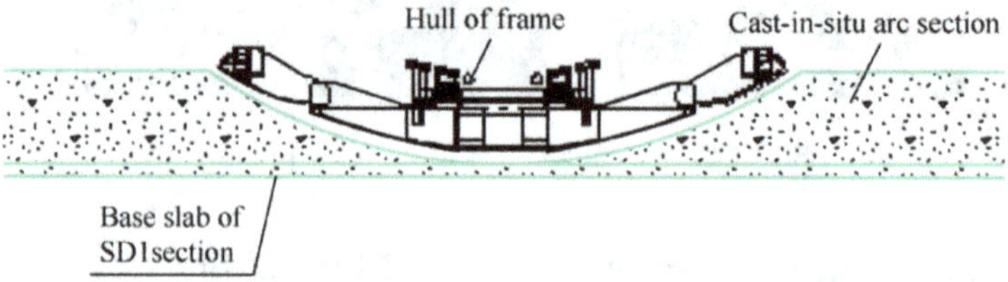

Fig. 2.19. Pouring of arch section of hull.

front hoisting hole is mainly used to hoist the No. 1 frame, and the rear one is used to hoist the Nos. 2 and 3 frames. After lifting with the rear hoisting hole, the No. 2 frame is placed on the steel structure and moved forward on the temporary rail for connection with the No. 1 frame. The No. 3 frame, installed with outriggers, is hoisted onto the temporary rail and moved forward to connect with the No. 2 frame.

The No. 1 frame is directly hoisted onto the hull. In order to place the hull, an arc section with a radius compatible with the inner diameter of the tunnel segment is poured at the corresponding position at the front end of the buried section. Moreover, it should be noted that the elevation and gradient should be consistent with the tunnel axis design (Fig. 2.19).

2.4 System functions and configuration

2.4.1 Main shield

A shield machine is generally composed of a main shield and a frame system. The main shield generally consists of a cutterhead, shield shell, main drive, thrust jack, erector, shield tail system, etc., which reflects the comprehensive capability of the shield machine, including the excavation diameter and the adaptability to geological conditions.

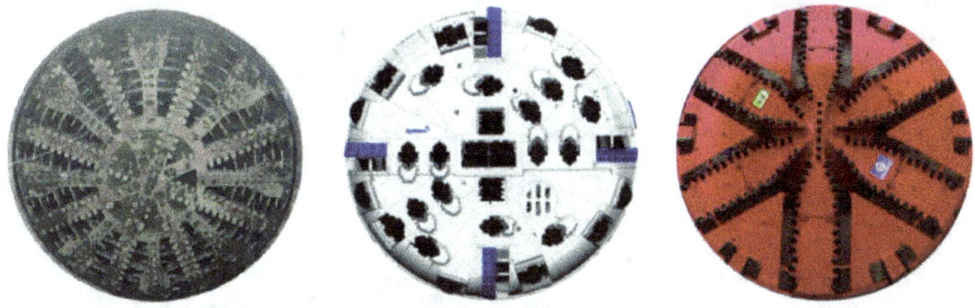

(a) Cutterhead for composite stratum (b) Cutterhead for hard rock stratum (c) Cutterhead for soft soil stratum

Fig. 2.20. Forms of cutterhead.

The cutterhead is specially designed for each tunnel. According to the type of shield machine selected, such as the earth pressure balanced shield machine or the slurry balanced shield machine, the geological conditions of the project site, such as soft soil, composite stratum, rock stratum, the size, structural form and opening ratio, of the cutterhead are designed, and the cutters are reasonably configured (Fig. 2.20).

The cutterhead structure should be sufficient in rigidity and strength to ensure no deformation or excessive wear during tunneling under adverse geological conditions. Some protective measures can be taken in the design to mitigate cutterhead wear, such as protective cutters in different numbers for the peripheral cutters and the outer edge of the cutterhead; a wear-resistant layer set on the surface and back of the cutterhead where fluidity is high, and along the circumference of the outer edge, and a corresponding wear detection device.

The structure and layout of the cutterhead, the design of the cutterhead opening, as well as the layout of cutters should fully consider the fluidity of muck to reduce the wear of the cutterhead (Fig. 2.21).

Different types of cutters can be installed on the cutterhead to adapt to excavation in different strata. The main cutters are classified according to their structure, including the common types of hobs, toothed cutters, parting cutters, scrapers, profiling cutters and center cutters (Fig. 2.22).

Among them, the advance cutter is mainly used to loosen sandy gravel, clay and other strata. It has a high cutting efficiency and low wear and plays a role in protecting the cutter and cutterhead.

The hobs are used to squeeze and break rock strata. In the composite formation with hard rock, the double-edged disc hob can be configured to break hard rock.

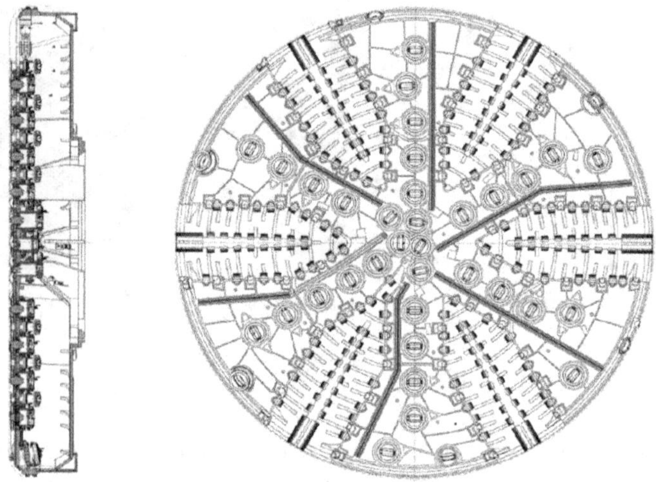

Fig. 2.21. Schematic diagram of cutterhead structure.

The parting cutter is mainly used to cut loosened soil and peel off soil on the working face. It is equipped with hard alloy blades with a high wear resistance, and the cutter body is protected by a wear-resistant welding layer.

The peripheral scraper is installed at the outer edge of the cutterhead opening to ensure the excavation diameter and indirectly prevent wear around the cutterhead body. The curve design of the peripheral scraper effectively prevents the collapse of the tunnel roof. The peripheral scraper is equipped with a double-layer carbide tooth edge to enhance the wear resistance of the cutter during tunneling.

Moreover, in order to deal with cutter wear, spare cutters are included in the general cutter configuration, which can be divided into replaceable tools under pressure and replaceable tools under atmospheric pressure according to different replacement environments. During the construction of cutter replacement under pressure, the staff should enter the front of the cutterhead for operation under pressure, which is risky and difficult to operate; the cutter replacement under atmospheric pressure is conducted in the cutter arm, which is easy to operate and has little risk. However, cutter replacement under atmospheric pressure requires a large installation space, which needs to be considered in cutterhead design.

Among the super-large-diameter shield tunnels in China, most were soft soil tunnels before 2011, and they were built using shield machines with cutter replacement under atmospheric pressure, even the hybrid

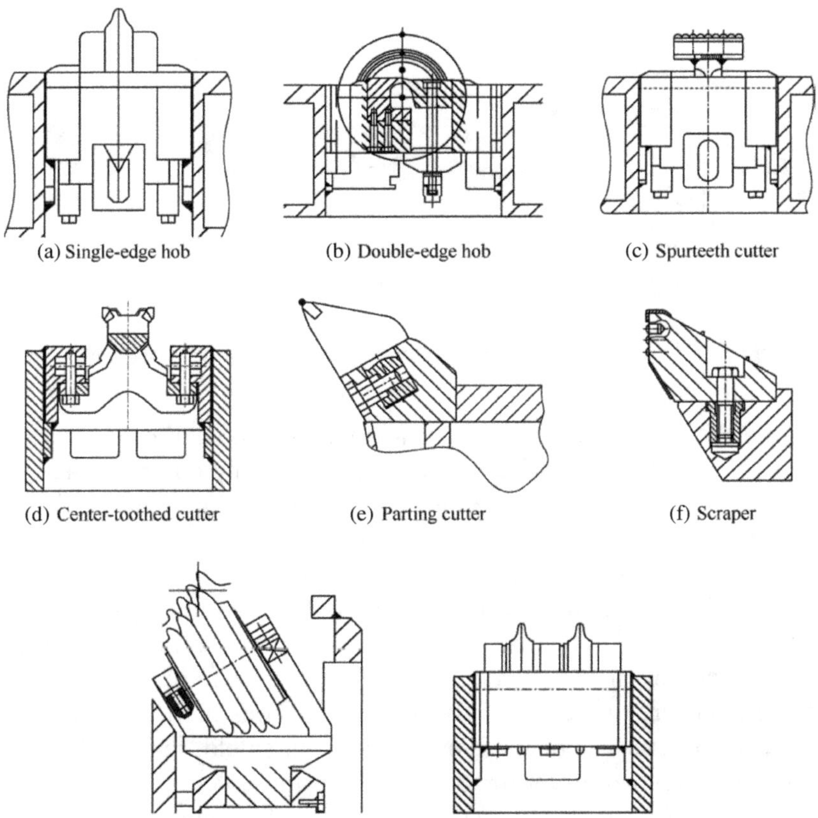

(a) Single-edge hob (b) Double-edge hob (c) Spurteeth cutter

(d) Center-toothed cutter (e) Parting cutter (f) Scraper

(g) Profiling cutter (h) Double-edge center cutter

Fig. 2.22. Schematic diagram of various cutters.

shield machine used for the Nanjing Weiqi Road Tunnel built in 2007. After 2011, super-large-diameter shield tunnels with a diameter of more than 11 m, especially over 14 m, were popularized in other regions across the country, from the Yangtze River Delta region to the whole country, and the geological conditions became more and more complex. Due to the rapid wear of cutters used in composite formation and frequent cutter replacement, in order to improve the efficiency and safety of cutter replacement, large-diameter hybrid shield machines' cutterheads are often equipped with cutter replacement under atmospheric pressure.

The cutterhead drive system of shield machines usually has two modes: hydraulic drive and electric drive. The ultra-large-diameter shield machines generally adopt the frequency conversion motor drive mode,

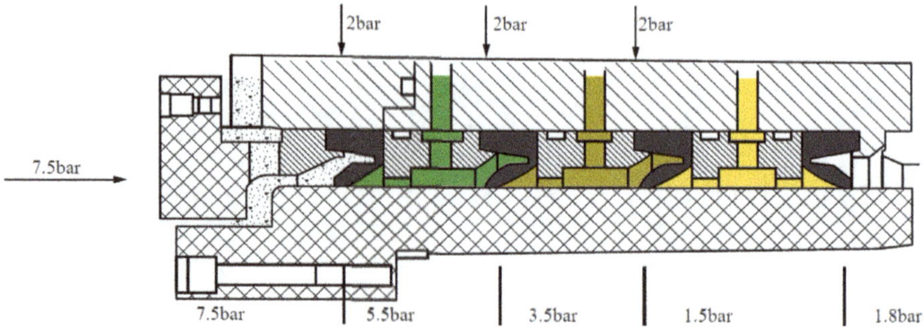

Fig. 2.23. Schematic diagram of working principle for main drive seal.

including the main bearing, large gear ring, pinion, gearbox, safety coupling, drive motor, main drive lip seal and other main components.

The cutterhead drive device must be designed with greater rigidity to optimize the service life of the main bearing and provide stable speed and sufficient torque for the cutterhead excavation.

The main drive seal is generally composed of multiple lip seals with high wear resistance. The cavity formed by each seal is filled with sealing grease, which, together with the lip seal, can resist external water and soil pressure to ensure the safety of shield machine construction (Fig. 2.23).

The shield consists of three parts: the front shield, the middle shield and the tail shield. In order to adapt to tunneling on curves, the shield is generally designed as an inverted cone; that is, the diameter of the shield tail is smaller than that of the middle shield, and the diameter of the middle shield is smaller than that of the front shield. The steel structure of the shield shell is designed to withstand specific earth pressure, hydrostatic pressure and dynamic loads. High-strength steel Q345B is generally used to provide sufficient rigidity and wear resistance.

The shield tail generally adopts a special "sandwich" structure composed of welded support bars sandwiched in two layers of steel plates to enhance the structural stiffness and withstand the working pressure.

Taking the shield machine with a diameter of 15.43 m used in the Shanghai Yangtze River Tunnel as an example, which was required for a single advance distance of 7.5 km, in order to ensure the safety of shield tail sealing, an emergency sealing method of a reserved freezing pipe plus emergency chemical grouting pipe was adopted for the first time in the design of the shield tail. The shield tail was a four-piece welded structure (for convenient transport), which was connected with the shield body

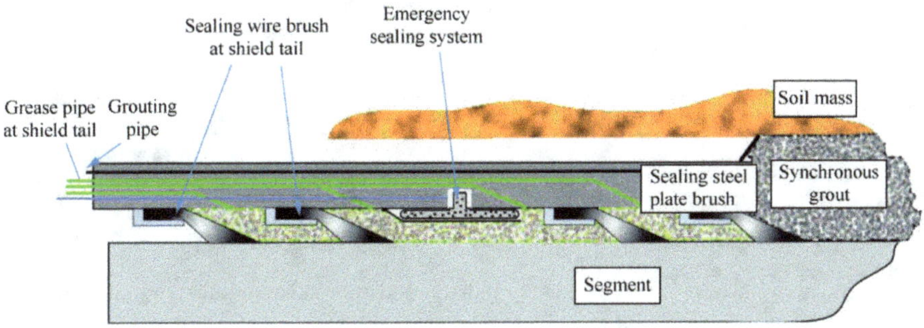

Fig. 2.24. Schematic diagram of shield tail structure.

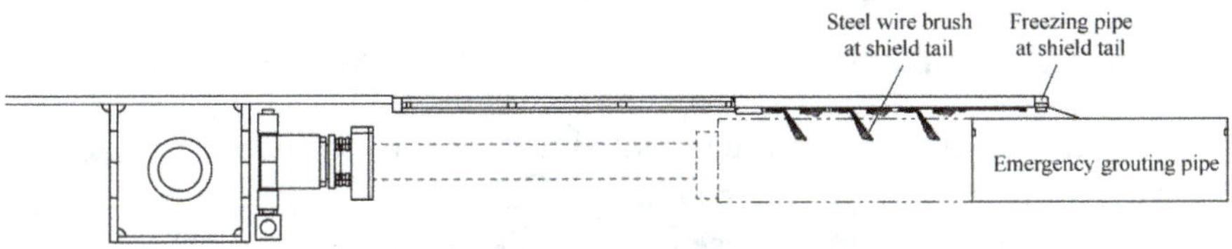

Fig. 2.25. Emergency grouting pipe.

by welding. The shield tail seal consists of three sealing wire brushes, one sealing steel plate brush and one emergency seal. The innermost two seal brushes were replaceable (Fig. 2.24).

The emergency seal was the inflatable expansion type. When compressed air is injected (10 bar), the airbag expands to play a sealing function. The function of emergency sealing is firstly to stop water in case of a large amount of water leakage at the shield tail, and secondly to protect the segment installation area and prevent water from flowing into the shield machine when replacing the first two wire brushes.

The reserved freezing pipe is a ring pipe along the circumference of the shield machine at the rearmost end of the shield tail. Before replacing the first two wire brushes, the soil and water in this part are frozen by injecting the freezing medium to ensure the safety of replacing the wire brushes. The chemical grout may also be injected via the emergency grouting pipe to stop water and leak in time (Fig. 2.25).

The grease injection system at the shield tail is an important component of the shield tail seal. According to the shield machine diameter and the shield tail seal form, it generally includes the front, middle and rear three grease pipelines, each of which is symmetrically distributed at multiple points of the shield tail. Each grease pipeline branch is pressurized

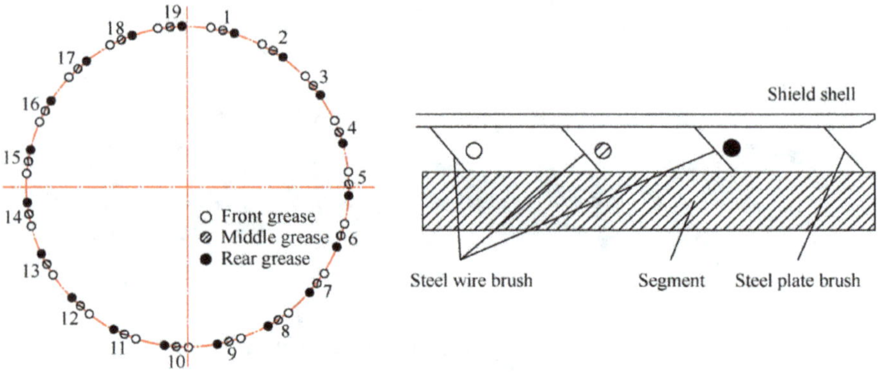

(a) Distribution of grease pipes at shield tail (b) Positions of front, middle and rear grease pipes

Fig. 2.26. Schematic diagram of grease injection system at shield tail.

by a shield tail grease pump connected to a header for conveying to the corresponding position of the shield tail seal.

The shield tail sealing grease plays a role in rust prevention, corrosion prevention and wear reduction on the wire brush and steel structure, and its lubrication and sealing effect can effectively protect the multiple wire brushes and spring steel sheets at the shield tail and together isolate the soil sediment and the grout backflow. The pressure injection of shield tail grease is generally controlled in both pressure and consumption, and too high a pressure should be avoided since it is required to match the advance construction. The synchronous pressure injection of shield tail grease is operated by the Programmable Logic Controller (PLC) automatic control system (Fig. 2.26).

The shield machine thrust system is composed of a motor, hydraulic pump, hydraulic jack, hydraulic oil pipe, filter element, etc. There are 57 hydraulic jacks in total, every three of which form a unit, numbered 1–19 clockwise in the shield advance direction. In the erection mode, Nos. 1–19 can be independently controlled for telescopic motion, which is convenient for segment erection. In the advance mode, Nos. 1–19 jacks are divided into six zones to facilitate the control of oil pressure in each zone to ensure the advance attitude (Fig. 2.27). In each zone, two jacks are equipped with built-in sensors (one for use and one for standby) to measure the travel of the jacks in real time. The jack power is mainly sourced by a 200 kW motor-driven duplex pump at a working pressure of up to 350 bar.

The slurry circulating system is a unique key system of the slurry balanced shield machine (Fig. 2.28). The slurry system is composed of

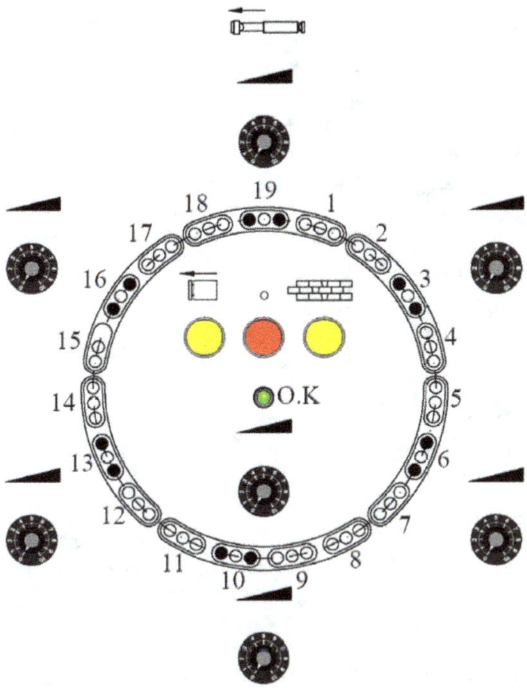

Fig. 2.27. Control panel of shield machine thrust jack.

various pipelines and slurry pumps, usually with three working modes of bypass, advance and backwash. Through slurry circulation, the adjusted fresh slurry is fed to the excavation face, mixed with the excavated soil and then transported to the ground slurry treatment site for further filtration and separation treatment before the final outbound transport of muck.

The air balance system is composed of a pressure measuring transformer, regulating station and control station, air supply and exhaust control valve and pipeline (Fig. 2.29). In order to ensure construction safety, there are generally two identical systems, subject to parallel switching. The pressure in the bubble chamber is adjusted and set by adjusting the air balance system. During the construction, the bubble chamber pressure is kept close to the pre-set value through automatic exhaust and air feed to achieve the purpose of real-time compensation.

Segments are generally erected using a segment erector, which is a kind of automatic manipulator with multi-dimensional movement capability, is controlled by a special remote controller and grabs segments with its vacuum suction cups and other forms. With forward and backward

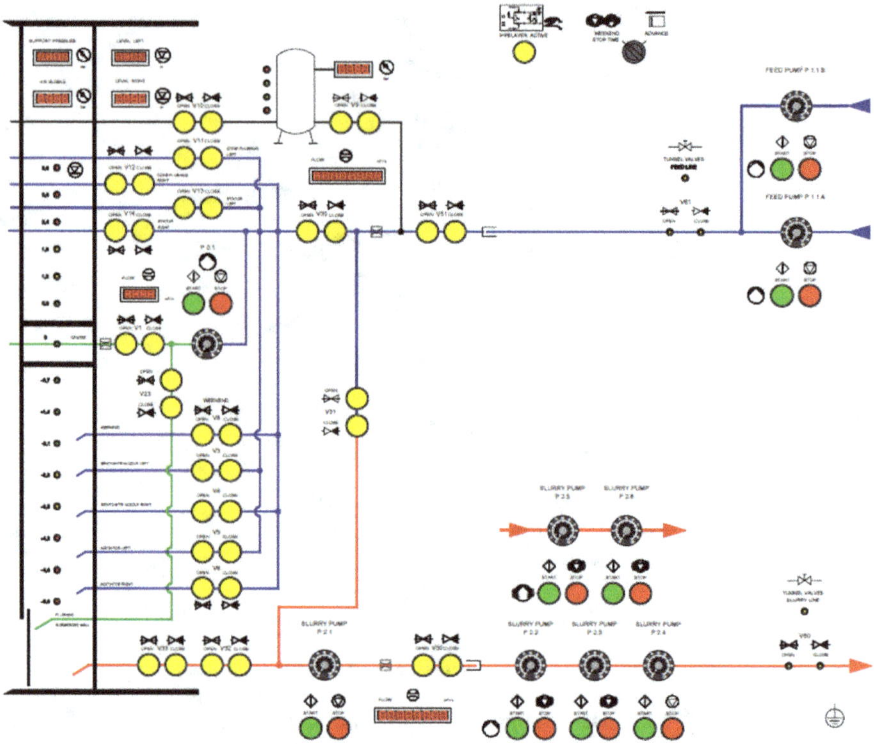

Fig. 2.28. Slurry circulating system.

movements, rotation and other adjustments, it transports segments to the designated positions, and it can make fine adjustments, such as inward and outward turns, to achieve accurate segment erection.

There is an erecting platform on the upper half of the erector to accommodate the construction staff fastening the segment bolts (Fig. 2.30).

During shield tunneling, in order to fill the structural voids generated due to the construction, grouting is carried out simultaneously at the tail of the shield machine. At present, single-liquid slurry is widely used. The special grouting pump is used to inject slurry into the structural space between the segment fed from the shield tail and the soil mass through the synchronous grouting pipeline branches at the shield tail, reaching a filling rate generally controlled at 110–120%.

The grouting pipeline of super-large-diameter shield machines generally branches into six or eight routes, or even more. The grouting pipelines

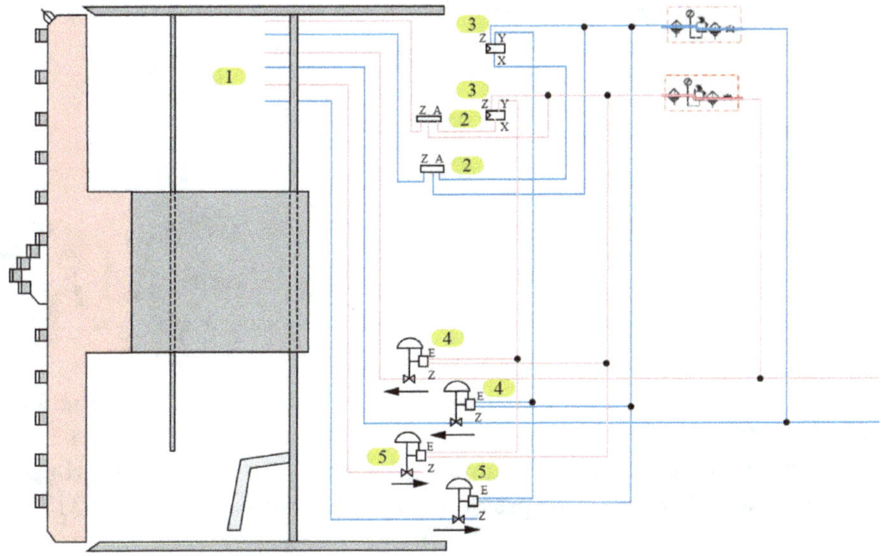

Fig. 2.29. Air balance system.

Note: 1 — Bubble chamber; 2 — Air pressure measuring sensor; 3 — Control regulator; 4, 5 — Valve.

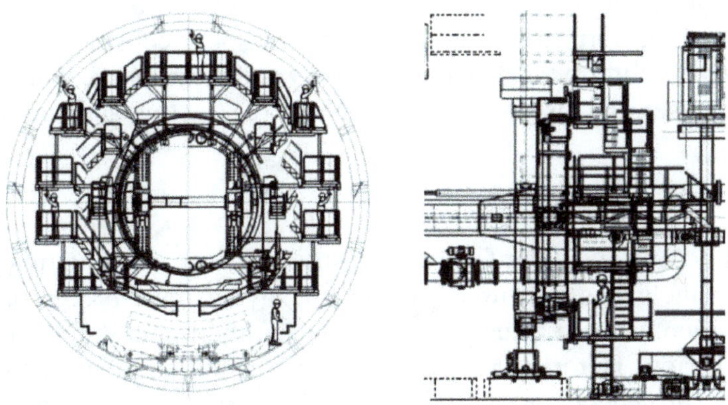

(a) Front view (b) Side view

Fig. 2.30. Structural diagram of erector and platform.

are symmetrically distributed. The grouting system is generally composed of grout barrels, grouting pumps, grouting pipelines and a corresponding control system. The system pressure is collected by the counting sensor and fed back to the construction staff, who controls the grouting pressure

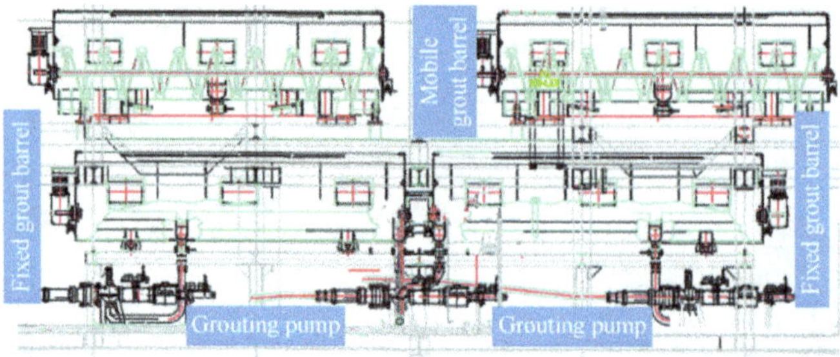

Fig. 2.31. Layout of fixed grout barrel and grouting pump.

and grouting amount with the control system according to the construction requirements. The grouting pressure and grouting amount should be set in real time by the construction technicians according to the data of ground settlement and water and soil pressures during construction (Fig. 2.31).

2.4.2 Rear supporting system of shield machines

The rear supporting system of shield machines plays an auxiliary role in shield construction, including component prefabrication, synchronous slurry mixing, slurry treatment and muck transport, construction material transport, pipeline extension, power supply and distribution, water supply, air supply, sewage discharge, synchronous construction and ventilation.

2.4.2.1.1 *Site layout*

With the development of urban construction, the construction site of shield tunneling is narrowed, so the site layout must be reasonable in combination with the actual construction conditions and in full consideration of various factors to improve the construction efficiency as much as possible. In general, prefabricated members, such as segments used in shield construction, are produced separately at other sites. The temporary stacking of prefabricated members, as well as the layout of slurry treatment, muck transport, substation, synchronous grout mixing plant and other systems, should be considered at the shield construction site. Taking the site layout of super-large-diameter slurry balanced shield machines as an example, in addition to the above supporting systems, a segment stacking yard, air compressor room, clean water tank, various storages, etc., should be considered (Fig. 2.32).

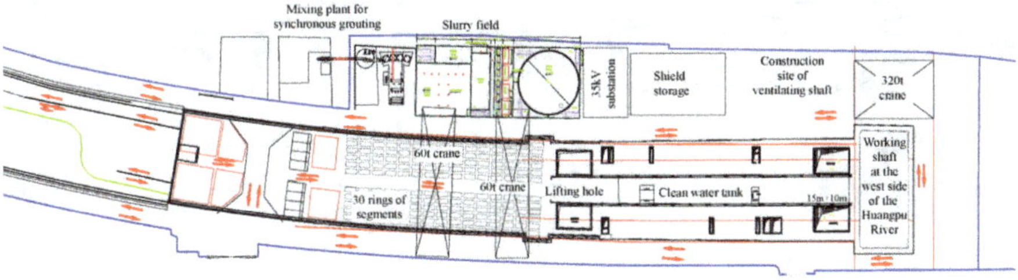

Fig. 2.32. Layout of shield construction site.

The slurry treatment site covering a large area should be designed from a professional view to accommodate numerous equipment and complex systems. Dry mucking should be considered if the conditions are available, or the muck may be transported by water in the form of slurry in combination with the actual construction conditions.

The segment site, clean water tank and air compressor room are arranged by fully utilizing the structure of the buried section to save space. The layout of the segment site should be designed taking full account of the construction progress, waterstop pasting, hoisting and transport, and construction in the winter and rainy seasons.

2.4.2.1.2 *Synchronous structural construction*

The constructions using super-large-diameter shield machines generally adopt the mode of synchronous shield tunneling and the construction of internal road structures.

Hollow rectangle prefabricated members are lifted to the double-headed trucks using a crane from the production site or temporary storage site, then transported to the shield machine frame through the buried section (or the launching working shaft) and the built tunnel pavement, and then lifted to the installation location by the lifting equipment on the frame for installation (Fig. 2.33). After horizontal transport to the working face, the special spreader on the No. 2 frame should be used to turn them over 90° before placement.

In order to ensure a smooth horizontal transport road in the tunnel and rapid tunneling of the shield machine, the synchronous construction of structures should lag behind the frame by 100 m, and the cast-in-place pavement structure construction should be carried out using the typified forms and the quick-release supports (Fig. 2.34).

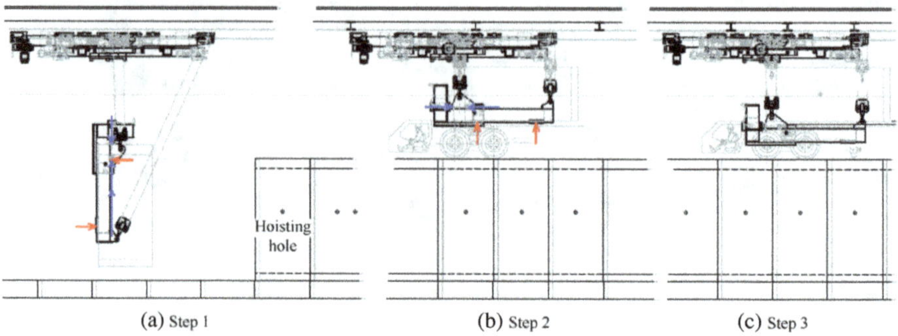

Fig. 2.33. Installation of hollow rectangle prefabricated members.

Fig. 2.34. Synchronous construction of pavement.

On the one hand, the synchronous construction of structures plays a role in anti-floating during the construction, and on the other hand, it serves as a pavement for transport vehicles so as to realize the rapid advancement of the shield machine. With the advance of the shield machine and the frame, the prefabricated pavement members are hoisted closely behind the front frame.

2.4.2.1.3 *Tunnel section layout*

The section layout of the tunnel mainly considers the rational use of space and safety factors. All pipelines from the wellhead to the shield machine and other facilities, such as lighting in the tunnel, are arranged on the retaining side wall of the pavement under synchronous construction,

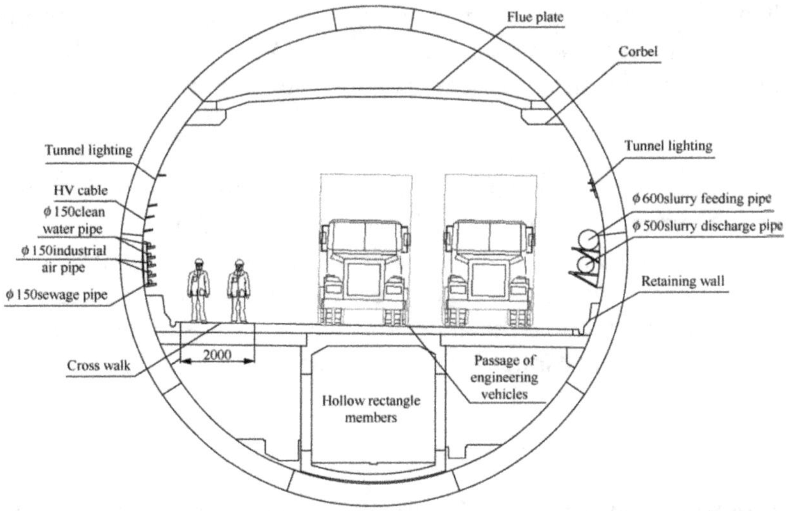

Fig. 2.35. Tunnel section layout (unit: mm).

without affecting to the synchronous road construction in the tunnel (Fig. 2.35).

(1) **Transport passage:** Along with the shield tunneling, the cast-in-situ pavement should be poured in sections and strips, and the passage for construction materials and footway should be divided on the pavement.
(2) **Tunnel lighting:** In general, one lamp holder is arranged every six rings at the upper left and right of the tunnel, and the lighting cable and lamp are fixed on it.
(3) **Piping:** One side of the tunnel is provided with the slurry feeding pipe and discharge pipe for slurry circulation in shield machine advance; the other side of the tunnel is provided with the cooling water pipe and air pipe to supply water and compressed air to the working face and a drainage pipe to discharge sewage from the working face.
(4) **Cabling and communication wiring:** High Voltage (HV) and Low Voltage (LV) cables, control cables, communication cables and other cables are arranged on the left and right sides of the middle and upper parts of the tunnel.
(5) **Pipeline extension:** With the advance of the shield machine, the pipelines erected in the tunnel should be continuously extended,

including slurry feed and discharge pipes, industrial water pipes, industrial air pipes, waste water pipes, HV cables, LV lighting wiring, communication optical cables, and Profibus remote control cables.

For the sake of convenience of transport and installation, the pipeline is generally extended every 10 m. The slurry feed and discharge pipes are equipped with special pipe connectors and mobile hoses, while the water and air pipes are provided with hose reels. Pipelines should be connected once every 10 m of shield machine advance. All kinds of cables are continuously unreeled and spliced when a whole reel is used up.

2.4.2.1.4 *Construction transport*

(1) **Inbound transport from the ground:** Generally, if the open section and buried section structures have been completed during shield machine construction, engineering materials such as tunnel segments, grout for synchronous grouting and prefabricated members. should be directly transported into the tunnel by special vehicles through the open and buried sections after being loaded on the ground.
(2) **Transport in tunnel:** In order to speed up the construction progress, during the synchronous construction of the road structures, the hollow rectangle prefabricated members are erected between the shield machine and the frame along with the shield machine tunneling as the transport passage. To ensure safety, stoppers for truck transport are installed on both sides of the hollow rectangle prefabricated members (Fig. 2.36).

After the prefabricated road structures are transported to the place ahead of the shield machine frame by a double-headed truck, they are lifted by a lifting appliance on the connecting beam between the front and rear frames.

The segments are transported by special trucks through the tunnel pavement to the connecting beam between the front and rear frames, transferred to the segment storage and transport mechanism by a lifting appliance on the connecting beam, and then conveyed to the erection area by a segment storage and transport mechanism (Fig. 2.37).

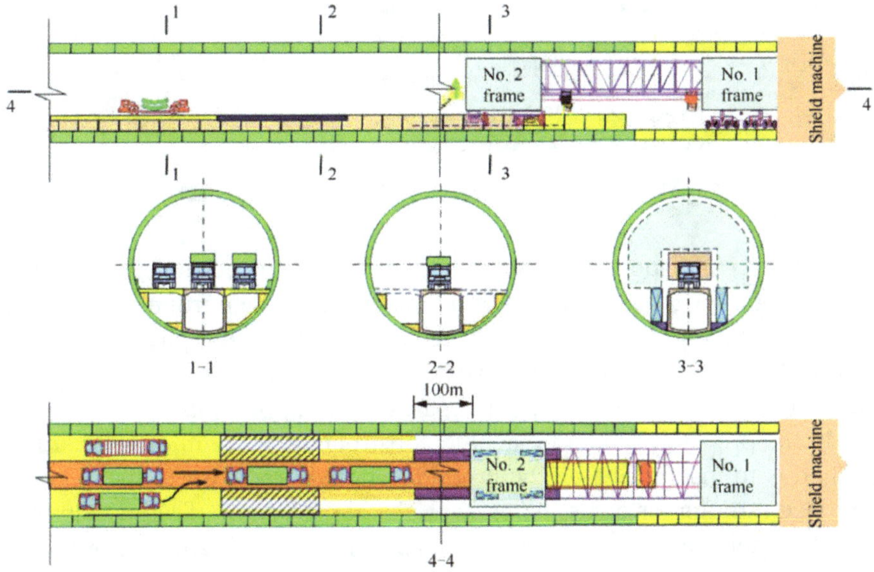

Fig. 2.36. Schematic diagram of construction transport in tunnel.

Fig. 2.37. Schematic diagram of storage and transportation of segments.

2.4.2.1.5 *Ventilation in tunnel*

In super-large-diameter tunnels, when the shield machine advances uphill, heat and moisture generated on the working face gather on the working face, resulting in a foggy state instead of naturally discharging. Moreover,

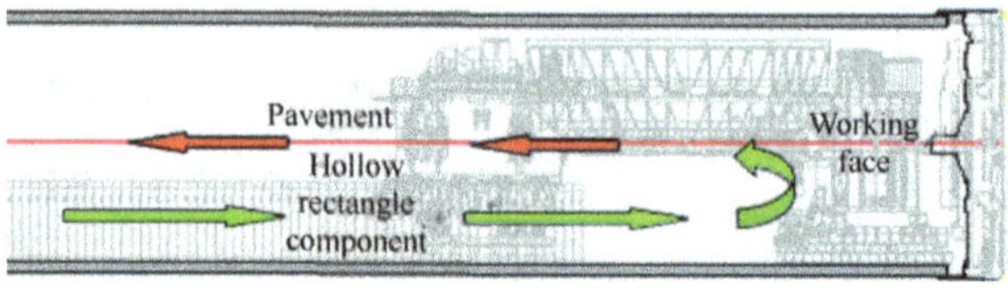

Fig. 2.38. Schematic diagram of ventilation in tunnel.

heavy trucks transporting construction materials in the tunnel also produce a lot of exhaust. In addition, the harsh air environment has an adverse impact on the shield machine equipment and the physical condition of the staff, and it also seriously affects the progress of the measurement, so ventilation in the tunnel is particularly important.

In the process of shield machine construction, a common ventilation method is to install fans on the ground to deliver fresh air to the excavation face through air ducts. At present, the construction of large-diameter shield tunnels generally adopts the synchronous construction of prefabricated members. Making full use of the passage in the middle of the installed hollow rectangle components, special tunnel axial flow fans are installed inside the hollow rectangle components at the launching shaft to transmit fresh air from outside the tunnel to the tail of the frame, and the fresh air is transmitted to the working face through the ventilation system equipped on the rear frame (Fig. 2.38).

Moreover, fans are installed at the left and right sides of the upper tail of the shield machine frame, which are connected to the machine head through the air duct and have adjustable blowing directions to realize forward air supply or outward air exhaust at the same time. This can also form a small cycle of two fans.

If the tunnel spans a long route with significant internal fluctuations, a certain number of large-diameter jet fans may be added to the pavement built under synchronous construction for outward air exhaust, forming a ventilation cycle with the air supply passage through the hollow rectangle components, so as to ensure the air quality in the tunnel.

Chapter 3

Slurry System

As an important part of construction using a slurry balanced shield machine, the slurry system functions to pump slurry that meets the requirements of the indicators to the excavation face, mix slurry with soil cut, and then transport the mixture to the ground for separation and adjustment; the treated slurry is recycled and then pumped to the excavation face again, repeating the cycle.

The slurry system of a large-diameter slurry balanced shield machine generally consists of a slurry treatment system and a slurry conveying system. The commonly used slurry treatment equipment is divided into pretreatment equipment, primary treatment equipment, secondary treatment equipment and drying treatment equipment according to the treatment process.

This chapter focuses on the design basis, the design of subsystems, the calculation of technical parameters and the equipment classification of the slurry treatment system. It introduces the composition of the slurry conveying system in detail and finally discusses the relevant indicators of slurry management monitoring.

3.1 Slurry treatment system

3.1.1 System design basis

The quality of the slurry affects the operation of the whole system and, subsequently, the shield tunneling process. Slurry circulates smoothly and quickly, just like the blood circulation of the human body. In order to ensure the smooth operation of the slurry equipment, it is necessary to deeply understand the stratum characteristics of the tunnel to be bored, customize the slurry equipment and make corresponding adjustments for stratum changes so as to ensure that the shield machine is fed with sufficient and qualified slurry at any time to maintain its smooth advance.

The design basis of a slurry treatment system mainly includes the following five aspects:

(1) technical parameters of the shield machine;
(2) engineering geological conditions;
(3) the construction scale and the relationship between the treatment site and the launching site;
(4) the scale of the construction site;
(5) surrounding environment.

Among them, the most critical is engineering geological conditions. Therefore, in the design stage, it is necessary to fully study and analyze the engineering geological conditions, understand the particle composition of the stratum, fill in the records (as shown in Table 3.1), and design and configure the corresponding treatment capacity in the treatment subsystems at all levels to ensure that the treated slurry meets the requirements of the shield machine construction.

After a full statistical analysis of the particle size distribution of all layers involved, the particle size distribution curve is drawn to guide the selection of separation points at all levels of the slurry separation station. The general pre-screening system treats coarse particles above 5 mm. The primary cyclone is designed with a smaller division point of 60–75 μm generally to reduce the treatment load of the secondary cyclone, provide slurry with an appropriate concentration for the secondary cyclone treatment and ensure the division point of the secondary cyclone. The particles smaller than 30 μm in slurry are generally slurryable. However,

Table 3.1. Main particle composition (sample).

Stratum code	Stratum designation	Contents of particle fractions (%)					Water content (%)	Unit weight (kN/m^3)	Density (g/cm^3)
		>5 mm	0.075–5 mm	0.05–0.075 mm	0.02–0.05 mm	<0.02 mm			

as the shield machine advances, the increasing content of this part of the particles also leads to an increase in slurry density and viscosity, thus affecting the muck-carrying performance and increasing the pump load. Therefore, 20–30 μm division points should be selected during the design and selection of secondary cyclones to slow down the rising rate of density and reduce consumption of fresh slurry.

In addition, the particle size grading of the primary and secondary cyclones can be adjusted by adjusting the feed concentration, pressure, overflow pipe diameter and underflow port size. In the process of actual construction, division points can be adjusted depending on different formations to retain more useful particles.

3.1.2 Overall system design

3.1.2.1 *Slurry treatment method*

According to the principle of slurry treatment, the commonly used slurry treatment methods include natural sedimentation and mechanical solid control separation. The natural sedimentation method is to transport slurry to the ground through the slurry discharge pipe and carry out sedimentation by gravity in the sedimentation tank. However, due to a high treatment demand for slurry and a large number of stabilizers added in the process of slurry preparation to improve the ability of slurry to carry muck and stabilize the excavation face, which affects the rate of sedimentation by gravity. In order to meet the treatment requirements, it is often necessary to prepare multiple sedimentation tanks, which increases the temporary land area for construction. In order to solve the problem of the poor sedimentation effect of fine particles, a large amount of flocculants is usually added to slurry to speed up sedimentation, and the addition of flocculants leads to the risk of environmental pollution from discharged muck.

The mechanical solid control separation method is widely used in shield machine construction due to its advantages of high-efficiency separation and compact structure.

The usual treatment method is to remove the large particles in the slurry through a vibrating screen or rotary screen, and then remove the small particles through a hydrocyclone and dehydration screen. The treated slurry is sampled and analyzed. If it meets the construction requirements, it can be recycled. If it fails to meet the requirements, it is

Fig. 3.1. General layout of combined slurry treatment system.

adjusted by adding fresh slurry before reuse. The waste slurry and muck from the treatment are transported for discharge. With the rising requirements for environmental protection in cities, the management of slurry discharge is becoming increasingly strict, and the traditional direct discharge of slurry will be prohibited. Therefore, it is imperative to introduce drying treatment measures such as centrifuges or filter presses to achieve dry mucking.

Although the natural sedimentation method and mechanical solid control separation method have different ways to treat slurry, their purpose is to recycle slurry discharged by shield machine equipment after removing muck. The natural sedimentation method requires a large construction site, and the mechanical solid control separation method imposes high equipment and maintenance costs. Therefore, a proper combination may be carried out according to their characteristics in the construction process to improve treatment efficiency and reduce construction costs. See Fig. 3.1 for a general layout of the combined slurry treatment system.

3.1.2.2 *Slurry treatment process*

The slurry treatment system mainly includes a slurry separation system, a slurry adjusting and mixing system, a clean water system and a waste slurry (emergency sedimentation) system. See Fig. 3.2 for the treatment process.

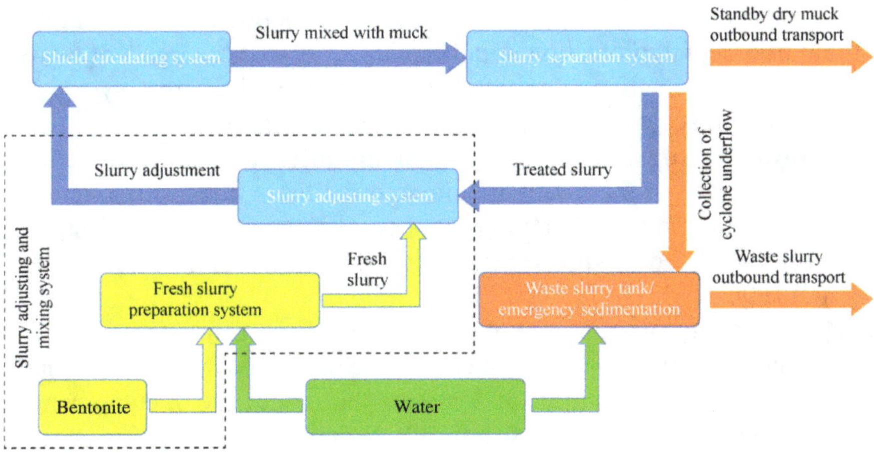

Fig. 3.2 Slurry treatment process.

3.1.2.3 *Design criteria for slurry treatment*

Constant improvement and innovations have been made to the slurry treatment design concept in order to make the slurry treatment system more compatible with shield machine construction, raise construction efficiency, facilitate construction management and make the construction cost more controllable.

(1) **Attach importance to the design of the "separation process":** The vibration motors with high-frequency vibration acceleration and two-stage cyclones with high sizing efficiency are used to not only reduce the water content of waste muck and facilitate the outbound transport of muck but also efficiently recover useful polymer materials for slurry adjusting and mixing of fresh slurry.
(2) **Design to be "best compatible" with shield machine construction:** Through remote data interaction and mutual monitoring, the slurry treatment control system is closely coordinated with the slurry conveying system and the shield tunneling system so as to achieve efficient tunnel construction and the best performance of the entire tunneling system.
(3) **Design to achieve "zero discharge" and "less discharge" of waste slurry:** Through two-stage treatment, "less discharge" of waste slurry

is basically achieved. At the same time, in order to meet higher environmental protection requirements, “zero discharge” of waste slurry can be achieved with the addition of a filter press system or centrifugal separation three-stage treatment system.

(4) **Equipment design in “integrated modules”:** In order to reduce the land area of the treatment site and facilitate transport, assembly and disassembly, the main components of the system are designed with integrated modules, making the functional zoning more clear and tidy.

(5) **Component “wear resistance” design:** Wear resistance design is adopted for key components such as sieve plates and cyclones to reduce the replacement frequency and prolong the service life of equipment.

(6) **Construction “sound insulation” design:** The separation station adopts a fully enclosed sound insulation design, is equipped with a low-noise vibration motor to minimize noise and has dust-proof and aesthetic functions.

3.1.3 Design of subsystems

3.1.3.1 *Slurry separation system*

The main design idea of the mechanical slurry separation system is that the slurry discharged from the shield machines is screened to remove soil particles of large diameter through a pre-screen, which are then conveyed to the desander cyclone; the overflow containing small solid particles from the desander cyclone is then subjected to secondary desalting cyclone treatment; finally, the underflow of the desander and desilter is concentrated into the solid phase using a dehydration screen and other equipment before waste disposal. The water generated in the concentration process flows into the sedimentation tank again. The treatment principle is illustrated in Fig. 3.3. Integrated sludge separation equipment in Fig. 3.4.

3.1.3.2 *Slurry adjusting and mixing system*

The slurry adjusting and mixing system has the function of “metabolism” in the circulation system of slurry balanced shield machines. In the process of tunneling, the slurry indicator performance and slurry volume always change in the slurry circulation, especially in formations with a high content of fine particles, and other indicators, including slurry

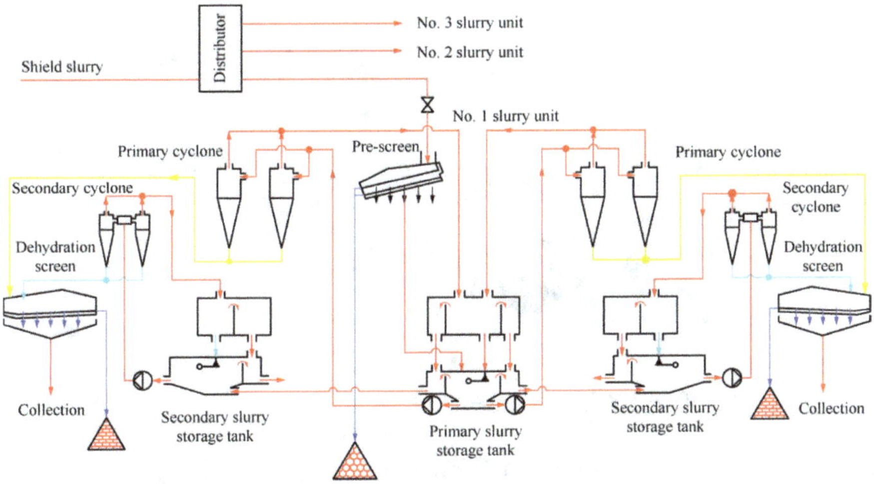

Fig. 3.3. Schematic diagram of sludge separation system.

Fig. 3.4. Integrated sludge separation equipment.

density and viscosity, change more rapidly. The control of these indicators and slurry pressure is the most important factor affecting the stability of the excavation face. Therefore, the excavation face of slurry balanced shield machines must be attached with a slurry membrane with appropriate indicators to maintain stability. It is particularly important to prepare fresh slurry and control and adjust the physical indicators of the circulation.

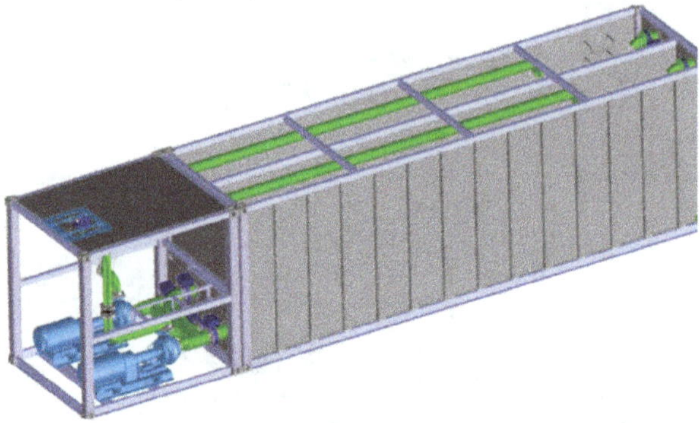

Fig. 3.5. Schematic diagram of integrated slurry adjusting and mixing equipment.

In order to ensure a smooth shield machine advance, slurry suitable for muck from different formations should be timely adjusted, prepared and fed according to the changes in geological conditions, and the slurry.

The slurry adjusting and mixing system consists of various slurry tanks, delivery pumps, pipelines, control equipment and valves of different functional units. In the design of the slurry preparation system, it is necessary to consider the demand and storage of fresh slurry mixed for the maximum number of advance rings of the shield machine in a single day when advancing in the most unfavorable stratum and reasonably position the civil structures, slurry pipelines and the mechanical and electrical installations of the slurry preparation system in combination with the fresh slurry materials, slurry mixing process and site conditions. The commonly used integrated slurry adjusting and mixing equipment is shown in Fig. 3.5.

3.1.3.3 *Clean water system*

The clean water system is mainly used to supply industrial circulating water, slurry preparation system water, slurry separation system water, site flushing and cleaning water, consisting of two parts, respectively, for water intake and water supply (Fig. 3.6). As the water consumption of large-diameter slurry balanced shield machines is very large, river water is usually selected as the water source. Considering the requirements of water quality, the clean water system is equipped with a filtering device.

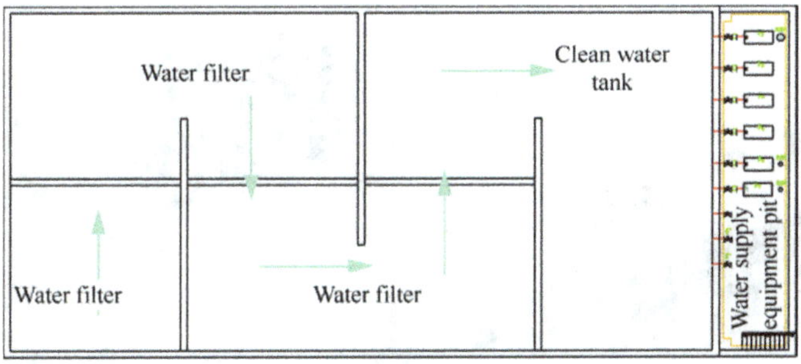

Fig. 3.6. Layout of clean water tank and water supply equipment pit.

3.1.3.4 *Slurry disposal system*

Generally, the slurry disposal system is divided into two parts:

(1) **Treatment system mainly for waste slurry:** waste slurry (thick slurry) precipitated in cyclone underflow → sand collecting tank → pumping → barge transport; waste slurry (excess thin slurry) in sedimentation tank → pumping → barge transport.
(2) **Treatment system mainly for dry mucking:** soil collection pit → muck truck assisted by excavator → outbound transport; excessive clean water → pumping (flow by gravity) → discharge to municipal piping network.

3.1.3.5 *Waste slurry drying system*

After secondary treatment of the main slurry separation system, the "less discharge" of waste slurry can be basically achieved. However, in order to meet higher environmental requirements, the "zero discharge" of waste slurry can even be achieved with additional three-stage treatment systems such as filter pressing or centrifugal separation.

Drying treatment mainly includes the filter pressing process and the centrifugal process. The filter pressing system is equipped with a box-type automatic plate pulling filter press, which can realize automatic control of multiple processes, including filter plate pressing, filtering, pressing, back blowing, washing, filter plate releasing and unloading. The system requires a large land area, but the energy consumption is relatively small.

(a) Filter press

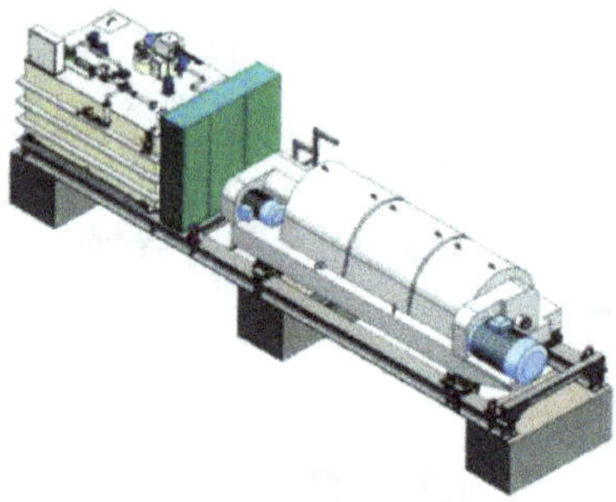

(b) Centrifugal treatment

Fig. 3.7. Schematic diagram of waste slurry crying treatment equipment.

The centrifugal treatment system uses a horizontal screw centrifuge to form an annular liquid flow in the high-speed rotating drum under the action of a centrifugal force field. The solid particles rapidly settle on the inner wall of the drum under the action of centrifugal force. Leveraging the differential speed between the drum and the screw conveyor generated by a differential, the screw conveyor pushes the sediment to the drying area at the cone end of the drum. After the two-way squeezing by the thrust action of the screw conveyor and the centrifugal component of the sediment, the sediment is further squeezed and dehydrated, discharged through the muck outlet at the small end of the drum, and the clear liquid from separation is discharged through the overflow port. The centrifugal treatment equipment has very high requirements for technical capacity, the selection and use of additives and high energy consumption, but it occupies a relatively small land area.

The waste slurry filter press and centrifugal treatment equipment are shown in Fig. 3.7.

3.1.3.6 *Emergency treatment system of sedimentation tank*

Any failure of the mechanical solid control equipment may lead to the suspension of the entire shield tunneling. In order to avoid this situation and reduce the impact of equipment failure, a special sedimentation tank can be set up for emergency treatment if the conditions are available.

In case of equipment failure, slurry can be directly discharged to the sedimentation tank for emergency treatment by the natural sedimentation method. The treated circulating slurry is then adjusted through the adjusting system to ensure that the slurry fed to the shield machine excavation face can meet the construction requirements.

3.1.4 Calculation of technical parameters

Parameter calculation of the slurry treatment system of the shield machine includes conservation of matter (capacity to be compatible with a slurry balanced shield machine), cyclone treatment capacity, motor power for the slurry pump, material consumption (water, fresh slurry materials, etc.), power consumption and total equipment power.

3.1.4.1 *Design calculation parameters*

Excavation diameter: D, tunnel sectional area: S_e, advance distance: L, maximum advance rate: $V_{\max}$, segment ring width: H, undisturbed soil density: ρ_{SM}, slurry feeding density: ρ_B. slurry discharge density: ρ_S, slurry feeding rate: Q_B, slurry discharge rate: Q_S, density of soil to be cut: ρ.

3.1.4.2 *Calculation of relevant circulation parameters*

(1) Calculation of mucking from circulation of a single shield machine:

(i) *Mucking volume per unit time*:

$$Q_{SM} = S_e V_{\max}$$

(ii) *Mucking mass per unit time*:

$$M_{SM} = S_e V_{\max} \rho_{SM}$$

(2) Calculation of circulating slurry in circulation of a single shield machine:

(i) *Slurry discharge rate*:

$$Q_S = Q_{SM} \cdot \frac{\rho_{SM} - \rho_B}{\rho_S - \rho_B}$$

(ii) *Sludge feeding rate*:

$$Q_B = \frac{Q_S \cdot \rho_S - Q_{SM} \cdot \rho_{SM}}{\rho_B}$$

3.1.4.3 *Design calculation of performance parameters*

(1) Calculation principle of treatment capacity of sludge separation equipment:

   (i) *The first stage*: The maximum throughput of the coarse screen must be greater than the maximum rate of the slurry discharge pump.
   (ii) *The second stage*: The maximum treatment capacity of the primary cyclone must be greater than the flow downstream of the pre-screen.
   (iii) *The third stage*: The maximum treatment capacity of the secondary cyclone must be greater than the flow downstream of the primary cyclone.

(2) Calculation of cyclone treatment capacity:

   (i) *Calculation of cyclone treatment capacity*:

$$Q = 2.69nDd_i\left(\frac{20}{\alpha}\right)^{0.2}\Delta P^{0.5}\left[\rho_m\left(\frac{1.5D}{d_0}\right)^{1.28} - 1\right]^{-0.5}, \quad (3.1)$$

where
Q — Cyclone treatment capacity (m^3/h);
n — Number of cyclones;
D — Cyclone diameter (cm);
d_i — Equivalent diameter of cyclone slurry inlet (cm);
α — Cyclone cone angle (°);
ΔP — Cyclone slurry inlet pressure (MPa);
ρ_m — Density of inflow slurry (g/cm^3);
d_0 — Diameter of overflow pipe of cyclone (cm).

(ii) *Calculation of separation particle size and grading particle size of cyclone*:

$$d_{50} = 1098.6 D^{0.18} d_0^{0.32} d_i^{0.25} \mu_m^{0.5} \left(\frac{\tan\alpha}{2} \right)^{0.5} (\rho_d - \rho_m)^{-0.5} (3D - 2d_0)^{-0.5} \Delta P^{-0.25} d_m = 1.65 d_{50}, \tag{3.2}$$

where

d_{50} — Cyclone separation particle size (μm);
μ_m — Plastic viscosity of slurry (Pa · s);
ρ_d — Density of dispersed phase in slurry (g/cm^3);
ΔP — Cyclone inlet pressure (kPa);
d_m — Grading particle size of cyclone (μm).

(3) Calculation of screening capacity of vibrating screen:

(i) Calculation of material screening capacity of unit vibrating screen. Theoretical calculation formula for the material running speed (v) of the linear vibrating screen:

$$v = K_V \lambda \omega \cos\delta \,(1 + \tan\delta \tan\alpha), \tag{3.3}$$

where

v — Material running speed (mm/s);
K_V — Comprehensive empirical coefficient, generally taken as 0.75–0.95;
λ — Single amplitude (mm);
ω — Vibration frequency (rad/s);
δ — Vibration direction angle (°);
α — Inclination angle of screen surface (°).

Theoretical calculation formula for the treatment capacity (Q) of linear vibrating screen:

$$Q = 3600 bvh, \tag{3.4}$$

where

Q — Treatment capacity of vibrating screen (m^3/h);
b — Effective width of screen (m);
h — Average material thickness (m);
v — Material running speed (m/s).

(2) Calculation of muck-discharge capacity of single-line vibrating screen: According to the material screening capacity of the unit vibrating screen, the corresponding number of units is configured to finally meet the treatment capacity of the vibrating screen in this stage of the pre-screening process, as shown in the following formula:

$$n = \frac{K_1 Q_{SM}}{Q}, \tag{3.5}$$

where

n — Number of unit vibrating screens;

K_1 — Treatment coefficient of pre-screening, generally taken as 0.4–0.45.

3.1.4.4 *Material balance calculation*

According to the engineering geological data, the primary cyclone and secondary cyclone are preliminarily selected, and the material balance table is calculated for working strata (taking the soft soil formation as an example).

(1) Slurry feeding:

(i) *Mass concentration*:

$$P_{D1} = \frac{G(\rho_1 - \rho_w)}{\rho_1 (G - \rho_w)} \times 100\%$$

(ii) *Slurry feeding volume*:

$$V_1 = \frac{Q_1}{v} \quad (\text{m}^3/\text{m})$$

(iii) *Slurry feeding mass*:

$$W_1 = \rho_1 V_1 \quad (\text{t/m})$$

(iv) *Solid mass*:

$$ww_{a1} = W_1 P_{D1} \quad (\text{t/m})$$

(v) *Solid volume*:

$$wv_{a1} = \frac{ww_{a1}}{G} \qquad (\mathrm{m^3/m})$$

(vi) *Water mass*:

$$ww_1 = W_1\,(1 - P_{D1}) \qquad (\mathrm{t/m})$$

(vii) *Water volume*:

$$wv_1 = \frac{ww_1}{\rho_w} \qquad (\mathrm{m^3/m})$$

(viii) *Summation and check*:

$$Ww_1 = ww_{a1} + ww_1 \qquad (\mathrm{t/m})$$

$$Wv_1 = wv_{a1} + wv_1 \qquad (\mathrm{m^3/m})$$

$$\rho_1 = \frac{Ww_1}{Wv_1} \qquad (\mathrm{t/m^3})$$

(2) Soil mass to be cut:

(i) *Density*:

$$\rho_3 = \frac{G(\omega + 1)}{G(\omega + 1)} \qquad (\mathrm{t/m^3})$$

(ii) *Volume*:

$$V_2 = \frac{\pi D^2}{4} \qquad (\mathrm{m^3/m})$$

(iii) *Mass*:

$$W_2 = V_2 \rho_3 \qquad (\mathrm{t/m})$$

(iv) *Solid mass*:

$$ww_{a2} = \frac{W_2}{1 + \omega} \qquad (\mathrm{t/m})$$

(v) *Solid volume*:

$$wv_{a2} = \frac{ww_{a2}}{G} \qquad (\mathrm{m^3/m})$$

(vi) *Water mass*:

$$ww_2 = \frac{w_2\omega}{1+\omega} \qquad (\mathrm{t/m})$$

(vii) *Water volume*:

$$wv_2 = \frac{ww_2}{\rho_W} \qquad (\mathrm{m^3/m})$$

(viii) *Summation and check*:

$$Ww_2 = ww_{a2} + ww_2 \qquad (\mathrm{t/m})$$

$$Wv_2 = wv_{a2} + wv_2 \qquad (\mathrm{m^3/m})$$

$$\rho_2 = \frac{Ww_2}{Wv_2} \qquad (\mathrm{t/m^3})$$

(3) Slurry drainage water:

(i) *Particle composition*:

(a) Mass and volume of coarse particles:

$$ww_{y3} = ww_{a2}y \qquad (\mathrm{t/m})$$

$$wv_{y3} = \frac{ww_y}{G} \qquad (\mathrm{m^3/m}),$$

where

y — percentage of coarse particles in soil volume.

(b) Mass and volume of fine particles:

$$ww_{s3} = ww_{a2}s \qquad (\mathrm{t/m})$$

$$wv_{s3} = \frac{ww_s}{G} \qquad (\mathrm{m^3/m}),$$

where

s — percentage of fine particles in soil volume.

(c) Mass and volume of clay micro-particles:

$$ww_{c3} = ww_{a2}c + ww_{a1} \quad \text{(t/m)}$$

$$wv_{c3} = \frac{ww_c}{G} \quad (\text{m}^3/\text{m}),$$

where

c — percentage of clay particles in soil volume.

(ii) *Water mass*:

$$ww_3 = ww_1 + ww_2 \quad \text{(t/m)}$$

(iii) *Water volume*:

$$wv_3 = \frac{ww_3}{\rho_W} \quad (\text{m}^3/\text{m})$$

(iv) *Summation and check*:

$$Ww_3 = ww_{y3} + ww_{s3} + ww_{c3} + ww_3 \quad \text{(t/m)}$$

$$Wv = wv_{y3} + wv_{s3} + wv_{c3} + wv_3 \quad (\text{m}^3/\text{m})$$

$$\rho_3 = \frac{Ww_3}{Wv_3} \quad (\text{t/m}^3)$$

(4) Separation treatment:

(i) *Pre-screened residue of vibrating screen/rotary screen*:

$$ww_y = ww_{y3} \quad \text{(t/m)}$$

$$wv_y = wv_{y3} \quad (\text{m}^3/\text{m})$$

(ii) *Secondary cyclone grit chamber*:

$$ww_s = ww_{s3} \quad \text{(t/m)}$$

$$wv_s = wv_{s3} \quad (\text{m}^3/\text{m})$$

(iii) *Slurry particles into adjusting tank*:

$$ww_c = ww_{c3} \qquad \text{(t/m)}$$

$$wv_c = wv_{c3} \qquad \text{(m}^3\text{/m)}$$

(iv) *Loss of water*:

(a) Vibrating screen/rotary screen area:

$$W_g = wv_y \frac{G - \rho}{\rho - \rho_W} \qquad \text{(m}^3\text{/m)}$$

(b) Secondary cyclone equipment area:

$$W_c = wv_s \frac{G - \rho}{\rho - \rho_W} \qquad \text{(m}^3\text{/m)}$$

(v) *Amount of slurry into adjusting tank*:

$$V_t = wv_c + (wv_3 - W_g - W_s) \qquad \text{(m}^3\text{/m)}$$

(vi) *Density of slurry into adjusting tank*:

$$\rho_t = \frac{wv_c G + \left(wv_3 \rho_W - W_g - W_s\right)}{wv_c + (wv_3 - W_g - W_s)} \qquad \text{(t/m}^3\text{)}$$

(vii) *Dilution water*:

$$W_s = \frac{V_t(\rho_t - \rho_1)}{\rho_1 - \rho_W} \qquad \text{(m}^3\text{/m)}$$

(5) Waste slurry:

(i) *Waste slurry in each ring*:

$$Q_b = \frac{\left[(V_t + W_s)V_{\max} - Q_B\right]H}{V_{\max}} \qquad \text{(m}^3\text{/r)}$$

(ii) *Amount of waste slurry from drum screen and sump*:

$$Qq = \frac{\left(wv_y + W_g + wv_s + W_c\right)H}{V_{\max}} \qquad (\mathrm{m}^3/\mathrm{r})$$

3.1.4.5 *Calculation of power of auxiliary motor for slurry pump*

$$N = \frac{N\rho_m QH}{102\eta_p\eta_m} \tag{3.6}$$

where

N — motor power of slurry pump (kW);
n_s — safety factor;
ρ_m — density of pumped slurry (g/cm^3);
Q — pump working flow rate (L/s);
H — working head of cyclone (m);
η_p — pump efficiency (%);
η_m — motor to pump belt drive efficiency (%).

When selecting the above parameters, it is necessary to fully consider some special working conditions that may occur during the construction stage and select sufficient safety factors to ensure the stable operation of equipment during the construction process.

3.2 Slurry treatment equipment

The commonly used slurry treatment equipment is classified according to the treatment process into those used for pre-treatment, primary treatment, secondary treatment and drying treatment.

3.2.1 Pre-treatment equipment

As the first step in the slurry separation process, the pre-treatment equipment is very critical. It consists of usually two types: vibrating screens and rotary screens. Its main function is to carry out slurry pre-separation in the treatment process, screening out soil and rock in large

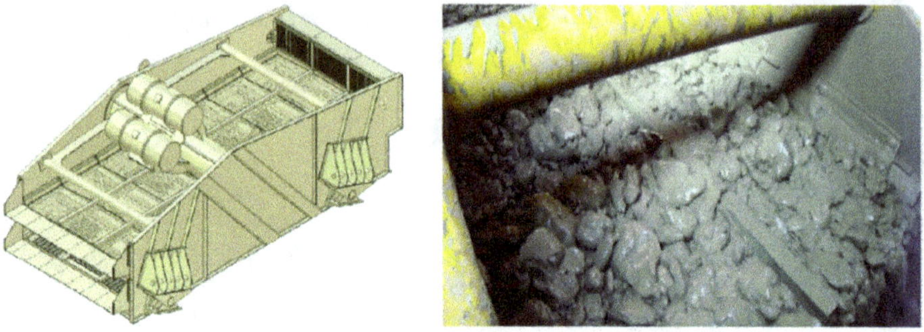

(a) Structural sketch (b) Treatment effect

Fig. 3.8. Pre-treatment vibrating screen equipment.

particles, and after screening, most of the particles above 3–5 mm can be removed. The screened slurry flows into the storage drum and waits for the next stage of treatment. The screened materials are transported to the muck yard for outbound transport.

(1) **Pre-treatment vibrating screen equipment:** The pre-treatment vibrating screen with double-layer polyurethane sieve plates and a high-frequency vibrating motor is used to screen solid particles, such as cobbles, gravel, clay lumps and coarse sand, discharged by shield machines. A set of high-pressure clean water washing systems should be designed to solve the blockage problem of sieve plates caused by clay agglomeration. According to the slurry circulating flow of large-diameter shield machines, the vibrating screening area of the sieve plate is configured. Generally, each set has a treatment capacity of 1,200–1,600 m^3/h. See Fig. 3.8 for pre-treatment vibrating screen equipment.
(2) **Pre-treatment rotary screen equipment:** The treatment capacity of slurry balanced shield machines is designed by defining and adjusting barrel diameter, inner layer diameter, length, rotating speed, power and screen mesh length, and each rotary screen can reach a capacity of 1,200–1,600 m^3/h. The equipment is chain-driven, with the outer layer made of simple replaceable stainless steel woven mesh, and equipped with an anti-impact pressure relief box and a water jet high-pressure cleaning system. See Fig. 3.9 for pre-treatment rotary screen equipment.

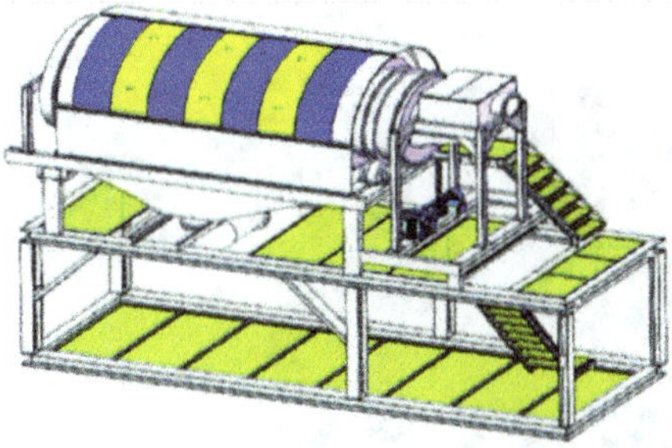

(a) Structural sketch

(b) Treatment effect

Fig. 3.9. Pre-treatment rotary screen equipment.

(a) Physical equipment

(b) Photo of partial equipment

Fig. 3.10. Primary treatment equipment.

3.2.2 Primary treatment equipment

The primary treatment equipment usually adopts a large-diameter cyclone with cyclone separation particle sizes of 60–75 μm, equipped with a full wear-resistant rubber lining, featuring a long service life and high separation accuracy. The cyclone is also equipped with a fishtail siphon device to ensure the stability of underflow concentration and improve the separation effect. The treatment capacity of a single cyclone is 180–220 m^3/h. See Fig. 3.10 for commonly used primary treatment equipment.

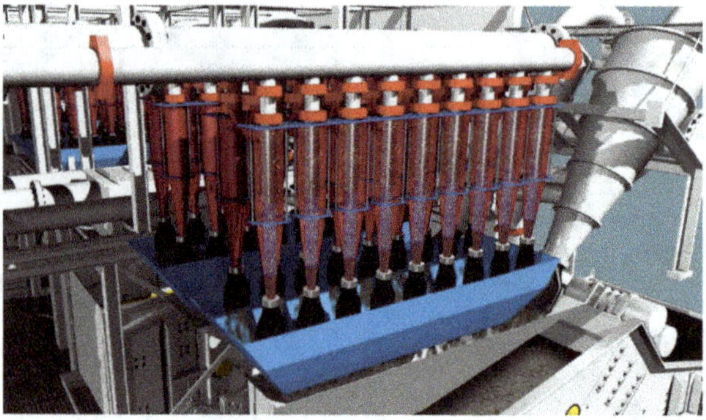

Fig. 3.11. Secondary treatment equipment.

3.2.3 Secondary treatment equipment

Small-diameter cyclones are mostly used for secondary treatment equipment, and the cyclone separation particle sizes are 18–25 μm, subject to the diameter of the underflow port and the slurry inlet pressure adjusted according to different formations. The treatment capacity of a single two-stage cyclone is 20–25 m^3/h. The commonly used secondary treatment equipment is shown in Fig. 3.11.

3.2.4 Drying treatment equipment

(1) **Dehydration screen:** The solid particles separated from the first- and second-stage cyclones are discharged to the dehydration vibrating screen for dehydration and then discharged through the muck discharge tank. The dehydration vibrating screen, similar to the pre-treatment vibrating screen, is equipped with a vibrating motor with high vibration acceleration and a large screen area, which has excellent dehydration performance. The mesh can be quickly replaced to change size according to different geological conditions. Due to the effect of high vibration acceleration, the dehydrated muck is suitable for truck transport without secondary environmental pollution such as slurry leakage. The capacity of the dehydration screen can be changed by adjusting the area of the dehydration sieve plate, generally reaching 200–300 m^3/h. In relative terms, the residues on the screen have a high moisture content, but they can basically be

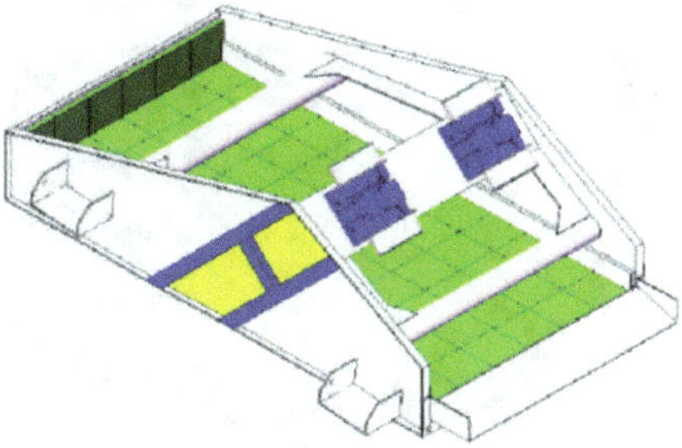

(a) Structural sketch

(b) Treatment effect

Fig. 3.12. Dehydration screen equipment.

directly transported. The commonly used dehydration screen equipment is shown in Fig. 3.12.

(2) **Filter press:** The filter press is usually used together with flocculant. By adding flocculant to the waste slurry and muck tank and stirring, the fine particles in the waste muck can be bonded to form floccules and promote sedimentation. The filter press discharges most of the pore water in floccules to form lumps that are convenient for handling. There are two common dehydration methods: pressurized dehydration and vacuum dehydration. Pressurized dehydration uses pumps and air compressors to pressurize floccules and dehydrate them with filter cloth; vacuum dehydration applies negative pressure into the rotating drum with the tightened filter cloth and realize dehydration under pressure difference. The treatment capacity of the filter press can be changed by adjusting the area and number of filter plates. Generally, a single machine has a treatment capacity of 8–15 m^3/h (dry muck). Relatively, the water content of the filter residues is extremely low, meeting the requirements of direct outbound transport.

(3) **Centrifuge:** The slurry centrifuge is used for solid–liquid dehydration and separation of the waste slurry based on the principle of centrifugal sedimentation. Under the action of centrifugal force, the solid particles of the waste slurry are pushed onto the inner wall of the drum and then to the muck discharge port of the drum by the blades on the screw pusher for discharge; the liquid overflows through the overflow hole at the big end of the drum. This cycle is repeated to achieve continuous separation. The centrifuge has a large treatment capacity. Generally, a single machine has a treatment

(a) Physical equipment (b) Photo of partial equipment

Fig. 3.13. Centrifuge equipment.

capacity of 50–70 m^3/h (dry muck). The muck from separation is relatively high in moisture content, but it can basically be directly transported. See Fig. 3.13 for commonly used drying centrifuge equipment.

3.3 Slurry conveying system

The slurry conveying system is composed of slurry feeding and discharge pumps, slurry pipes, piping sections and auxiliary equipment. The system uses the $P_{1.1}$ main slurry pump to send the high-quality slurry reaching a preset density in the adjusting tank of the slurry treatment system into the sealed compartment of the shield machines. The soil cut by the cutterhead is discharged from the tunnel through the $P_{2.1}$ slurry discharge pump or a series of slurry discharge relay pumps to the ground and then transported to the ground slurry treatment system for treatment and adjustment, and part of the treated slurry is disposed of. The remaining is delivered to the adjusting tank, mixed with fresh slurry and then delivered to the shield machine excavation face through the slurry feeding pipe. This cycle is repeated to help the shield machine advance.

3.3.1 System calculation basis

The calculation basis of the slurry conveying system mainly includes the geological conditions, the environmental conditions of tunnel construction and the technical parameters of the selected shield machines. See Table 3.2 for a brief summary.

Table 3.2. Parameters of slurry conveying system.

S/N	Name	Key	S/N	Name	Key
1	Shield machine cutting diameter	D	11	Distance between slurry treatment site to working shaft	L_1
2	Shield machine tunneling rate	S	12	Depth of working shaft	H
3	Minimum hydraulic pressure at cutting ring	$P_{c\min}$	13	Elevation difference between slurry treatment sit and working shaft	h
4	Maximum hydraulic pressure at cutting ring	$P_{c\max}$	14	True density of slurry feeding fluid	ρ_1
5	Fluctuation in hydraulic pressure	P_{cs}	15	Density of slurry feeding fluid	ρ_2
6	Efficiency at slurry feeding pump head	δ_s	16	Water content of stratum	ω
7	Efficiency at slurry discharge pump head	δ_p	17	True density of soil particle	ρ_3
8	Diameter of slurry feeding pipe	d_1	18	Slurry density of slurry feeding fluid	δ_1
9	Diameter of slurry discharge pipe	d_2	19	Density of mother liquid	δ_0
10	Tunnel length	L			

3.3.2 Calculation of slurry feeding and discharge flow rates

The flow mode of slurry in pipelines varies with fluid particle size, density and flow rate. If the particle size and density increase, the concentration difference between the upper and lower parts will be caused by the gravity difference in the pipeline, forming a heterogeneous flow; if the flow rate is low, particle precipitation will occur; if the flow rate is high, particles will jump and form a mixed flow. Therefore, the flow rate of slurry in the slurry pipeline must be kept above the critical value because when it is lower than the critical value, the particles in the slurry will precipitate, causing pipeline blockage and even serious blockage in the scenario of slurry discharge. Therefore, after calculating and determining the critical flow rate in the slurry conveying pipeline, the pump with the appropriate slurry feeding and discharge flow rates and power should be selected from the slurry pump series. The critical flow rate can be determined according to Durand's equation for the critical sedimentation flow rate:

(1) Flow rate in slurry feeding pipe (d_1):

$$v_1 = \frac{Q_1}{\frac{1}{4}\pi d_1^2} \tag{3.7}$$

(2) Flow rate in sludge discharge pipe (d_2):

$$v_2 = \frac{Q_2}{\frac{1}{4}\pi d_2^2} \tag{3.8}$$

(3) Critical sedimentation rate in sludge discharge pipe:

$$V = F_L \sqrt{\frac{2gd_2(\rho_3 - \rho_2)}{\rho_2}}, \tag{3.9}$$

where

F_L — slurry concentration determination coefficient, generally taken as 1.5;
g — gravity acceleration (m/s^2), taken as 9.8 m/s^2.

The critical flow rate values of common formations are 3.8 m/s for the sand layer, 3.5 m/s for the clay layer and 4 m/s for the rock layer.

If the flow rate in the sludge discharge pipe is greater than the critical sedimentation rate, the preset slurry discharge flow is feasible.

3.3.3 Determination of diameter of slurry feeding and discharge pipes

The diameter of the slurry feeding and discharge pipes is determined by the maximum particle size of the soil, slurry concentration, flow rate, etc. The relationship between the shield machine diameter and the pipe diameter is shown in Fig. 3.14.

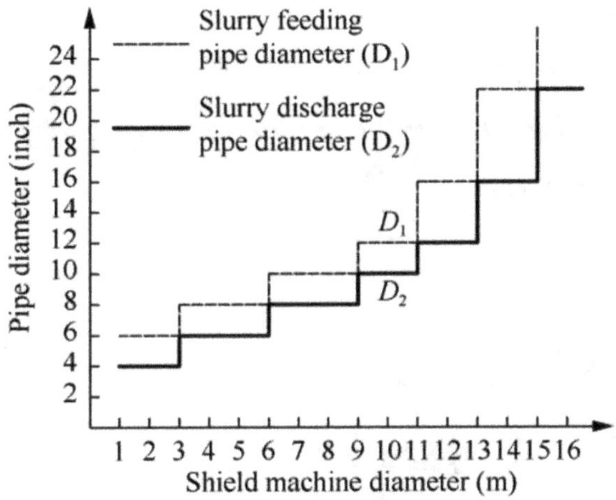

Fig. 3.14. Graph showing the relationship between shield machine diameter and pipe diameter (1 inch = 2.54 cm).

3.3.4 Selection of pump shaft power and motor power

When selecting motors for various pumps among different products by professional manufacturers with varied features, the following common formula is followed:

$$\begin{cases} M = \dfrac{0.163\gamma QH}{\eta_0 \tau}, \\ M' = 1.25M \end{cases} \tag{3.10}$$

where

M — pump shaft power (kW);
η_0 — pump efficiency;
M' — pump motor power (kW);
τ — efficiency reduction rate.

During the selection of pumps, one should also pay attention to the shape and number of impeller blades, as well as the pump speed, head and other indicators.

3.3.5 Determination of the full head for slurry feeding and discharge

3.3.5.1 *Determination of full head for slurry discharge*

(1) Cutting sectional area A:

$$A = \frac{\pi}{4} \times D^2$$

(2) Sediment content in soil, K:

$$K = \frac{100}{100 + \rho_3 \omega} \times 100\%$$

(3) Sand content in excavated soil, q:

$$q = \frac{AS}{100}$$

(4) Dry sand content in excavated soil sand, G:

$$G = A \times \frac{S}{100} \times \frac{K}{100}$$

(5) Slurry discharge and feeding flow rates, Q_2 and Q_1:

$$Q_1 = Q_2 - q$$

(6) Slurry feeding concentration, C_1:

$$C_1 = \frac{\delta_1 - \delta_0}{\rho_1 - \delta_0} \times 100\%$$

(7) Slurry discharge concentration, C_2:

$$C_2 = \frac{C_1 (Q_1 + 100G)}{Q_2}$$

(8) Slurry discharge density, δ_2:

$$\delta_2 = \delta_0 + \frac{\rho_2 - \delta_0}{100}$$

(9) Head loss coefficient of slurry discharge pipe J_{m2}:

$$J_{m2} = 10.666 \times G_2^{-1.85} d_2^{-4.87} \left(\frac{Q_2}{60} \right)^{1.85} \delta_2$$

(10) Equivalent straight pipe length of slurry discharge pipe, L_2:

$$L_2 = \left(L + H - \frac{D}{2} + 1 + h \right) \alpha,$$

where

α — length equivalent coefficient of slurry discharge pipe, taken as 1.05.

(11) Total resistance loss of slurry discharge pipe, H_{f2}:

$$H_{f2} = J_{m2} L_2$$

(12) Suction head, H_{s2}:

$$H_{s2} = P_{c\min} \frac{10}{\delta_2}$$

(13) Full head, H_2:

$$H_2 = H_{f2} + H - \frac{D}{2} + h - H_{s2} + \left(h_{f0} - 2 \right) + (h_{f0} - 4)$$

(14) Total head necessary for slurry discharge pump, H_{T2}:

$$H_{T2} = \frac{H_2}{\delta_P},$$

where

δ_P — efficiency at slurry discharge pump head, taken as 0.93.

3.3.5.2 *Determination of full head for slurry feeding*

(1) Head loss coefficient of slurry feeding pipe, J_{m1}:

$$J_{m1} = 10.666 G_1^{-1.85} d_1^{-4.87} \left(\frac{Q_2}{60} \right)^{1.85} \delta_1$$

(2) Equivalent straight pipe length of slurry feeding pipe, L_1:

$$L_1 = \left(L + H - \frac{D}{2} + 1 \right) \alpha$$

(3) Total resistance loss of slurry feeding pipe, H_{f1}:

$$H_{f1} = J_{m1} L_1$$

(4) Suction head, H_{s1}:

$$H_{s1} = p_{c\max} \frac{10}{\delta_1}$$

(5) Full head, H_1:

$$H_1 = H_{f1} - \left(H - \frac{D}{2} \right) + H_{s1}$$

(6) Total head necessary for slurry feeding pump, H_{T1}:

$$H_{T1} = \frac{H_1}{\delta_s},$$

where

δ_s — efficiency at slurry feeding pump head, taken as 0.98.

3.3.6 Determination of positions for slurry feeding and discharge relay pumps

(1) **Determination of position for the slurry discharge relay pump:** In the process of shield tunneling, the slurry feeding and discharge pipelines need to be frequently extended, causing an increase in pipe resistance. In order to ensure that the constant flow rate in the pipeline is greater than the critical flow rate, the speed of the slurry discharge pump $P_{2.1}$ should be changed accordingly at any time. When pump $P_{2.1}$ reaches the maximum head, intermediate relay pump $P_{2.n}$ is added.

The position of the slurry discharge relay $P_{2.n}$ pump (the suction head is taken as the minimum value of 10 m) is

$$L_{P2.n} = \frac{\delta_P (H_{P2.1} - 10)}{J_{m2}}. \tag{3.11}$$

According to the calculation results obtained using the above formula, the position of the slurry discharge relay pump can be determined. On this basis, the position of the slurry discharge relay pump can be determined according to the actual application of the pump.

(2) **Determination of position for the slurry feeding relay pump:** Similarly to the slurry discharge relay pump, when pump $P_{1.1}$ reaches the maximum head, the intermediate relay pump $P_{1.n}$ is added to the slurry feeding pipeline.

The position of the slurry feeding relay pump $P_{1.n}$ (the suction head is taken as the minimum value of 10 m) is

$$L_{P1.n} = \frac{\delta_s (H_{P1.1} - 10)}{J_{m1}}. \tag{3.12}$$

According to the calculation results obtained using the above formula, the position of the slurry feeding relay pump can be determined. On this basis, the position of the slurry feeding relay pump can be determined according to the actual application of the pump.

3.4 Slurry management monitoring

3.4.1 Slurry indicator management

Slurry indicator management is mainly to adjust the slurry indicator according to the slurry indicator changes during the shield tunneling process to meet the requirements of shield tunneling.

3.4.1.1 *Slurry indicator*

(1) **Density:** The density of the slurry is a main control indicator. The density of slurry in tunneling should not be too high or too low. A too high slurry density affects the slurry conveying capacity, whereas a too low slurry density affects the stability of the excavation face.

The slurry density is controlled at 1.15–1.25 (1.30) g/cm^3, with the lower limit of 1.15 g/cm^3. The upper limit depends on the special requirements of construction (shield machine construction in sandy soil, protection of ground buildings, shield machine crossing with shallow overburden, etc.), and the maximum is 1.30 g/cm^3.

(2) **Viscosity:** The viscosity of the slurry is another main control indicator. As far as the suspension of soil particles is concerned, the higher the viscosity of the slurry, the better the performance will be. Considering the slurry-making capacity of the slurry treatment system, with the increase in advance rings, slurry becomes thicker and thicker, with the density rising sharply. However, the increase in viscosity does not mean a higher quality of slurry; viscosity even declines for tunneling in sandy soil. Therefore, the viscosity of the slurry should be kept between 18 and 25 Pa · s.

Considering that the viscosity adjustment takes time, when the slurry viscosity is 18 s (in the adjusting tank), fresh slurry may be gradually fed at a dosage depending on the trend of viscosity decline.

(3) **Sand content:** The purpose of slurry treatment is to remove most of the large particles of soil above 20 μm. Therefore, the sand content in the working slurry is also an important indicator in slurry treatment. Generally, the sand content indicator should be controlled within 20%.

(4) **Bleeding and pH value:** Bleeding and pH values are comprehensive indicators in slurry management. They are related to the viscosity of slurry to a great extent. Slurry with a good suspension property means small bleeding, and vice versa.

The bleeding of the slurry must be less than 5%, and the pH value must present an alkaline result. Reducing sand content and increasing the viscosity of the slurry are the main means to ensure acceptable bleeding.

When tunneling in sandy and silty soil, due to the continuous deterioration of working slurry, it is necessary to constantly adjust the parameters of the slurry. However, when tunneling in clay and muddy clay, the concentration of discharged slurry rises due to the continuous increase in the number of cohesive particles. A commonly adopted method involves appropriate dilution by feeding water.

3.4.1.2 *Slurry mix proportion*

During construction, the site must be equipped with a slurry laboratory and a slurry control room. The indicators of the working slurry in the adjusting tank should be tested and adjusted before advance to the next ring, and the adjustments should be promptly made to meet the construction requirements and recorded. After tunneling for several rings, the change trend of the slurry indicator can be obtained, and minor adjustments can then be made on the basis of the guiding mix proportion. Therefore, slurry monitoring is a dynamic change process, and attention should be paid to the change trend of slurry indicators to make them stable in certain ranges.

Before shield launching, a certain volume of initial advance circulating slurry should be prepared to meet the shield launching requirements. Fresh slurry is fed into the adjusting tank, and the slurry indicators are adjusted to meet the construction requirements after launching.

3.4.2 Slurry control and monitoring system

The slurry treatment system usually adopts a programmable logic controller (PLC) to integrate and control the working processes of the whole slurry treatment system and the slurry adjusting and mixing system and monitor the working status of each system in real time, such as the start and stop of the vibrating screen, the slurry feeding pumps of the primary and secondary cyclones, the liquid level of each slurry tank of the slurry treatment system, the flow rate of the fresh slurry system and the real-time slurry feeding and discharge density. The PLC system and monitoring system are integrated into the operation room of the slurry treatment system.

The operation room of the slurry treatment system is usually designed on the slurry treatment site to facilitate the staff in monitoring the abnormal vibration or sound of the slurry treatment system in real time through the view port or by patrol inspection and facilitate the timely identification and handling of any problems.

The slurry treatment system supports selection among the automatic, manual and maintenance modes for operation.

When the equipment operates automatically, the slurry treatment system equipment is associated with the shield machine. After receiving the shield machine start signal, the operator instructs the starting or stopping of the equipment according to the specified sequence and interlocking relationship, as well as the requirements of the production process.

When the manual mode is indicated, the treatment equipment at all levels of the slurry treatment system can be started and stopped by manual control.

In the maintenance mode, some remote functions of the equipment are locked, and single pieces of equipment must be started and stopped in the local mode. For the sake of safety, after the maintenance is completed, the maintenance personnel need to unlock the control room locally and reset it before proceeding to the next step.

In addition, the operation screen can display the process flow of the entire production system and the operation status of equipment; multiple functions are available in case of any fault in the system operation, including giving an immediate alarm, displaying the fault location, locking the fault equipment and recording fault processing.

The slurry treatment system, slurry adjusting system, fresh slurry mixing system, and liquid levels of each slurry tank are clearly displayed at different interfaces. The slurry treatment site is equipped with a number of HD CCTV monitoring points for operators to monitor each working point in real time.

Chapter 4

Segment Fabrication

Segments are an important part of tunnels since their quality is directly related to the overall tunnel quality and safety. Different from the ordinary segment, the segment of large-diameter shield tunnels has the characteristics of large volume, large weight and multiple blocks. It is of great engineering significance to study the fabrication process, materials and production technology of large-diameter segments. Therefore, this chapter introduces the design, fabrication and inspection process of high-precision steel formwork for large-diameter shield machines, systematically analyzes and verifies the high-performance concrete (HPC) used to make segments and elaborates on the quality management of large-diameter shield segments throughout the whole life cycle in combination with engineering practice.

4.1 High-precision steel formwork

The steel form for large-diameter segments is generally of the double-taper universal design, and the inclined straight mandrel and embedded inner nut boss are used to construct the segment hand hole. Due to their super-large diameter, the arc length of blocks is large, and the whole side plate of the steel form is a long and narrow structure, so attention should be paid to deformation control during the fabrication process. At present, the steel forms of major manufacturers have their own characteristics in structural form, vibration mode and opening and closing modes. Most foreign manufacturers adopt flexible designs, enabling the form width to be adjustable during use; in China, the rigid design is basically adopted without width adjustment. Forms are opened mainly around the hinge joint and by translation motion.

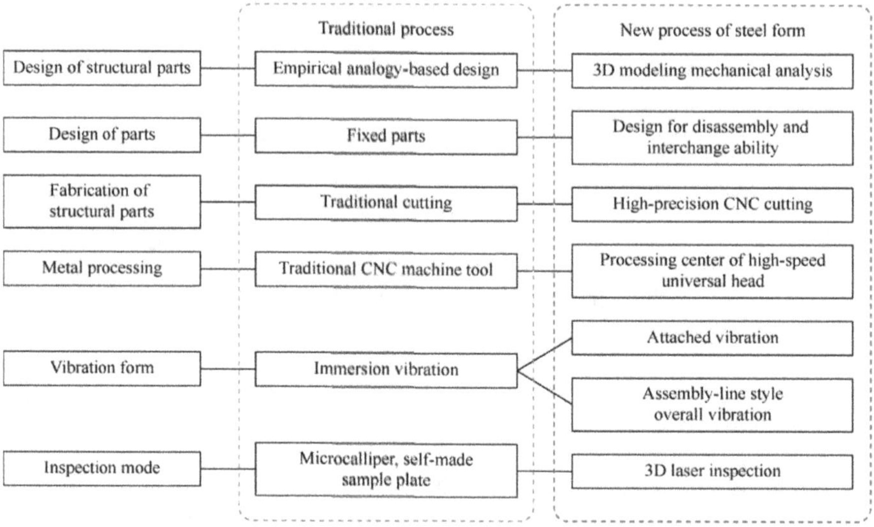

Fig. 4.1. New process flow of large-diameter and high-precision steel form.

See Fig. 4.1 for the new process flow for large-diameter and high-precision steel forms.

4.1.1 High-precision steel form design

4.1.1.1 *Structural form of segment*

The lining ring is generally of the double-taper universal ring form with a thickness of more than 500 mm. A convex surface is set at the segment ring face to be in contact with the jack to improve its local compression resistance. The segments are generally connected with inclined bolts circumferentially and longitudinally.

Figure 4.2 shows the segment ring formation of the Shanghai Yangtze River Tunnel with a diameter of 15 m. The lining ring is divided into 10 blocks, namely 7 standard blocks (B), 2 adjacent blocks (L) and 1 closure block (F). A 4 mm high convex surface is set at the point of contact between the ring face of the 650 mm thick segment and the jack. Thirty-eight M30 longitudinal inclined bolts are used to connect the segment rings, and two M39 circumferential inclined bolts are used to connect the blocks. Segment rings are assembled with staggered joints.

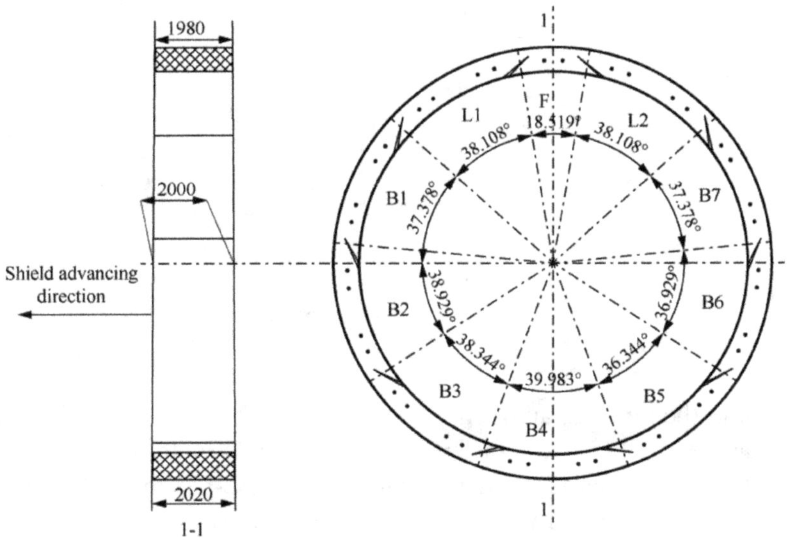

Fig. 4.2. Schematic diagram of segment ring formation of Shanghai Yangtze River Tunnel (unit: mm).

4.1.1.2 *Structural form of steel form*

The steel form is composed of a base, side plate, end plate and auxiliary parts (embedded device, opening device, drive mechanism, locking mechanism, core mandrel device and cover plate) (Fig. 4.3). The base, end plate and side plate (commonly known as the three major pieces) are connected into a whole through a locking mechanism, and the inner cavity space formed is deemed the actual segment shape and size. The high precision of steel forms is a necessary prerequisite to ensuring the dimensional accuracy requirements of concrete segments. The functional requirements of steel forms are as follows:

(1) The base, end plate and side plate are all steel structures welded by the non-annealing process. The four sides of the base are subject to metal processing, except the arc face; the end and side plates are all subject to metal processing.
(2) The base, end plate and side plate are connected in the form of bolt locking.
(3) The opening and closing of the side plate are realized by pushing and pulling rollers on a guide rail; the end plate is opened and closed

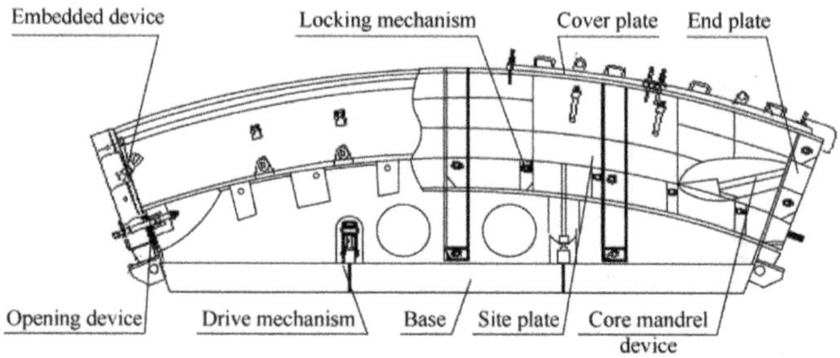

Fig. 4.3. Structural diagram of steel form.

around the hinge at an opening angle of 15°–20°; and the cover plate, composed of blocks, is fastened to the end/side plate by fixtures.

(4) The embedded parts ,such as shear pins and pre-tensioned steel pipes, are selected according to the engineering requirements.

4.1.1.3 *Key points of steel form design*

Compared with ordinary steel forms, the steel form of large-diameter segments has a longer arc length and increased weight, which puts forward higher requirements for the design to solve the problems of a long and narrow side plate structure of the steel form due to a large arc length, which results in instability and easy twist deformation. It is also difficult to ensure the arc face accuracy of the base due to its large arc length, and the reliability of the seal between the base and the side plate is also affected. With the increase in segment weight, the stress of the steel form base changes, which puts forward higher requirements for the strength and stiffness of the steel form base.

(1) **Deformation and precision control of steel form:** It is necessary to carry out 3D mechanical analysis on the structure of large-diameter steel forms to determine the fabrication process, for the sake of reducing structural deformation. The steel form structure should be designed in detail to ensure the steel form accuracy, with a width tolerance of ±0.5 mm, an arc length tolerance of ±0.7 mm for the cavity, an angle tolerance of ±0.02° for the ring face and an angle tolerance of ±0.01° for the end face. The key dimensions of the segment

form should be detected accurately and comprehensively by the 3D laser scanning technology, such as the ring width, arc length and depth after erection, providing reliable data support for controlling the deformation of the steel form, improving the overall accuracy and ensuring the quality of segment fabrication.

(2) **Sealing form of steel form structure:** The sealing design of the steel form structure should meet the sealing reliability at the fit positions of the side plate, end plate and base.

(3) **Interchangeable design of steel form fittings:** Due to the requirements under different working conditions of tunnel construction, it is often necessary to produce segments to several special specifications. If a new steel form is designed separately, the fabrication cost will increase. Therefore, interchangeable fittings should be considered in the steel form design, including shear pin and pre-tensioning device, so that the same set of steel forms can produce multiple types of segments.

4.1.1.4 *Structural design and 3D mechanical analysis of three major parts*

The base, side plate and end plate are the three most important parts of the steel form. Scientific and reasonable structural design is required to meet the requirements in the aspects of cost and function of the steel form. Mechanical analysis should be carried out by establishing a 3D calculation model.

(1) **Base design:** The base serves as a benchmark of the steel form, mainly consisting of (Fig. 4.4) an arc plate, end and side plates, inner and outer plates and the bottom steel structure. The sizes of all the base parts determine the internal cavity width, arc chord length, end face angle and arc face angle of the entire segment steel form, so they are deemed critical to ensuring the overall accuracy of the steel form, among which controlling the arc face size and the angle between the two end faces α become the focus in fabrication.

The base height is designed according to the height of the arc face plus an allowance to adapt to machining, generally not exceeding 1,000 mm. While meeting the strength and stiffness requirements, the arc plate should also match the side plate sealing strip.

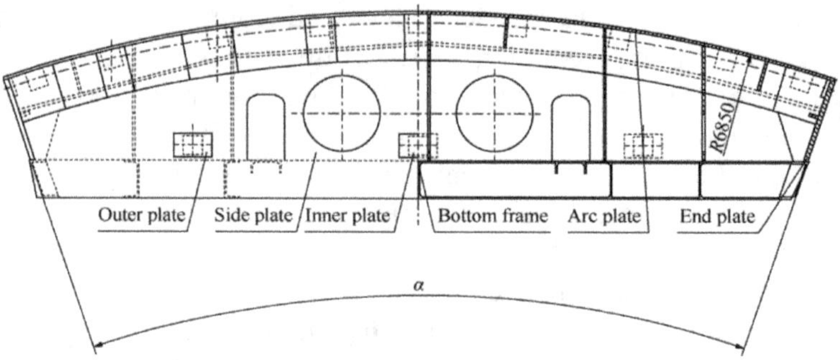

Fig. 4.4. Schematic diagram of base (unit: mm).

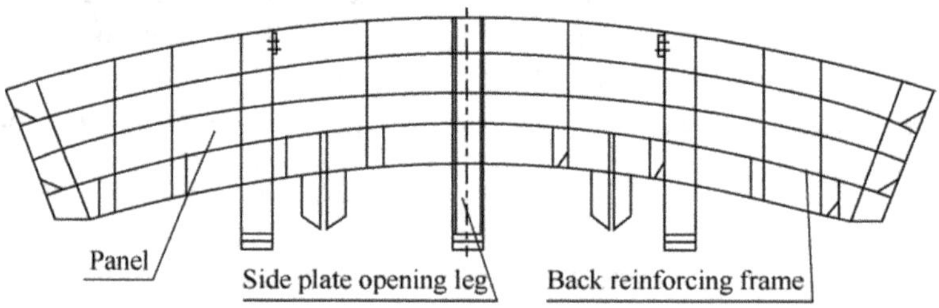

Fig. 4.5. Schematic diagram of side plate structural parts.

The bottom frame of the base is a steel structure made of channel steel, which is required to be in steady contact with the ground. Moreover, to ensure overall accuracy, the overall twist deformation of the base must be strictly controlled within 1 mm.

(2) **Side plate design:** With the main function of controlling the side shape and arc chord length of the segment, the side plate (Fig. 4.5) is mainly composed of a panel, back reinforcing frame, side plate opening leg, etc. The structural parts of the side plate are thin and long, bringing potential structural instability and deformation. Therefore, a considerable number of reinforcing ribs are designed at the appropriate positions on the back of the panel.

The side plate cavity surface (Fig. 4.6) should not only ensure the flatness of the entire side of the steel form but also effectively ensure the actual size after segment casting. Both ends of each mortise-and-tenon on the cavity surface is beveled 45° to facilitate assembly with the end plate.

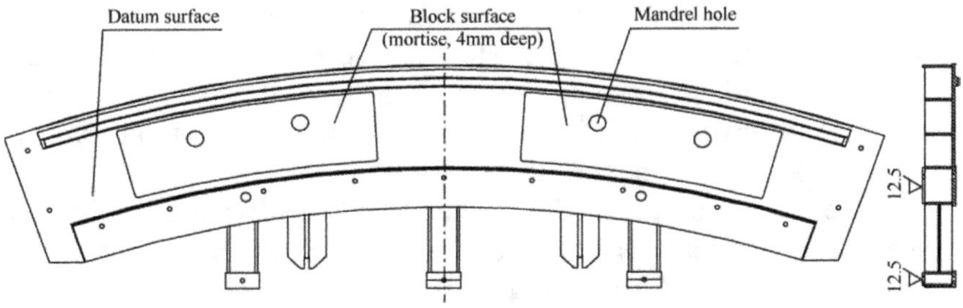

Fig. 4.6. Schematic diagram of side plate cavity surface.

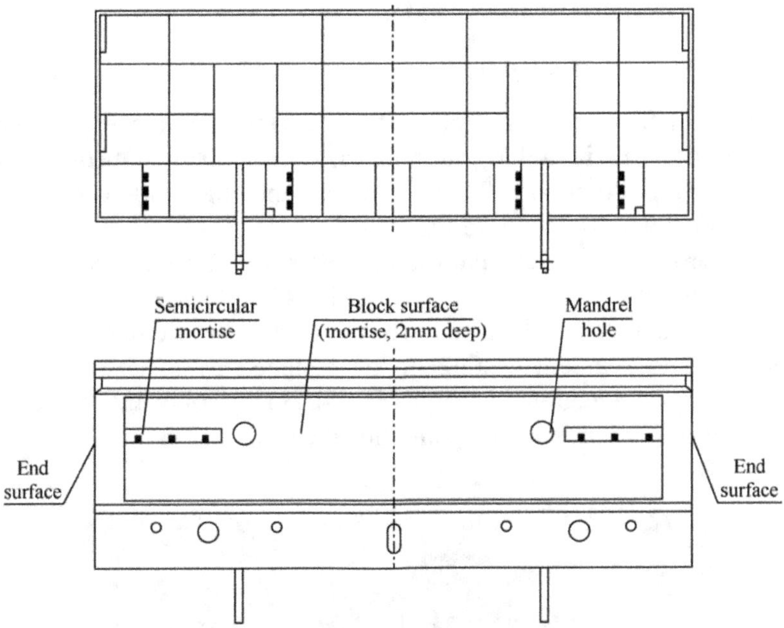

Fig. 4.7. Schematic diagram of end plate.

(3) **End plate design:** The end plate (Fig. 4.7) is mainly used to control the end face shape and arc chord length of the segment. The end plate is designed similarly to the side plate, which is mainly composed of a panel, a back reinforcing frame and a connecting leg with the base.
(4) **Safety analysis:** The influence of concrete pouring on the strength, stiffness and fatigue life of the steel form should be analyzed through

computer-based 3D numerical simulations so as to realize the optimal design of super-large-diameter steel forms. The base, side plate and end plate are the most important parts to form the main structure of the steel formwork, so it is necessary to conduct a safety analysis on the strength and stiffness of the three major parts.

The 3D numerical model of the structural parts is established to analyze their safety performance. See Fig. 4.8 for the safety analysis results of the steel form of segments for the Shanghai Yangtze River Tunnel. The analysis results show that the safety performance of the base, side plate and end plate of the steel form meets the requirements.

4.1.1.5 *Design of other parts*

(1) **Design of form core and embedded parts:** The adjacent segments are connected with a diagonal bolt, fastened with an embedded nut at one end and provided with a hand hole at the other end to connect the embedded nut, as shown in Fig. 4.9.

According to the segment requirements, the form core and embedded parts are designed, as shown in Fig. 4.10.

The hexahedral embedded parts are used to carry embedded plastic nuts during segment casting, to be fastened by a hexahedron and spacer in an integrated design (Fig. 4.11) to solve the positioning problem of the inclined hole and hexahedron.

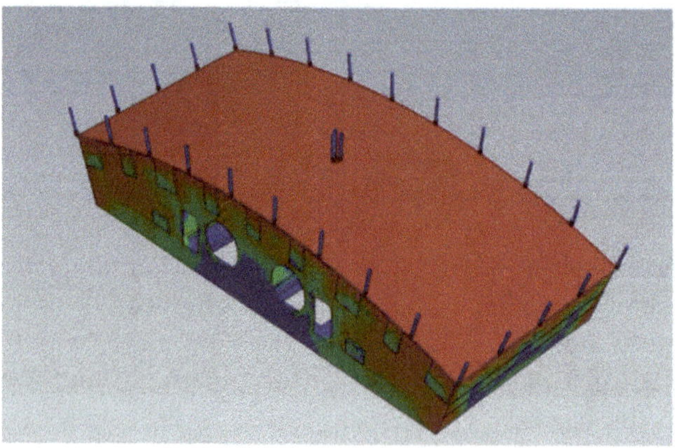

Fig. 4.8. Safety analysis of base at 70°C.

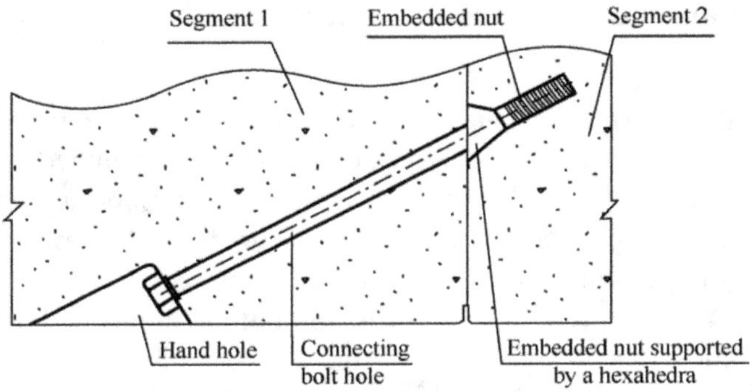

Fig. 4.9. Schematic diagram of segment connection.

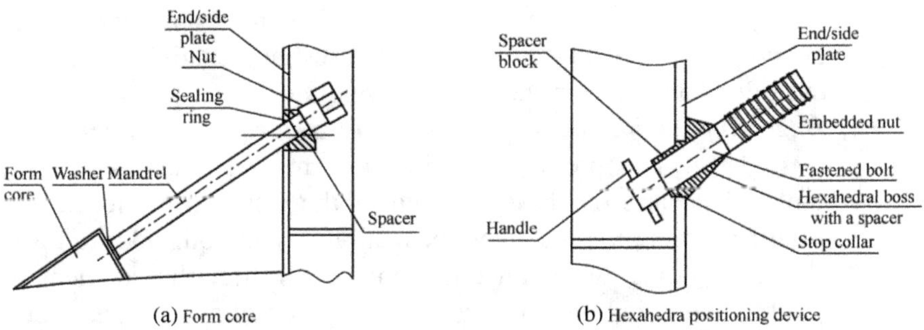

Fig. 4.10. Schematic diagram of form core and embedded parts.

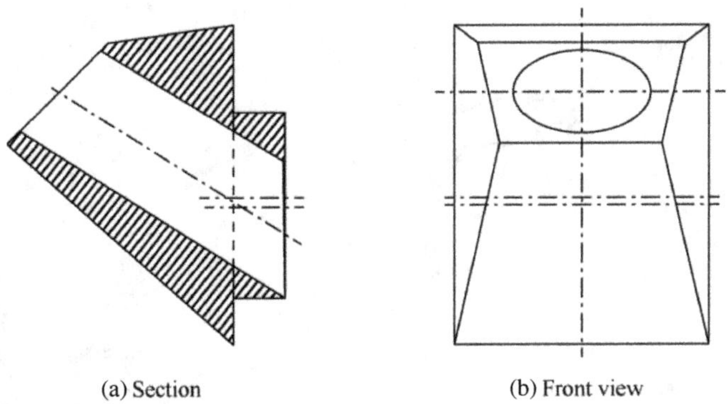

Fig. 4.11. Schematic diagram of hexahedral embedded parts.

(2) Interchangeability design of steel forms for profile-varying segments

(i) *Interchangeability between shear pin steel form and universal steel form*: The shear pin steel form is used to add the embedded parts on the side of the steel form during the segment pouring and mount a shear pin into the corresponding shear pin hole formed after the segment is cured so as to meet the design requirements of the segment. The universal steel form, without the embedded parts, is flat on the panel. A $\varphi 60$ hole is designed in the side plate, and the outer diameter of the pedestal of the shear pin steel form and the cylindrical bulkhead of the universal steel form is also $\varphi 60$ (Fig. 4.12). A splice is added on the outer side of the side plate, which is directly welded to the side plate for fixation. These two forms are interchangeable, with the same welding position on the splice and without contact with the side plate so as to prevent local deformation of the side plate cavity surface during welding.

(ii) *Interchangeability between pre-tensioned steel form and universal steel form*: The pre-tensioned steel form is to add hand hole embedded parts on the side plate of the steel form and embed seamless steel pipes between the hand holes on both sides to form the hand holes and connecting bolt holes after the segment is poured and cured, facilitating the connection of special segments. The hand-hole embedded parts are fixed to the side plate with nuts. The hand-hole embedded parts on both sides fix the seamless steel pipe at the closure of the side plate, as shown in Fig. 4.13.

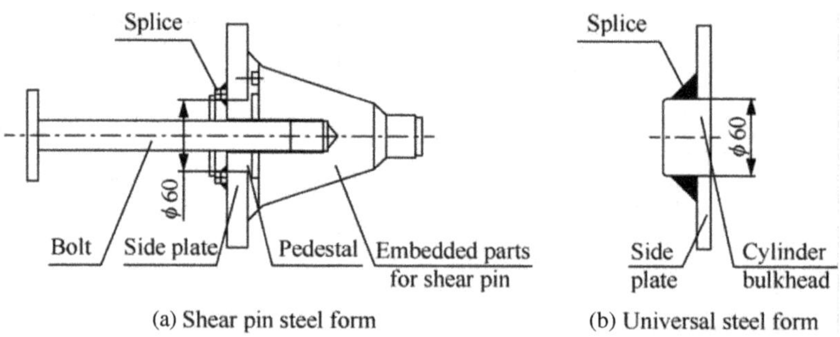

Fig. 4.12. Schematic diagram of interchangeability between shear pin steel form and universal steel form.

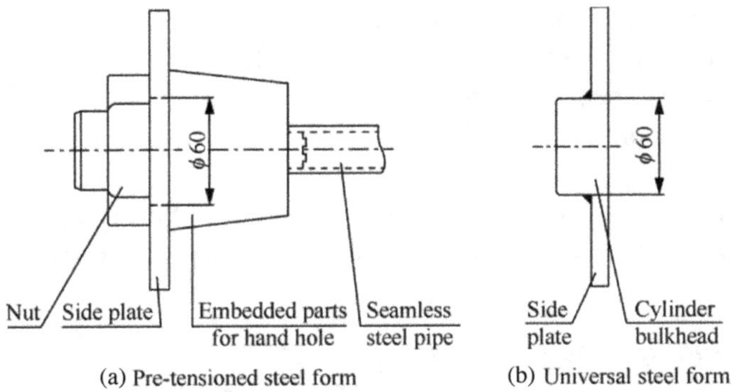

Fig. 4.13. Schematic diagram of interchangeability between pre-tensioned steel form and universal steel form.

4.1.2 Fabrication of high-precision steel form

4.1.2.1 *Fabrication and metal processing of the structural parts of the three major parts of the steel form*

(1) **Fabrication and metal processing of the base structural parts:** The two side plates at the front and back of the base are cut by a high-precision CNC cutting machine. The base arc surface should be controlled with accuracy to maintain a zero clearance within 1,000 mm at both ends and less than or equal to 0.5 mm clearance in the middle.
(2) **Fabrication and metal processing of side plate and end plate structural parts:** A CNC flame cutting machine is used for the blanking of the panel to control the flatness of the panel within 1 mm.

The non-annealing stress relief metal processing is adopted for the fabrication of end/side plate structural parts, which can effectively avoid large internal stress on the workpiece due to frequent welding, thus affecting the metal processing accuracy of the end/side plate and even causing deformation or cracking of the workpiece. The fabrication steps for structural parts are as follows:

(i) Make the solid modeling of the end/side plate using computer-based digital 3D modeling software.
(ii) Produce a datum bottom plane by machining.
(iii) Produce a plate cavity surface by machining.

(iv) Use special tools for roughing.
(v) Finish the cavity surface and pin hole.
(vi) Use the special fast polishing tool in cooperation with the new jet equipment for ultra-fine finishing.

Finally, the flatness on all planes and between parallel planes is controlled within 0.05 mm, and the surface roughness reaches Ra1.6.

4.1.2.2 *Final assembly of steel form*

The final assembly is key to ensuring the accuracy of the steel form, and the combined method of assembly platform and sample plate is usually adopted. The two sides of the end plate and the base of the double-taper steel form are processed at a CNC machining center, and the vertical installation technology is directly adopted. After the end plate and base are processed as a whole, the final assembly process is started, and the steps are as follows:

(1) Use a sample plate to measure the arc chord length of the inner diameter of the steel form, with an error less than 0.2 mm acceptable.
(2) Fit the side plate. The side plate is kept in a height position consistent with that of the end plate, subject to measurement by a depth micrometer with an acceptable error of 0–2.5 mm, and is then positioned and locked. The angle tolerance accuracy at the ring face of the end/side plate must be ±0.02°, and that at the end face must be ±0.01°.
(3) Install the side plate drive mechanism and end plate opening device, which must be agile without jamming.
(4) Install the form core, embedded parts and related parts.

After the final assembly, the width and height of the inner cavity of the steel form, the arc chord length, the distance, angle and height of the form mandrel, the angle and position of the hexahedral embedded parts, the joint clearance, the form closing condition and other items must be rechecked.

4.1.3 Inspection of high-precision steel form

In the process of final assembly, a high-precision sample plate is used to detect the arc chord length, a micrometer with an accuracy of 0.01 mm is used to measure the inner cavity width, and a depth gauge with an accuracy of 0.02 mm is used to inspect the depth of the steel form. Due to the particularity of the double-taper steel formwork, the values measured at any width of a steel form are unequal, so the width is measured at 10 positions selected on each steel form to ensure the correctness and validity of the data.

In order to ensure the overall accuracy of the steel form, the accuracy requirements are increased for the angles on the steel form ring face and end face, so clearance is measured with a detection sample plate made based on the study on angle detection on the ring face, and the measurement result is converted into angle.

As shown in Fig. 4.14, β refers to the angle at the end face, that is, the included angle between the end plate and the base, with an acceptable requirement of ±0.01°. α refers to the angle at the ring face, that is, the included angle between the side plate and the base, with an acceptable tolerance of ±0.02°. There is no tool for directly measuring angle on the steel form, so length must be measured for conversion. As shown in Fig. 4.15, when the height is 351.5 mm and the included angle is 0.02°, $x = 351.5 \times \tan 0.02° = 0.12$ mm.

A ring-face angle sample plate is designed as shown in Fig. 4.16. The sample plate is processed and fabricated by a numerical control machine tool to ensure that the four contact points are parallel or perpendicular to each other in pairs, and its effective width should be less than the inner cavity width of the steel form.

A theoretical width difference can be calculated by placing the sample plate at the center line of each steel form. As shown in Fig. 4.17, the width at one measuring point on the steel form is smaller than that at the next; the theoretical value is 0.4 mm, and the single-side value is 0.2 mm. Within the angle tolerance of 0.02°, the clearance dimension is (0.2 ± 0.12) mm.

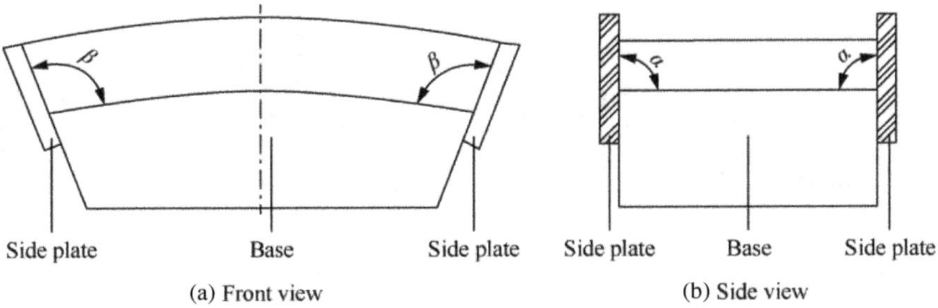

Fig. 4.14. Schematic diagram of end face angle and ring face angle of steel form.

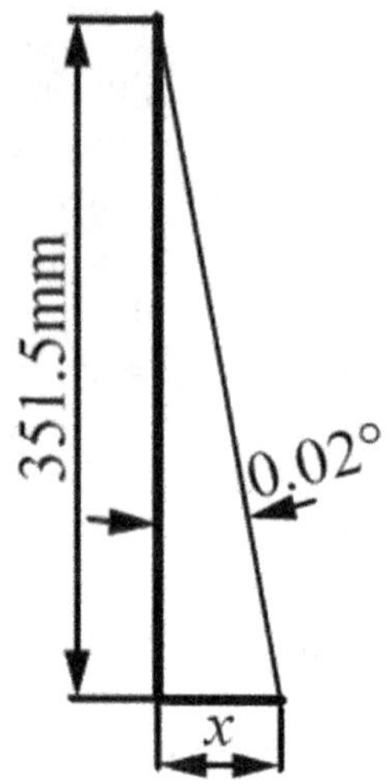

Fig. 4.15. Length conversion from height and included angle.

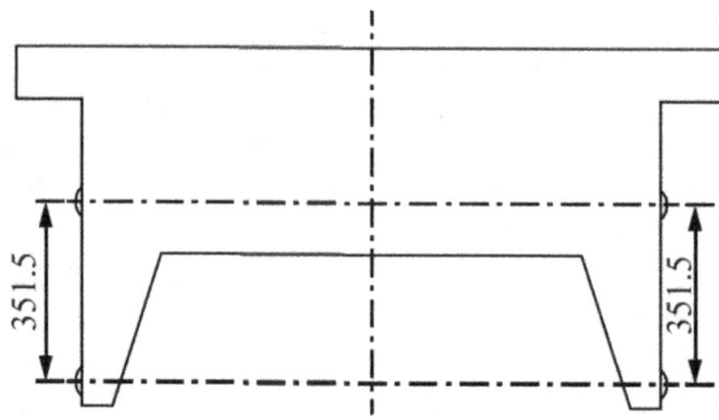

Fig. 4.16. Schematic diagram of ring face angle sample plate (unit: mm).

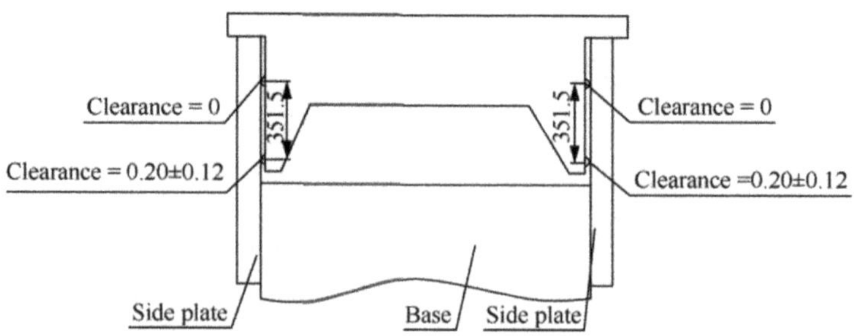

Fig. 4.17. Schematic diagram of ring-face angle measurement (unit: mm).

If the detected clearance is within this range, the ring face angle meets the design requirements.

For the detection of end face angle β, the arc length detection sample plate and spacer block are used together in the inspection process, and the same clearance measurement method is used for the ring face angle detection.

The final acceptance is tested again with a VMT measuring instrument (see Section 4.3.2 for the testing method).

4.2 High-performance concrete

Compared with other concrete structures, the factors that affect the durability of segment concrete are more complex, including the erosion from the air environment inside tunnels and from the external soil and water environment. In particular, groundwater contains a large number of corrosive media, which are adsorbed and deposited on the concrete surface, causing serious local erosion and damage to the concrete. Therefore, reasonable preparation of HPC is key to improving the durability of segments and the service life of tunnels.

HPC is generally used for large-diameter concrete segments. HPC has comprehensive technical indicator requirements such as high strength, high permeability resistance, high crack resistance and high durability. Through a large number of experimental research and engineering applications of super-large-diameter tunnel segment fabrication, certain experience has been gradually accumulated in the preparation technology of high-performance segment concrete. With the production process of large-diameter tunnel segments becoming mature, as well as the deepening of raw material research and the development of Internet technology, HPC preparation technology has also been constantly optimized, with a large number of new technologies emerging, as shown in Fig. 4.18.

4.2.1 Indicators of high-performance concrete

At present, the self-waterproofing design of segment concrete for large-diameter shield tunnels requires the use of HPC to achieve a 100-year service life of the structures generally, so the segment concrete is required to have good durability. Therefore, segment concrete is commonly prepared using the double-mixing technology of high-quality fly ash and ground slag to reduce the hydration heat and improve the performance of hardened concrete.

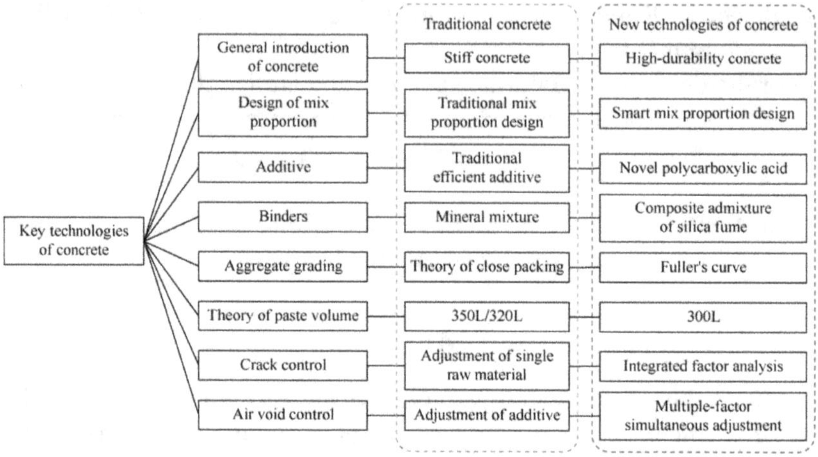

Fig. 4.18. Key technologies of large-diameter segment concrete.

In order to meet the construction progress and quality requirements, the segment concrete is required to be possessing high early strength, convenient for rapid vibration and highly plastic for construction. According to the traditional concrete design route, that is, considering the characteristics of concrete such as crack resistance, chloride ion erosion resistance, high-pressure water permeability resistance and carbonation resistance, high requirements are also put forward for the compactness of concrete. The specific indicators are as follows:

(1) Strength grade: C60.
(2) Impermeability grade: P12.
(3) Consumption of binders: 400–500 kg/m^3.
(4) Water-binder ratio: not greater than 0.35.
(5) Total alkali content: not more than 2.5 kg/m^3.
(6) Chloride ion content: not more than 0.06% of the total amount of binders, free of antifreeze and other additives containing chloride.
(7) Chloride ion diffusion coefficient: 3.0×10^{-12} m^2/s.
(8) Single super-large segment, with uniform texture, smooth surface and free of cracks.

According to the above indicators, segments are generally fabricated with stiff concrete prepared with conventional materials, at a low water–binder ratio, and with a large amount of composite mineral admixtures as the main technical measures, as shown in Table 4.1.

Table 4.1. Mix proportion design of concrete.

	Binder (kg)				Additives	Aggregate	Workability
Water-binder ratio	**Cement**	**Mineral power**	**Fly ash**	**Total binders**	**Water reduction rate (%)**	**Sand ratio (%)**	**Slump (mm)**
0.28–0.32	250–350	90–200	45–100	450–500	>20	34–38	60 ± 20

Table 4.2. Test items of high-durability concrete.

No.	Test item	Test method	Test standard
1	Crack resistance	Crack resistance test	L–V
2	Chloride ion erosion resistance	RCM method	$<3.0 \times 10^{-12}$ m²/s
3	High-pressure water permeability resistance	Impermeability test	>P12
4	Carbonation resistance	Quick carbonation test	<1.0 cm

In order to achieve a certain degree of compactness, the stiff concrete must be liquefied to an apparent density close to compactness so that its mechanical properties can be brought into play and meet the mechanical properties and durability required by structural design. In the process of practical application, higher vibration requirements may also be proposed.

4.2.2 High-durability concrete

Concrete durability refers to the ability of concrete to resist the action of environmental media and maintain its good service performance and appearance integrity for a long time, so as to maintain the safety and normal service of concrete structures. Generally, tunnel segment diseases are classified into harmful salt erosion, high water pressure erosion and concrete carbonization. Therefore, segment concrete is imposed with a high durability requirement, embodied in four indicators: concrete crack resistance, chloride ion erosion resistance, high-pressure water permeability resistance and carbonation resistance, as shown in Table 4.2.

4.2.2.1 *Crack resistance*

Cracks inevitably occur in the precast process of concrete segments, mainly due to an uneven concrete hydration rate, non-uniform internal temperature stress or uneven local water loss of concrete. In addition, the steam curing process of segment concrete accelerates the hydration reaction of binders and the generation of concrete hydration heat, and the sharp rise in internal temperature leads to cracking in the segment. According to their causes, cracks can be classified into shrinkage cracks and temperature cracks.

The measures adopted to deal with the above two kinds of cracks include: reducing the concrete hydration heat and slowing down the concentrated release of concrete hydration heat using the technology of composite admixtures; reducing the temperature stress caused by the temperature difference between the segment and the outside by optimizing the steam curing process; reducing the probability of internal cracks in concrete by improving the homogeneity in the concrete pouring process. Moreover, the paste control technology can also be used to reduce the paste on the concrete surface, or additives can be added to improve the moisture retention ability of the concrete surface so as to reduce micro cracks on the surface.

4.2.2.2 *Chloride ion erosion resistance*

The most dangerous corrosive medium that reinforced concrete structures may encounter during their service life is chloride ion (Cl^-). The segment structure is in contact with groundwater, soil layer and air and is in the dry and wet states alternatingly. The corrosion of reinforcement caused by chloride ion erosion is the main factor affecting the durability of the segment structure. The characteristic of chloride ion erosion is that the chloride ion diffuses from the surface to the internal reinforcement and then penetrates the oxide film on the reinforcement surface to produce electrochemical corrosion. Although the chloride ion does not produce corrosion products and is not consumed in corrosion, it plays a catalytic role in corrosion as a carrier of corrosion medium.

The durability of concrete structures in the chloride-ion environment mainly depends on the compactness of the concrete, the thickness of the concrete cover, reinforcement and environmental conditions. In order to prevent reinforcement from being penetrated by chloride ions and subsequent rusting, it is necessary to maintain concrete in high alkalinity and the passivation state of reinforcement. Therefore, improving the compactness of concrete can effectively reduce the harm of chloride ion erosion. By adopting the technical route of a low water–binder ratio, the grading of coarse and fine aggregates is adjusted to reduce the porosity of concrete, and high-performance additives are added to optimize the pore structure to gradually improve the compactness of concrete.

4.2.2.3 *High-pressure water permeability resistance*

High-pressure water permeability refers to the difficulty level of concrete in resisting the seepage, diffusion or migration of confined water, which reflects the size, number and connectivity of the internal pores of concrete to a certain extent. The penetration of high-pressure water not only causes certain damage to the concrete structure but also acts as a carrier of other harmful ions to accelerate the corrosion of concrete.

The interconnecting pores in concrete are gradually formed, mainly due to the hydration process of water in concrete. A low water–binder ratio can reduce the amount of free water in concrete. In addition, the additives can optimize the structure of the internal pores of concrete, reduce the pore size of concrete pores and improve the impermeability of concrete.

4.2.2.4 *Carbonation resistance*

Carbonation of concrete refers to the process in which alkali in concrete reacts with CO_2 in the environment to produce $CaCO_3$. The carbonation reaction neutralizes the alkaline substances in the concrete materials, leading to a decrease in the concrete pH value, destroying the potential difference formed by the passivation film on the reinforcement surface, and causing electrochemical corrosion. Aggravating corrosion will lead to concrete cracking, structural strength reduction and other problems. The carbonation resistance of concrete is closely related to its own compactness. The higher the compactness, the fewer the channels (such as pores) available for CO_2 diffusion in concrete, and the stronger the carbonation resistance.

4.2.3 New concrete technology

4.2.3.1 *Mix proportion design method*

Through analysis and research on the data of the concrete mix proportion design test of large-diameter tunnel segments, HPC mix proportion design has become more reasonable. There are many factors to be considered in

HPC mix proportion design, such as durability, strength, workability, volume stability and economic efficiency, and the numerous and complex constituents of the raw materials involved.

The relevant calculation program and database have been established with the help of computers to make the mix proportion design of HPC programmed and systematic. The test or design mix proportion is generated according to input on raw material standards, key design parameters and expected effects.

Taking the measured bulk density of coarse aggregate as the starting point, through the data processing and analysis on the concrete coarse and fine aggregates, water, various admixtures and cement constituents in the database and defining an appropriate water consumption and the dosages of various mineral admixtures, a closed HPC mix proportion design program with feedback and regulation capabilities is established, and the test results are fed back to the programmed design system for fitting. The process of determining the benchmark mix proportion of HPC is shown in Fig. 4.19.

With the gradual maturity of HPC mix proportion design and programmed design technology, the approach, first through reasonable data modeling and parameter design, including water–binder ratio, proportions of composite admixtures, aggregate grading and paste proportion, and then test fitting, can avoid the artificial influence factors relying on expert judgment or analogical experience. Any deviation revealed in the process

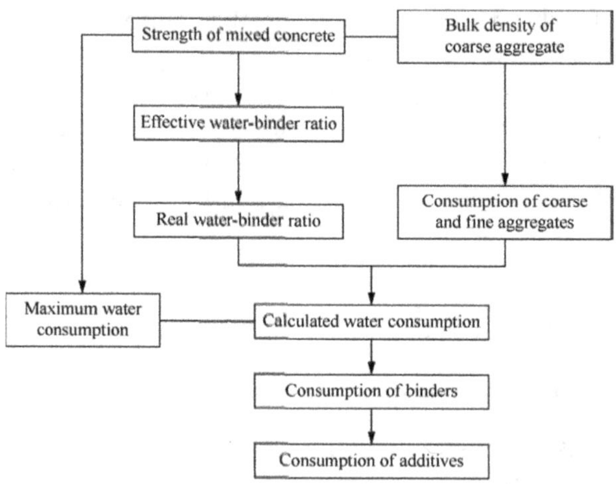

Fig. 4.19. Flowchart of the process of benchmark mix proportion of HPC for segments.

of test fitting is added to the database for quality correction so as to achieve consistency between the HPC test results and mix proportion design.

4.2.3.2 *High-performance admixture*

Due to the large volume and the large area of the outer arc face of a single segment of super-large-diameter tunnel segment, a high camber plasticity is required for the segment concrete. Moreover, considering that the arc face of segments is subject to finishing, the workability of the segment concrete is deemed the main consideration for the selection of additives.

In order to ensure the high strength, low water–binder ratio and low water consumption of concrete, the water reduction rate of the admixture must exceed 20%. Due to the limitations of the performance of the admixture master batch, most segment concrete is mixed with aliphatic admixtures and polycarboxylic acid admixtures.

Aliphatic admixtures are widely used in segment concrete because of their special flow pattern performance, low sensitivity and small change in fluidity, while polycarboxylic acid admixtures are not conducive to stable production due to their high sensitivity, leading to large fluctuations in application despite the performance advantages of high water reduction rate, good concrete slump, low concrete shrinkage, etc. In recent years, the requirements for high strength and high fluidity of concrete have become increasingly strict, which has promoted the technical development of polycarboxylic acid water reducers. The market share of aliphatic admixtures has gradually decreased due to their poor solubility of main constituents and low feasibility of compounding.

With in-depth research on reducing the sensitivity of polycarboxylic acid-type admixtures and the types of polycarboxylic acids on the market, polycarboxylic acid admixtures have developed into the third generation of high-performance admixtures.

Polycarboxylic acid admixtures have good adaptability to binders in concrete, which enables higher concrete performance than expected. The existing polycarboxylic acid admixtures can not only avoid the problems of insufficient slump retention and strong temperature sensitivity in high-temperature environments but also solve the problems of high viscosity of concrete with a high content of admixtures and a low water–binder ratio and strong sensitivity to the clay content of sand and stone aggregates.

Polycarboxylic acid concrete segments have the following characteristics:

(1) Admixtures can be uniformly blended in solution with a low proneness to precipitation.
(2) It is more adaptable than binders with different constituents and proportions.
(3) At the same dosage, the water reduction rate of the polycarboxylic acid high-performance admixture is far higher than that of other admixtures.
(4) On the premise of ensuring various properties of concrete, thixotropy is improved to reduce construction difficulty.

4.2.3.3 *Compound mixing technology of admixtures*

In binders, the admixtures other than cement are called composite admixtures. The addition of composite admixtures to segment concrete is conducive to improving the cohesiveness and hydration of concrete (reducing the hydration heat of cement and the temperature of concrete), improving the late strength of concrete, improving the internal structure, compactness and workability of concrete, raising the density of particles, reducing porosity, improving the pore structure, improving the corrosion resistance of concrete and delaying the performance degradation of concrete. In particular, fine mineral admixtures are important to inhibit alkali aggregate reactions.

Composite admixtures mainly include fly ash and ground slag micropowder. Research experience shows that pozzolanic materials, such as fly ash and ground slag micro powder, mixed synchronously will produce a large amount of CSH gel in the process of hydration. The gel contains many tiny gel pores and capillary pores, which can play a better role in impermeability, crack resistance and chloride ion diffusion resistance than single mixing. However, there are also dosage limits for such admixtures. Any dosage beyond these limits causes a reduction in the early mechanical properties of concrete instead of further improvement in its performance.

In order to quickly determine the impact of composite admixtures on the chloride ion diffusion resistance of concrete and accordingly determine its appropriate dosage, the chloride ion diffusion is evaluated by using the AC impedance spectrum method with seven-day concrete samples. The advantage of this method is that the concentration gradient causing diffusion is generated by the electrode reaction itself, rather than any artificial factor. The AC electric field produces a very small concentration difference without causing a large disturbance to the whole measuring system.

4.2.3.4 *Fuller's curve fitting technology*

Continuously graded crushed stone is generally used as coarse aggregate in segment concrete. Coarse aggregates with different particle sizes are combined in different grades to form the basic skeleton of concrete. The particle shape, particle composition and other properties of coarse aggregate have a very important impact on the mechanical properties, shrinkage and creep, durability, etc. of concrete. Among them, good particle grading of aggregate can minimize the skeleton accumulation porosity, reduce the consumption of mortar and binders, thus reducing the drying shrinkage of concrete, reducing the hydration heat, and improving the volume stability and durability. In addition, good aggregate grading can improve the workability and service performance of concrete with the same water consumption. Therefore, it is very important not only to seek the best particle size of aggregate but also to quickly determine the reasonable mix proportion of different size fractures of aggregate in concrete production and to maintain the stability of aggregate grading.

In the concrete of segments for super-large-diameter tunnels, the close packing theory is generally used to calculate the aggregate consumption and grading.

According to the principle of maximum compactness, 5–25 mm and 16–31.5 mm gravel concrete (the proportion is 3:7) are used to form continuous grading. The mixture grading is conducive to improving the segment's workability. The mixture grading test values are shown in Table 4.3.

In recent years, aggregate grading based on the traditional close packing theory has often been affected by the fluctuation of raw materials in the practical application process, which leads to the failure to achieve the desired results. The maximum packing density and minimum porosity can

Table 4.3. Grading test values of mixture.

	Particle size (mm)							
Mesh size	**40**	**31.5**	**25**	**20**	**16**	**10**	**5**	**2.5**
Cumulative residue rate (%)	0	2	27	71	88	96	98	99

be obtained by exploring different combinations of coarse aggregates according to Fuller's curve theory. Taking Fuller's graded particle composition as the standard, the proportion of crushed stone can be calculated by using the dominant particle size content of coarse aggregate in different specifications, and under the condition of its inherent particle grading, the particle composition close to Fuller's grading can be obtained; compared with Fuller's grading, the smaller the sum of absolute deviation values of each size fraction, the greater its compactness.

Fuller's grading curve is ideal, and the actual mineral aggregate should be allowed to fluctuate to some extent. In the actual application, the indicator should not be a constant (Fuller's formula indicator is 0.5), so Fuller's maximum density curve is changed to the general formula of *n*th power, expressed in the following formula:

$$P_i = 100\left(\frac{d_i}{D}\right)^n, \tag{4.1}$$

where

P_i — cumulative throughput of sieve with mesh size of i mm (%);
d_i — particle size of aggregate at each size fracture (mm);
D — maximum particle size of aggregate (mm).

According to a large number of tests, it is found that $n = 0.3$–0.5 contributes to good compactness.

For the 5–25 mm continuous graded aggregate commonly used in segment concrete, crushed stones in four single-size fractures of 5–10 mm, 10–16 mm, 16–20 mm and 20–25 mm can be sieved, respectively, and mixing samples can be prepared by using the calculation results of Fuller's formula with an indicator of 0.5 (Table 4.4).

Calculations based on Fuller's curve fitting technology can be conducted with the help of computers. As long as the coarse aggregate sieve analysis data is input, the mix proportion can be generated quickly. Minor

Table 4.4. Size composition of 5–25 mm Fuller's graded coarse aggregate.

Indicator of Fuller's formula	Proportion of size fracture (%)				Bulk		Close packing	
	20–25 mm	16–20 mm	10–16 mm	5–10 mm	Density (kg/m³)	Porosity (%)	Density (kg/ m³)	Porosity (%)
0.5	19.10	17.08	30.31	33.51	1,564	42.50	1,682	38.18

adjustments can be made with accurate crushed stone data from sieve analysis, and the need for adjustments can be determined by the increase, decrease or change in the range of the sum of the absolute values of deviations in size fractures. In order to achieve the best combination of coarse aggregate, the reasonable particle size grading of crushed stone itself is critical. The control standard for coarse aggregate particle size grading used in the production process is formulated according to Fuller's grading requirements.

In order to ensure compliance with the particle grading requirements, sand and stone aggregates are supplied by grades in some developed countries, and fed according to Fuller's curve fitting technology for mixing. At present, the technology of double-graded aggregate mixing is gradually maturing. Optimized as per Fuller's curve, 5–25 mm and 16–31.5 mm are used for mixing at a ratio of 7:3, which not only ensures the strength grade of concrete but also guarantees the stability of concrete quality in the production process.

4.2.3.5 *Paste control technology*

Concrete paste refers to the sum of the volumes of binders and water in concrete, which can be calculated by the following formula:

$$V_p = V_w + V_c + V_{ma}, \tag{4.2}$$

where

V_p — paste volume (m^3);
V_w — water volume (m^3);
V_c — cement volume (m^3);
V_{ma} — admixture volume (m^3).

The workability of fresh concrete is mainly determined by the volume and rheological properties of the concrete paste. The paste volume plays a key role in the impermeability, volume stability and even surface cracks of concrete. A too large concrete paste volume increases the hydration heat and uneven shrinkage inside concrete, raising the probability of large volume deformation and subsequent shrinkage stress, which is unfavorable to the crack resistance of concrete; however, a too small concrete paste volume results in insufficient wrapping, which is not conducive to construction with fresh concrete, in addition to the problem of insufficient cohesion of aggregates. Therefore, in the segment fabrication industry, there are two basic properties of HPC that are contradictory but must be considered concurrently: high impermeability and high volume stability (small shrinkage and creep). To ensure the optimization of these two properties, there is an optimal paste volume.

For the segment fabrication for large-diameter tunnels, through the comparison and selection of binders and the application of new additives, the constraint of 350 L optimal paste volume was successfully broken, and the paste is controlled below 320 L to meet the requirements of high impermeability and high volume stability of concrete. Moreover, with the reduction in paste volume, the surface cracking of the finished segment products is significantly reduced during the construction process.

With the continuous expansion of the polycarboxylic acid admixture market in recent years, the corresponding compounding technology has matured, and the polycarboxylic acid admixtures have played an increasingly more important controllable adjustment role in segmenting concrete. According to the indicator requirements, the maximum total consumption of binders for large-diameter segment concrete should not exceed 500 kg/m^3, and it is also necessary to consider the adverse impact of high and concentrated hydration heat on large volume concrete, so the control constraint of 320 L paste volume has also been gradually challenged. The concrete paste volume of segments for some large-diameter tunnels has been controlled within 310 L.

By improving the rheological property of fresh concrete paste, the workability and thixotropy of concrete can be improved to meet the requirements of on-site construction. The improvement of rheological properties directly affects the requirements for the volume of concrete paste, so that the wrapping requirements of concrete aggregate can be met at a lower volume of concrete paste, and in consideration of better fluidity concurrently, control of concrete paste below 300 L can also meet the requirements for segment fabrication in the future.

4.2.3.6 *Crack control technology*

In the production process of tunnel segments, due to the influence of materials, processes, equipment, environment and other factors, shrinkage cracks occur, greatly affecting the compressive strength and impermeability strength of the segments and causing much trouble during the subsequent repair work. In addition, cracks in segments will directly affect the safety and service performance of the tunnel structure, so measures must be taken to control the cracks within the allowable limit.

The segments for large-diameter tunnels, due to the large concrete consumption and cement hydration, are more prone to cracking, so stricter crack control is needed.

(1) **Analysis of causes of cracks in segments:**

(i) *Loads*: The loads acting on segments are classified into direct load and deformation load, and the load-induced crack is called a load crack. The former acts directly on concrete segments, causing concrete cracks and damages due to the stress reaching or exceeding the concrete strength value; the latter causes cracks in the structures under the stress beyond the specified limit, due to changes in external environmental factors, including humidity and pH. According to relevant research and analysis, the proportion of cracks caused by direct load is about 20%, while cracks caused by deformation load account for about 80%.

(ii) *Temperature change*: Temperature change causes cracks in the concrete segment, namely temperature cracks, which mainly include the temperature change in the environment and that inside the concrete. The change in environmental temperature is mainly reflected in the seasonal temperature difference. For example, during construction in the summer, any steep descent in temperature will lead to uneven solidification inside and outside of the concrete segment, and the subsequent temperature stress will lead to concrete cracking. In addition, rapid temperature change leads to rapid water loss on the concrete surface, which causes the shrinkage of cracks and finally leads to cracks that are difficult to repair. The internal temperature change of concrete is reflected as a large amount of hydration heat released from concrete hardening inside the segment, which makes the internal temperature rise and causes expansion. However, the segment surface is in a cooling shrinkage state because of fast heat dissipation. In this way, the too-large internal and external temperature difference results in large temperature gradient changes, leading to tensile stress on the surface and cracks. Especially, the segments for large-diameter tunnels, due to their large volume and poor heat transfer effect, are more likely to cause cracks due to changes in internal and external temperature differences.

(iii) *Equipment and process*: In the production process, due to improper selection or operation of equipment and processes, it is also easy to cause cracks. In segment fabrication, when concrete is poured into the form, a large amount of heat will be generated, and the water vapor will disperse. Concrete bleeds on the surface during the vibration construction, which is concentrated at the junction of the segment side and the side plate. After steam curing, the excessive water loss is more likely to cause cracks.

(iv) *Mix proportion of raw materials*: Concrete is the basic material for segment production, so the quality of its raw materials greatly affects the quality of segments. For example, any change in water content in concrete has a very obvious impact on segments. If the water content is too low, concrete will rapidly solidify, making the vibration construction inoperable; if the water content is too high, when concrete is solidified in its form, a large number of air voids are concentrated on the side plate, and the water loss is serious after steam curing, resulting in a large number of shrinkage cracks. In addition, excessive water–cement ratio, concrete segregation, improper mix proportion, poor vibration and other reasons during concrete construction may cause plastic deformation and plastic shrinkage after concrete pouring, up to 1% of the concrete volume; the hydration of cement produces $Ca(OH)_2$, which absorbs CO_2 in the air to form $CaCO_3$ and causes shrinkage. The combined action of carbonation shrinkage and drying shrinkage leads to concrete surface cracking; if the alkali content in concrete aggregate exceeds the national standard, cracks will also occur to damage the concrete structure.

(2) **Technical measures for crack control**

(i) *Strictly control the quality of concrete raw materials*: The raw materials for concrete segment production mainly include cement, water, sand, gravel and certain additives. Cement with low hydration heat and stable indicators, such as No. 425 Portland cement, should be used to prevent cracks caused by excessive hydration heat. In addition, in order to ensure the workability of concrete, coarse sand with an aggregate size of 5–25 mm is used to facilitate the pumping and mixing of concrete. Additives should be high-efficient retarders and water-reducing agents to ensure good cohesiveness and workability.

(ii) *Adjust the mix proportion of concrete*: The mix proportion of concrete directly affects the quality of cured segments, which is an important condition for controlling concrete cracks from the following aspects. First, the cement consumption should be controlled. On the premise of meeting impermeability and strength, cement consumption should be reduced to lower the heat of hydration. Secondly, active admixtures should be used, such as fly ash, which can reduce the probability of temperature cracks. Fly ash with a cement content less than 20% should be added to concrete to extend the setting time and improve concrete strength, durability and impermeability. The water–cement ratio should be strictly controlled. Engineering practice has proven that each 10% increase in water consumption reduces the segment strength by about 20%, reduces the adhesion by about 10%, and increases the dry shrinkage by about 25%. Controlling the water–cement ratio and appropriately reducing water consumption can effectively improve the impermeability and avoid concrete bleeding and drying shrinkage. In addition, the concrete slump should be regulated. A large slump causes rapid aggregate settlement. When the settlement becomes stable, the mortar settlement will be suspended, causing external bleeding, and the excess free water will gradually evaporate, forming capillary channels in the concrete and subsequently causing cracks.

(iii) *Improve the construction and curing process*: The vibration process should be improved to avoid quality problems caused by uneven vibration and reduce the probability of cracks. In addition, it is also necessary to optimize and adjust the curing process to ensure the rapid hardening of concrete, ensure that the strength of the segment meets the design requirements and prevent the segment from shrinkage cracking. When the strength of the segment meets the design requirements, it is placed into water for one-week curing, which should be included in the engineering design.

(iv) *Reduce the impact of environmental factors*: Reducing the impact of environmental factors mainly involves temperature and humidity. Ice cubes are added to the concrete to keep the temperature low so as to avoid rapid water loss; the construction should be carried out in a low-temperature environment as far as possible to ensure good solidification of segments. Attention should be paid to the initial curing of segments. During the period from the end of segment pouring to the time before form release, a large amount of heat is released to solidify concrete. Frequent watering is carried out to ensure that the concrete is solidified uniformly.

4.2.3.7 *Air void control technology*

For the segments used in the large-diameter tunnels, the existence of air voids will seriously affect the safety and service performance of the tunnels, mainly including:

(1) **Appearance:** The presence of air voids seriously affects the appearance of the segment.
(2) **Strength:** Air voids greatly reduce the compactness and strength of concrete. Relevant research shows that the strength of concrete is proportional to the number of pores; that is, the more air voids in the segment, the worse the overall strength of the segment.
(2) **Durability:** The existence of air voids on the surface of the segment reduces the effective thickness of the concrete cover of reinforcement and the corrosion resistance of the concrete segment. Especially in tunnels with rich groundwater, the segment's durability is seriously threatened.
(4) **Water seepage channel:** When there are many air voids on the segment side, even if the waterstop is used for waterproofing, the concave air voids are likely connected into leakage channels, which will cause water leakage in the tunnel and endanger the safety of the tunnel structure under serious circumstances.

(1) **Causes of air void formation on segment surface**

(i) *Raw materials*: Cement, water-reducing agents and the shape and size of aggregate particles have a great impact on the generation of air voids. If the cement contains calcium lignosulphonate, diethylene glycol, propylene glycol and other mixed grinding aids with an air-entraining effect, it is difficult to eliminate the uneven air voids introduced by mixing; the chemical substances in the concrete water reducer, such as polycarboxylic acid, will introduce a large number of tiny air voids that are uniformly distributed, stable and closed in the concrete. If the supplier of the water-reducing agent uses a cheap and low-quality air entraining agent to save costs, large air voids will be formed in the concrete segment, or even large interconnected air voids will be formed. Any unreasonable thickening constituents in the additives will make concrete too viscous, and it is difficult to discharge air voids during vibration.

(ii) *Concrete mix proportion*: Any change in the proportion of the concrete mix will greatly affect the slump. When the concrete slump is too large and there is relatively more free water, because of the curved surface characteristics of the segment, the concrete cannot be fully vibrated after being poured into the form, and a large number of air voids in the segment cannot be discharged; if the slump is too small, it is difficult to vibrate and discharge air voids.

(iii) *Production process*: The form and vibrating system have the most significant influence on air voids. Concrete generates heat when it is solidified in a steel form, and water in the concrete is also discharged under the squeezing action of a vibrating rod. The discharged water is concentrated on the side of the form. Because the steel form has good sealing performance, the side is provided with waterstop groove and positioning rod groove, and the inner wall of the form has many special-shaped convex structures, resulting in air void accumulation which cannot be easily discharged. Air void accumulation or residue occurs in concrete under the conditions of insufficient vibration and vibration omission.

(2) Prevention and treatment of air voids on the segment surface

(i) *Control slump reasonably*: Slump has a great influence on the strength of concrete after solidification and the generation of air voids. When preparing concrete, the slump should be strictly controlled. In particular, it takes some time for concrete to be transported from the mixing plant to the pouring and vibrating positions. During this process, the concrete will partially solidify, causing slump loss. Therefore, it is necessary to add a certain amount of retarder when preparing concrete to slow down the loss of slump.

(ii) *Control vibrating quality strictly*: At present, manual vibration is mostly used in China. During manual operation, due to the small gap between the reinforcement framework and the segment form, improper operation is likely to cause inadequate vibration and air void accumulation. The vibration treatment is carried out in a hybrid manner, that is, the combination of attached vibrator and manual vibration, with attached vibrator as the main method and manual vibration as the auxiliary method, which can achieve good results in the prevention and control of air voids in segments.

In addition, multiple rounds of tests have been conducted to explore the influence of cement type, sand ratio, water–binder ratio, release agent, etc., on air voids in concrete segments, revealing that cement type has little influence on the air voids on the concrete surface; the sand ratio has obvious influence on the air voids on the concrete surface; the water–binder ratio has obvious influence on the air voids on the concrete surface; and the form release agent has an obvious improvement effect on air voids on concrete surfaces, better than hydraulic oil.

4.3 Latest technologies for segment fabrication

The segments for large-diameter tunnels are characterized by their large size and weight and high precision requirements. Taking the segment of the Shanghai Beiheng Passageway as an example, its diameter reaches 15 m, and the concrete consumption per ring is about 58 m^3, while the diameter of the metro segment is only 6.6 m, and the concrete consumption per ring is only about 8 m^3. It can be seen that it is difficult and demanding to produce large-diameter tunnel segments in the automatic assembly line mode, and the traditional production process needs to be innovated to meet the fabrication requirements. See Fig. 4.20 for the technologies of segment fabrication developed in recent years.

4.3.1 Assembly-line-style technology of segment fabrication

With the assembly-line-style technology of segment fabrication centered on the intelligent control system, forms are controlled in circulation on the production line, and locomotives or translation trolleys are used to pass through all the positions on the track in order to realize the whole process of segment fabrication. The production process is shown in Fig. 4.21. The assembly-line-style fabrication is divided into five major systems: reinforcement framework delivery system, concrete feeding system, concrete

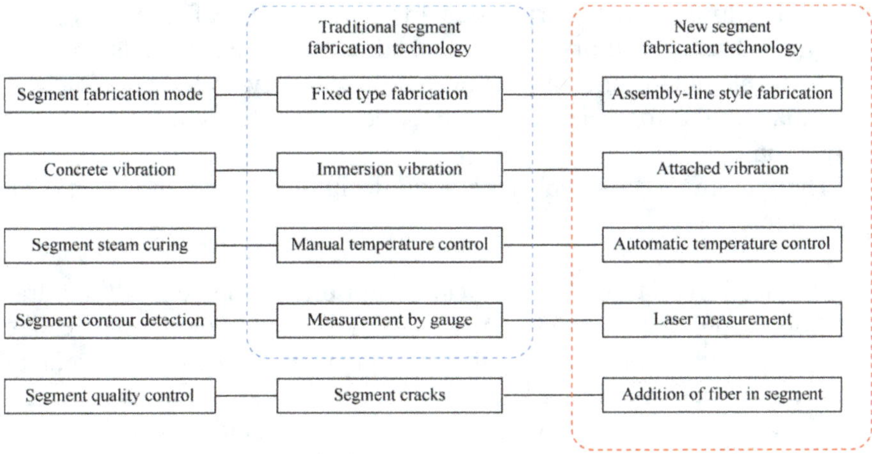

Fig. 4.20. Latest technologies for segment fabrication.

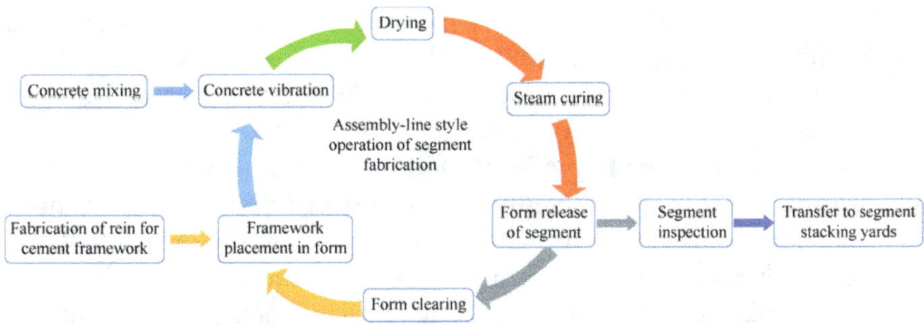

Fig. 4.21. Schematic diagram of assembly-line-style segment fabrication.

pouring and vibrating system, form delivery system and steam curing system. Under the cooperative operation of the above systems, an automatic assembly-line-style process can realize the standardized fabrication of large-diameter tunnel segments and improve production efficiency.

In the assembly-line process of segment fabrication, the setting and cooperation among different positions are important guarantees for smooth production. After the reinforcement framework is processed, it is transported to the form feeding position by the delivery system and lifted into the form after clearing and maintenance; concrete is transported to the concrete pouring and tamping position by the feeding system and then transported to the steam curing kiln for steam curing after pouring and tamping, vibration and drying; after steam curing, the segments are subject to form release and inspection before delivery. After clearing and maintenance, the form enters the next cycle for the assembly-line-style fabrication.

The automatic assembly-line-style technology has the following advantages:

(1) **Reduced production energy consumption:** Compared with the traditional production mode, the assembly-line technology can realize the fixed point operation of the reinforcement framework placement in the form, form release and other processes, and a small lifting area. The concrete pouring does not require crane operation, which greatly reduces crane operation time and effectively reduces production energy consumption.
(2) **Improved segment quality:** Under the assembly-line-style production mode, the workers work at fixed points, which is conducive to proficiency improvement and quality control. Moreover, the steam curing process is more precise and reliable in the heating and cooling modes through the control system.
(3) **Improved universality of automatic assembly line for segment production:** On the assembly line, forms of different specifications can be placed on the universal form trolley according to the engineering requirements, and the form positions are spaced at preset intervals so as to realize a share of the production line among segments of different specifications. In addition, the universal form trolley is used to facilitate form maintenance. When a form is faulty, the normal operation of the assembly line can still be ensured.
(4) **Improved productivity:** In the traditional production mode, each set of forms can only be used for production once or twice a day. Moreover, because the production site occupies a large area, the energy consumption of lighting and steam curing is high, increasing the cost. Each set of forms can be used for production 2–3 times a day by adopting the assembly-line-style production mode, with the output 1.5 times that of the traditional production mode, so as to improve the production efficiency. Moreover, through reasonable planning of production areas and scientific layout of steam curing areas, the land use rate can be effectively improved.

4.3.1.1 *Design and layout of positions*

In the segment production assembly line, the working time of different positions is the key to affecting the operation efficiency of the whole production line, so it is very important for the design and layout of the positions in the segment production line. According to the characteristics of large-diameter tunnel segments and different requirements, the production line is divided into eight procedures. Each procedure is arranged with a different number of positions according to the time required.

Procedure 1: form release and lifting of concrete segments, generally with 1 position.
Procedure 2: form clearing, application of release agent, generally with 1–2 positions.
Procedure 3: reinforcement framework placement in form, generally with 1 position.
Procedure 4: embedded parts installation, form closing, generally with 1 position.
Procedure 5: pouring and vibrating concrete, generally with 1 position.
Procedure 6: first finishing, generally with 1 position.
Procedure 7: natural curing, generally with 2–3 positions.
Procedure 8: secondary finishing and polishing, generally with the position in the pre-curing kiln.

After the secondary surface finishing, the semi-finished products are transported to the steam curing kiln for steam curing treatment. Different positions need to cooperate with each other to complete the assembly-line-style production efficiently. The assembly-line-style segment production is shown in Fig. 4.22.

4.3.1.2 *Control system*

The computer software-based control system controls each drive motor to perform corresponding actions by the programmable logic controller (PLC). As the brain of the whole assembly line, the control system sends all the action commands to coordinate all the steps of the production line, including the movement of forms, the transport of the reinforcement framework, the lifting and temperature control of the steam curing kiln, and the lifting and lowering of the vibration table. When each action is in place, a corresponding sensor or proximity switch will feed back the

Fig. 4.22. Assembly-line-style production of large-diameter tunnel segments.

action to the PLC, and the PLC will make a judgment and then execute the next action, thus realizing the intelligent control process.

The application of the intelligent control system greatly facilitates the management of the production process, making the production process of segments under control, realizing standardized management and avoiding errors caused by manual operation. According to the requirements of the production process, the control system is divided into the PLC automatic control system, monitoring system, the temperature control system, and the form-type intelligent identification system. Different systems cooperate with each other to enable the intelligent production process of the assembly line. The control system for the segment assembly line is shown in Fig. 4.23.

4.3.1.3 *Steel form translation mechanism*

The steel form translation mechanism mainly realizes the horizontal movement of steel forms, to optimize the design of the assembly line, and enable close coordination among different positions. For example, when the form cleared and cured is transported to the form feeding position of the reinforcement framework, the route coincides with the transport route of the reinforcement framework. With the steel form translation mechanism, the reinforcement framework and the form can be transported from different directions to the same position for the form feeding of the reinforcement framework. The thrust system and translation mechanism of the steel form are controlled by the PLC system, and the form is accurately positioned using proximity switches and photoelectric switches,

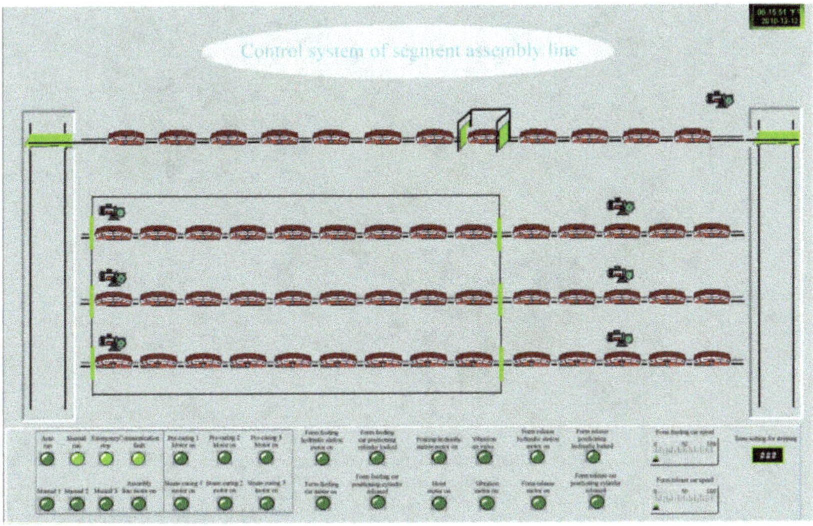

Fig. 4.23. Control system of segment assembly line.

effectively improving the efficiency and reliability of the assembly-line-style operation.

In the production process of the assembly line, any fault in any procedure will cause the entire production line to be suspended, causing operation delays at other positions and reducing the efficiency of the production line. By introducing a steel form translation mechanism, guide rails or universal form trolleys are added at the necessary positions to move the failed steel form out of the assembly line, avoiding any effect on the operation of the assembly line and facilitating form maintenance. The translation mechanism of the segment assembly line is shown in Fig. 4.24.

4.3.1.4 *Concrete pouring and vibrating system*

The concrete pouring and vibrating system is an important unit in the complete segment pouring and forming, mainly including the closed pouring chamber, pouring console, lifting table, concrete hopper, automatic door, hydraulic system and vibrating system.

In the process of concrete pouring and vibrating, the vibrating system plays a very important role in eliminating defects such as poor compactness and air voids in concrete. At present, the mainstream vibration methods include integral vibration, immersion vibration and attached vibration.

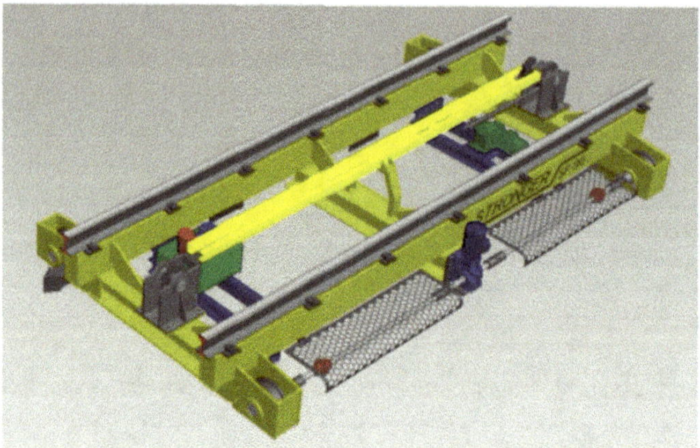

Fig. 4.24. Translation mechanism of segment assembly line.

The effects of the three vibration methods are different. The integral vibration and the immersion vibration often cause some defects because the corners are difficult to compact; the attached vibration is widely used for the purpose of compacting concrete through the uniform distribution of vibrators and the adjustment of speed and frequency. In addition, the combination of multiple vibration modes can effectively reduce slump, reduce the air voids inside and on the surface of the segment and further improve the quality of the segment. The pouring and vibration of the segment assembly line are shown in Fig. 4.25.

4.3.1.5 *Curing system*

The curing system includes a fully enclosed steam curing chamber, a walking rail and transmission system, and an automatic temperature control system. Compared with the traditional production mode, the assembly line intelligent curing system introduces several changes the traditional curing method that used the steam curing cover, overcoming the shortcomings of poor sealing, easy escape of steam, difficulty in controlling the steam curing temperature, etc., and through computer intelligent control of the temperature and humidity of the standstill curing area and steam curing kiln, effectively control of the steam consumption and production cost saving are achieved. A temperature sensor is set in the steam curing chamber, through which the temperature signal is fed back to the central

Fig. 4.25. Pouring and vibration of segment assembly line.

control computer to adjust the opening of the steam electronic valve. The application of the intelligent steam curing system in the assembly line can improve the operation efficiency of the assembly line as well as the product quality.

4.3.2 3D inspection technology

The 3D inspection technology is applied to complete the real-time 3D coordinate measurement of the product under computer-based control and conduct data processing to generate the inspection results with various measuring instruments (total station, laser tracker, scanner, etc.).

With the help of high-precision instruments to obtain a large number of measuring points, this technology can approach the surface shape of the model as much as possible and finally generate the size data of the steel form and segment through a series of calculations so as to achieve quality control. The all-around geometric information of steel form and segment is helpful to evaluate the segment quality objectively and comprehensively and effectively control the links that should be paid attention to in production.

4.3.2.1 *Principle of 3D inspection technology*

Laser tracking 3D inspection technology is a polar coordinate measurement system based on laser interference length measurement and angle

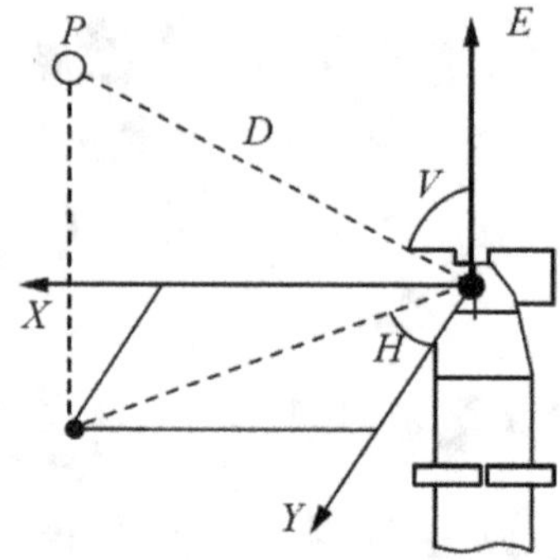

Fig. 4.26. Schematic diagram of coordinate measurement principle of laser tracker.

precision measurement. Its basic working principle is to place a reflector on the target point and use a tracking head to emit a laser to the reflector and receiving the laser reflected. When the reflector moves, the tracking head can adjust the laser emission direction to aim at the target. In the above process, the reflected beam is received by the detection system to measure the spatial position of the target.

In short, laser tracking 3D inspection technology is used to track a moving point in space, determine its spatial coordinates and draw a 3D model according to its motion trajectory. The 3D model drawn by laser tracking is compared with the segment design model. The spatial coordinates of the measuring points are determined by the polar coordinate measurement principle, that is, the horizontal angle *H*, the vertical angle *V* and the linear distance *D* between the point and the tracker. The measurement principle is shown in Fig. 4.26.

4.3.2.2 *Composition of 3D inspection system*

The laser tracking and measuring instrument (Fig. 4.27) consists of a spherical fixed reflector, support, chuck, tracker measuring head, cable for information transmission, microcontrol unit (MCU) for comprehensive information processing, temperature sensors for temperature compensation, network cable connecting the control unit and the computer to transmit information, CAM2 software working interface and other components.

For the geometric parameters of large tunnel components, the continuous data processing software (MPS-STEC) is used to directly output the geometric evaluation report of segment accuracy.

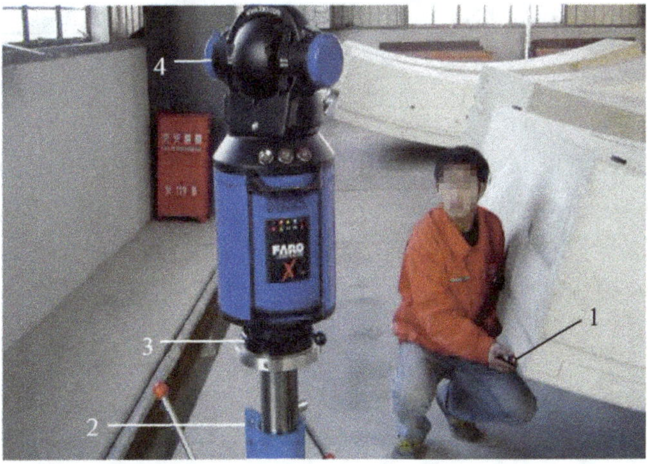

Fig. 4.27. Physical image of laser tracker.

Notes: 1 — Spherical fixed reflector, reflecting the measured object back to the tracker with laser; 2 — Support, to be adjusted according to the measured object height; 3 — Chuck for fixing the tracker; 4 — Tracker measuring head, used for emitting laser, processing information, etc.

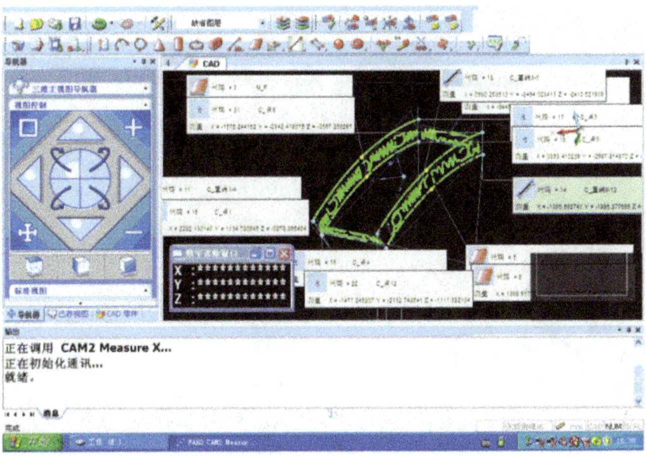

Fig. 4.28. CAM2 software work interface.

(1) CAM2 software performs the plane and arc measurements by collecting segment point data, generates eight measurement angular points through calculation, conducts iteration and outputs a measurement value report, as shown in Fig. 4.28.

Table 4.5. Statement of MPS-STEC software.

Item		Measured value	Design value	Difference	Precision
Longitudinal best plane (mm)	Left	0.16	0.00	0.16	±0.50
	Right	0.13	0.00	0.13	±0.50
Circumferential best plane (mm)	Front	−0.10	0.00	−0.10	±0.50
	Rear	−0.09	0.00	−0.09	±0.50
Parallelism (mm)	Inner radius	0.18	0.00	0.18	±2.00
	Outer radius	0.17	0.00	0.17	±2.00
Width (mm)	Line segment 1–7	1,494.47	1,494.35	0.12	±0.3
	Line segment 4–10	1,495.09	1,495.14	−0.05	±0.3
	Line segment 2–8	1,497.54	1,497.40	0.14	±0.3
	Line segment 5–11	1,497.79	1,497.77	0.02	±0.3
	Line segment 3–9	1,501.55	1,501.33	0.22	±0.3
	Line segment 6–12	1,501.04	1,501.14	−0.10	±0.3
	Average of line segment 1–7 and line segment 4–10	1,494.78	1,494.75	0.04	±0.3
	Average of line segment 3–9 and line segment 6–12	1,501.30	1,501.24	0.06	±0.3
	Average	1,497.91	1,497.86	0.06	±0.3
Width (mm)	Line segment 1–4	480.16	480.00	0.16	−0.15–+1
	Line segment 7–10	480.61	480.00	0.61	−0.15–+1
	Line segment 3–6	480.43	480.00	0.43	−0.15–+1
	Line segment 9–12	480.42	480.00	0.42	−0.15–+1
	Average	480.41	480.00	0.41	−0.15–+1
Arc length (mm)	Line segment 1–4	480.61	480.00	0.61	±0.40
	Line segment 7–10	480.43	480.00	0.43	±0.40
	Line segment 3–6	480.42	480.00	0.42	±0.40
	Line segment 9–12	480.41	480.00	0.41	±0.40

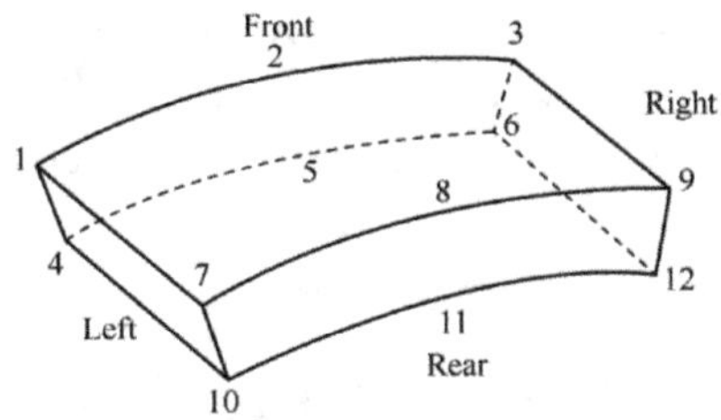

Fig. 4.29. Schematic diagram of generated model.

(2) The MPS-STEC software consists of modules such as data reading, analysis and operation, optimal adaptation and report generation. The data reading module reads the 3D coordinates of the contour points; the analysis and operation module constructs geometric elements such as lines and planes from points and calculates the measured plane through the operation among geometric elements; the best adaptation module compares the actual measurement model with the CAD model; the report generation module outputs various measurement results in both digital and graphical modes, in the general PDF format, and prints out the report (Table 4.5). The generated model is shown in Fig. 4.29.

4.3.2.3 *3D inspection method*

Whether the contact surfaces between concrete segments can closely and effectively fit directly affects the shield machine advance and tunnel quality. Factors affecting fitting between segments include:

(1) inner and outer radii of segments,
(2) angle and flatness at the ring face,
(3) angle and flatness at the end face,
(4) diameter and location of bolt hole,
(5) misalignment between bolt hole and nut,
(6) depth and width of the waterproof gasket and sealing strip groove.

All the above items can be accurately measured with the laser tracking measuring instrument. The segment is simplified into 4 planes of contact and 4 planes and 4 arcs of a ring segment composed of 2 internal and external arc faces, totaling 12 points.

4.3.2.4 *Application of 3D inspection technology*

The industry has begun to apply the laser tracking 3D inspection technology to the 3D measurement of steel forms and segments to achieve high-precision quality control of segments.

When the selected segment is inspected using the laser tracking 3D inspection technology, the data are output by items in a statement (Tables 4.6 and 4.7) to reflect the deviation between the structural model and the theoretical mathematical model, as shown in Fig. 4.30.

4.3.3 Vibration technology for segment fabrication

4.3.3.1 *Vibration mode*

The control of the vibration process is the most important key link in the whole production process, which is related to the internal and external

Table 4.6. Angular deviation between planes (unit: °).

Graphics6	Item	Difference	Precision
	Central angle	-0.01	±0.01
Angle between edges (°) 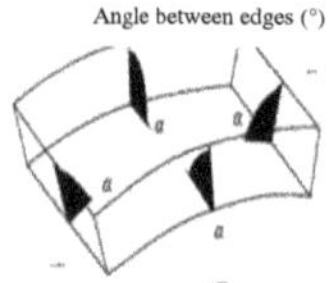	α_L	0.01	±0.02
	α_F	-0.01	±0.02
	α_R	0.01	±0.04
	α_B	0	±0.04
Longitudinal angle 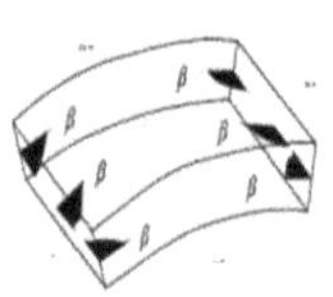	β_{LF}	0	±0.01
	β_{LB}	0.01	
	β_L	-0.01	
	β_{RF}	-0.01	
	β_{RB}	0	
	β_R	0	

Table 4.7. Circumferential width, thickness and arch length of segment (unit: mm).

Item		Difference	Precision
Longitudinal best plane	Left	−0.05	±0.10
	Right	−0.03	
Circumferential best plane	Front	0.12	±0.50
	Rear	0.13	
Parallelism	Inner radius	0.42	±2.00
	Outer radius	0.18	±3.00
Width	Line segment 1–7	0.40	±0.40
	Line segment 4–10	0.26	
	Line segment 2–8	0.34	
	Line segment 5–11	0.32	
	Line segment 3–9	0.35	
	Line segment 6–12	0.14	
	Average of line segment 1–7 and line segment 4–10	0.37	
	Average of line segment 3–9 and line segment 6–12	0.25	
	Average	0.37	
Thickness	Line segment 1–4	1.81	±2.50
	Line segment 7–10	1.71	
	Line segment 3–6	1.86	
	Line segment 9–12	1.20	
Arc length	Arc length 1–2–3	0.16	±0.50
	Arc length 4–5–6	−0.44	
	Arc length 7–8–9	0.26	
	Arc length 10–11–12	−0.02	
Twist		0.03	±3.00

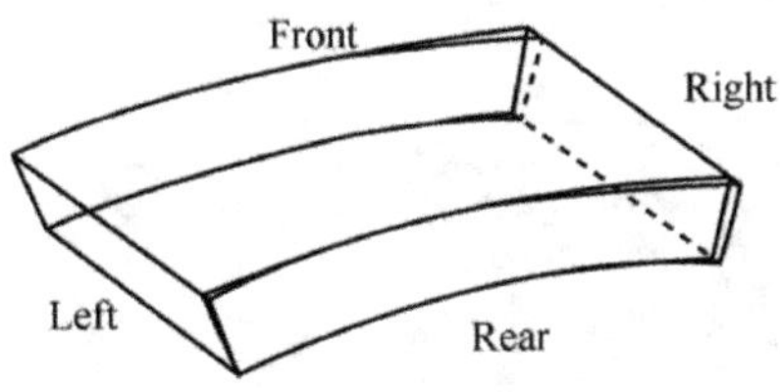

Fig. 4.30. Schematic diagram of deviation between structural model and theoretical mathematical model.

quality of segments, such as compactness, smoothness, flatness and air void conditions. It has the most direct influence on the appearance quality of segments, that is, the segment cracks, air voids, cavities and honeycombs, and the smoothness of the outer arc face.

The main difference between large-diameter and small-diameter tunnel segments lies in the selection among different choices of vibration technology. At present, there are three main vibration processes for segment fabrication: immersion vibration, attached vibration and integral vibration of the vibration table.

(1) **Immersion vibration mode:** The flexible shaft and motor are used to drive the eccentric block to vibrate inside the vibrating rod, and the exciting force can be transmitted to the concrete in a range with the vibrating rod as the center and a radius of 350 mm. Due to the limited range of vibration action, the operator needs to insert vibrators at different positions to overlap the ranges of concrete subjected to excitation action near each insertion point so as to achieve the effect that the concrete is basically evenly vibrated (Fig. 4.31).
(2) **Attached vibration mode:** Driven by the attached vibrator installed under the inner arc face steel plate, vibration occurs similar to that when a piece of drum skin is knocked and is transmitted to the concrete in contact for diffusion. Its vibration diffusion action distance is about 700 mm, and multiple vibrators must be arranged below the arc face in the form so that the action ranges of adjacent vibrators are overlapped (Fig. 4.32).

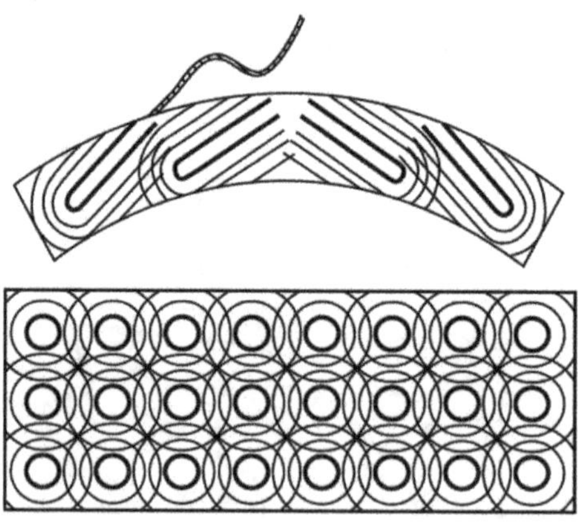

Fig. 4.31. Schematic diagram of energy transmission under immersion vibration.

(3) **Vibration mode of integral vibration table:** The high-rigidity form is clamped on the floating table of the integral vibration table. The variable frequency control motor drives the vibration motor to drive the floating table to perform variable frequency and amplitude vibration in stages. The concrete in the form is forced to vibrate up and down. Since the arc face arch and end plate in the form generate a horizontal component force on the concrete during the up and down vibration process, and the concrete paving speed is significantly faster than other methods, the internal homogeneity of concrete is also obviously better than other methods (Fig. 4.33).

Fig. 4.32. Schematic diagram of energy transmission under attached vibration.

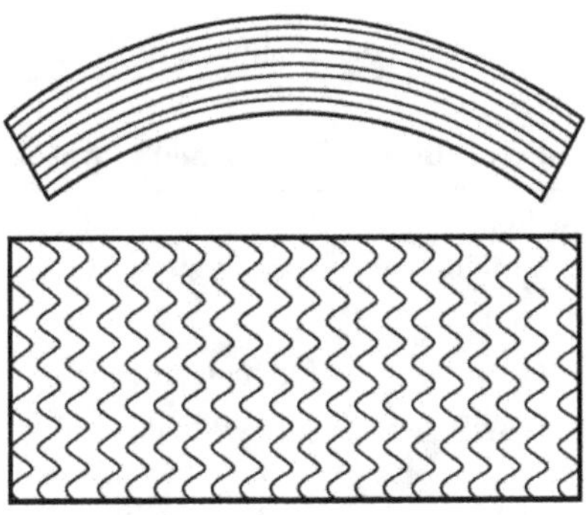

Fig. 4.33. Schematic diagram of vibration energy transmission of integral vibration table.

4.3.3.2 *Characteristics and application of different vibration modes*

The advantages and disadvantages of three vibration modes for large-diameter tunnel segments are shown in Table 4.8.

Through the comparison of the advantages and disadvantages among the three segment vibration modes, their applicability is known as follows:

(1) The immersion vibration mode is applicable to the fixed pedestal production process with low requirements for consistency of concrete pouring conditions and a low production rate.
(2) The attached vibration mode is applicable to projects with low requirements for form life, environmental protection and concrete homogeneity.
(3) The integral vibration table mode is applicable to projects with high requirements for form life, production strength, segment accuracy, environmental protection and concrete homogeneity.

Due to the large volume of large-diameter tunnel segments, their steel form is also required to be large in volume. If an integral vibration table is used for vibration, the form cost will be too high, and the power required for vibration is also extremely high, which is not suitable for the fabrication of large-diameter tunnel segments. The attached vibration can produce segments of good and stable quality and low requirements for operators, which is suitable for mass production. The immersion vibration is suitable for fixed pedestal production and is adaptive to the large

Table 4.8. Advantages and disadvantages of three vibration modes.

Item	**Immersion vibration**	**Attached vibration**	**Integral vibration table**
Advantage	Low construction investment, high economical efficiency	High segment quality, low crew requirements	High segment quality, low crew requirements
Disadvantage	Low segment quality, high crew requirements	High construction investment, high form requirements and low wear resistance	High construction investment, high form rigidity requirement

distribution area of forms in such a production process. Therefore, the immersion vibration is preferred in the fixed pedestal production process; the attached vibration is preferred in the assembly-line-style production process.

4.3.4 Intelligent steam curing technology for segments

Steam curing is a common production process for precast concrete members in order to improve the early strength of the concrete, speed up the form turnover, and improve the production efficiency during the segment production process.

After pouring, concrete placed in a hot climate and dry air environment, without curing in time, suffers from quick evaporation and subsequent dehydration, which prevents the cement particles already in the gel state from full hydration, resulting in failure in conversion into stable crystals, and a lack of sufficient adhesion, and flaky or powdery falling off on the concrete surface. In addition, the premature evaporation of water when concrete does not reach enough strength also produces large shrinkage deformation, resulting in dry shrinkage cracks. Therefore, the curing at the initial stage after concrete pouring is very important. The curing is carried out immediately after the final setting of concrete, and the curing of stiff concrete is carried out immediately after pouring.

The steam curing of concrete is divided into four stages: standstill, temperature rise, constant temperature and temperature drop. The steam curing of concrete should meet the following requirements:

(1) The ambient temperature should not be lower than 5°C during the standstill stage. The temperature can be raised after 4–6 h of pouring and the final setting of the concrete.
(2) The heating rate should not be greater than 10°C/h.
(3) During the constant temperature stage, the internal temperature of concrete should not exceed 60°C, and the maximum temperature should not exceed 65°C. The constant temperature curing time should be determined through tests according to the requirements for the strength of components upon form release, the concrete mix proportion and environmental conditions.
(4) The cooling rate should not be greater than 10°C/h.

4.3.4.1 *Intelligent steam curing system for segments*

In China, the temperature control in the steam curing process of tunnel segments is generally conducted by manually adjusting the switch on the steam piping for temperature rise, constant temperature and temperature drop of the segments. This method has many disadvantages, such as a heavy workload, low temperature control accuracy, serious energy waste and high cracking proneness of segments (especially in winter). The intelligent temperature control system mainly consists of an intelligent temperature controller. According to the process requirements, the temperature control program for temperature rise, constant temperature and temperature drop is compiled by the computer to control the steam curing of segments and support the computer in printing records, querying the login management of different users and batch control. The operation is more convenient and intuitive, and the system is more user-friendly (Fig. 4.34).

4.3.4.2 *Application of intelligent steam curing system for segments*

The intelligent steam curing system of the segments has been put into use. The system can automatically select the intelligent setting mode on a

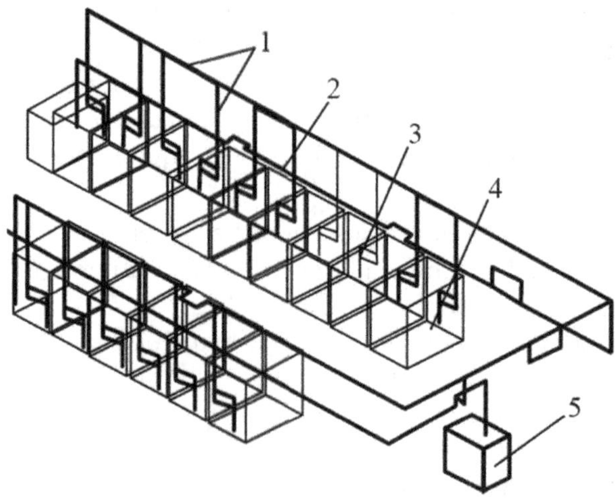

Fig. 4.34. Elevation of intelligent control system pipeline.

Notes: 1 — trunking; 2 — Steam curing header; 3 — Steam curing branch; 4 — Steam curing cover; 5 — Steam curing control chamber

variety of temperature curves, which conforms to the actual characteristics of the segment production and meets the process requirements; it is superior to the traditional steam curing process in terms of workload, temperature control accuracy, internal quality of segments and energy efficiency.

(1) **Steam supply device of steam curing cover:** The steam curing cover is mobile and sized depending on the size of the segment steel form. One temperature sensor (or multiple sensors, as appropriate) is installed in each steam curing cover. The sensor should be installed in the middle and upper parts of the steam curing cover. A steam supply device is installed in the front of a single steam curing cover, as shown in Fig. 4.35. Under normal conditions, the upper stop valve (standby) is usually closed, and the lower stop valve is normally open. The solenoid valve is normally closed. After receiving the command from the controller, the valve is opened to initiate steam supply.

(2) **Layout of steam curing chamber**

(i) *Heat insulation of air curtain machine*: An industrial air curtain machine (Fig. 4.36) is installed over the entrance and exit of the steam curing chamber for thermal insulation. The continuous high-speed airflow generated over the steel form by the air curtain machine is used to form an air door, isolate the air in the steam curing chamber from the outside and form two independent temperature zones, thus achieving the effect of thermal insulation.

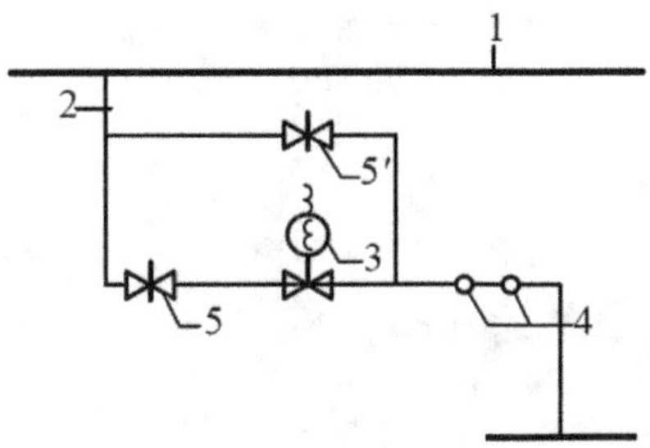

Fig. 4.35. Schematic diagram of steam supply device installed at the front end of steam curing cover.

Notes: 1 — Steam header (φ 50); 2 — Steam pipe (φ 25); 3 — Solenoid valve; 4 — Steam pipe (φ 15, connected to steam curing cover); 5 — Stop valve; 5' — Stop valve (standby).

(ii) *Dry steaming of fan blade equipment*: According to the conditions of the steam curing chamber, fan blade heating pipes (Fig. 4.37) are arranged, and steam is supplied through the pipes to heat the radiators so as to uniformly heat the air in the steam curing chamber. Moreover, the steam is controlled inside the steam curing pipes so as to reduce the steam condensing area and eliminate waste of heat due to the large amount of steam condensing in the steam curing chamber.

(iii) *Thermal insulation material*: The fire-resistant thermal insulation rock wool is used to fill the gap between plastic steel plates, and the thermal insulation foam is used to improve the thermal insulation effect of the wall.

Fig. 4.36. Air curtain machine at the entrance and exit of steam curing chamber.

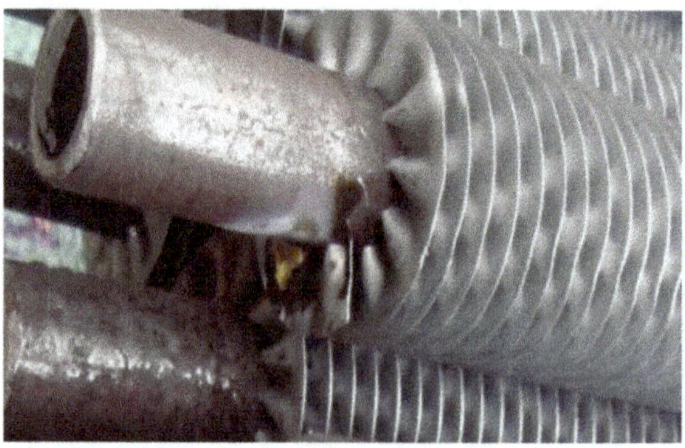

Fig. 4.37. Fan blade heating pipe.

4.3.5 Application technology of fiber reinforced concrete

As a new type of concrete composite, fiber-reinforced concrete can improve and enhance the performance of concrete. The application of fiber reinforced concrete in large-diameter tunnel segments can improve the quality of the segments, enhance safety performance indicators and improve the tunnel quality.

4.3.5.1 *Advantages of hybrid fiber reinforced concrete segments*

Hybrid fiber-reinforced concrete is a new type of composite building materials which has excellent mechanical properties. It can not only improve corrosion resistance and fatigue resistance but also have good fire resistance and excellent crack resistance.

Using steel fiber and polypropylene fiber (Fig. 4.38) to partially or entirely replace the steel bars in the concrete segment, the hybrid fiber-reinforced concrete segment is prepared, which has many advantages, including the following:

(1) The mechanical properties of the segments are improved, and the effects on toughness, bending strength and shear strength are obvious.
(2) The damage resistance performance of the segments is improved to avoid damage during transport and erection.
(3) With less or no reinforcement processing workload, the labor cost is reduced, and thanks to the short curing cycle of hybrid fiber-reinforced concrete, the production energy consumption is reduced.

(a) Polypropylene fine fiber (b) Polypropylene crude fiber (c) Steel fiber

Fig. 4.38 Fiber picture.

(4) The production process of hybrid fiber-reinforced concrete segments can reduce the loss of steel form, improve the industrial production rate of segments and the output can be more than doubled.
(5) Due to good waterproofing and durability, the later maintenance costs are significantly reduced.

To sum up, hybrid fiber-reinforced concrete has excellent performance, to comprehensively make up for the deficiencies of ordinary reinforced concrete segments. At present, there are two main ways to make fiber-reinforced concrete: (i) short fiber-reinforced concrete with high elastic modulus, represented by steel fiber; (ii) short fiber-reinforced concrete with low elastic modulus, represented by polypropylene fiber and nylon fiber. The addition of fiber can prevent the generation of internal and macro cracks in concrete, play a great role in improving the crack resistance of matrix concrete and greatly improve the mechanical properties compared with ordinary concrete.

4.3.5.2 *Application of hybrid fiber segments*

In European countries, the hybrid fiber-reinforced concrete segment technology has become more mature and has been widely used in tunnel engineering. Table 4.9 summarizes some foreign engineering practices.

Through numerous studies on steel fiber-reinforced concrete segments, China has developed effective theories and relevant specifications and carried out application research in engineering practices in key cities. However, the research on the application of steel fiber and polypropylene fiber reinforced concrete to large-diameter tunnel segments has just begun, so the research and development of new types of hybrid fiber high-strength concrete segments, as well as complete segment design and

Table 4.9. Foreign engineering practices of hybrid fiber reinforced concrete.

Tunnel	**Country**	**Length (m)**	**Segment width (m)**	**Segment thickness (cm)**	**Construction year**
Metrosud Naples	Italy	2,600	1.2	30	1998–2010
Meteor	France	4,500	1.6	40	1993–1995
Eole	France	3,400	1.4	35	1992–1998
Essie	Germany	2,400	1.5	40	1990–1996
CRTL	Britain	2,000	1.2	30	2001–2005

construction solutions, has a good industrialization prospect to further improve the mechanical properties and durability of tunnel segments.

4.4 Life-cycle quality management

China has begun to vigorously promote industrialized building, among which prefabricated buildings have gradually drawn public attention and experienced rapid market expansion by virtue of high quality assurance, fast construction, less environmental pollution and other advantages. However, the current prefabricated building cannot really realize industrialization because an all-round platform for sharing information has been absent for a long time. In the future, the combination of information technology, automation technology, modern management technology and manufacturing technology can improve or even transform the operation, management, product development and production of component manufacturing enterprises so as to realize real construction industrialization.

Digitalization of tunnel segments refers to the establishment of a digital model covering all information in the production process of tunnel segments and the establishment of a digital information system for precast concrete segments in five aspects: design optimization (BIM), production (RFID), logistics tracking (GPS), tunnel construction (VR, AR) and tunnel operation (RS, GIS).

4.4.1 Segment life-cycle management system

The segment life-cycle management system includes segment design, production, storage, transport, erection, operation and maintenance. The system is a product of the organic integration of building industrialization and informatization, with enterprise resource planning as the basic object, the building information model (BIM) and RFID in the Internet of Things as the key technologies. Relying on the RFID technology of the Internet of Things, the system records and tracks the key information of each link in the life cycle of PC components in real time, and integrating BIM technology with the management software platform of the system, establishes the database covering the whole life cycle of components, monitors and guides the quality and progress of the whole process of segment production, stacking, maintenance, transport and use to reduce engineering risks through intelligent early warning and optimization and improve construction efficiency and quality.

4.4.2 System composition

The segment life-cycle management system is established in a three-tier architecture: application layer, transmission layer and perception layer, and has three functions: basic information collection, data transmission and sharing, and intelligent management and decision-making (Fig. 4.39).

(1) The application layer includes the application system, software and middleware, which interact with the front-end functional layer through elastic compute service (ECS) and data exchange plug-ins.
(2) The transmission layer is mainly divided into an access network and an access unit. With the data exchange plug-ins, the information and user demands collected by the handset are uploaded to the cloud, and the query structure and decision conclusions are transferred from the cloud to the user.
(3) The perception layer is mainly centered on the RFID handset and guided by information recording software to perform the reading and collection of basic information about components. Managers at all levels of the decision-making platform log in to the management decision-making system via a browser, view the records and obtain various decision-making suggestions.

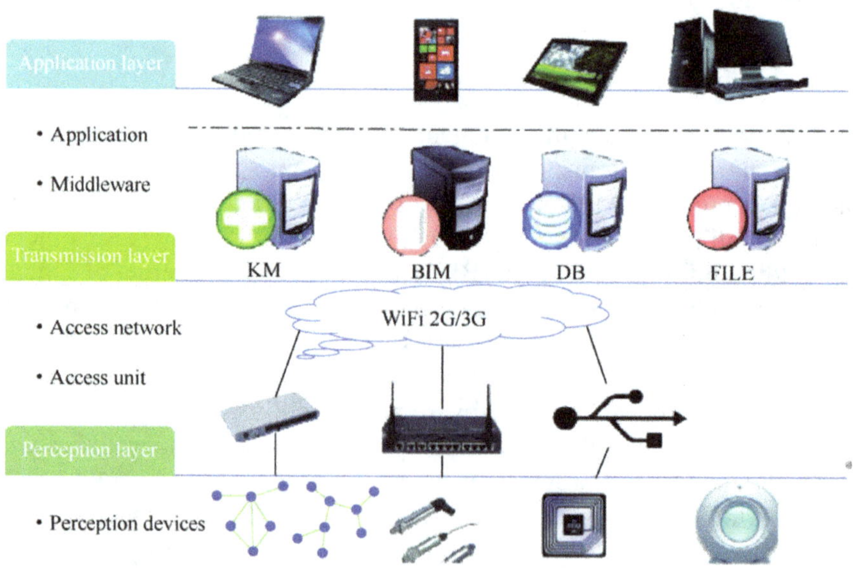

Fig. 4.39. Schematic diagram of segment life-cycle management system architecture.

4.4.3 Functions of management system

4.4.3.1 *Hardware equipment*

(1) **Special chip for tunnel segment (Fig. 4.40):** The information chip is installed in the segment grouting hole. The chip can store segment codes and has read and write functions. When the segment is lifted, it is screwed into the segment grouting hole, which is characterized by convenient installation and maintenance, strong adaptability, convenient reading, safe and reliable data, etc.
(2) **Segment information acquisition equipment (Fig. 4.41):** The segment information acquisition and tracking equipment is composed of a handset, a dedicated Bluetooth module and a camera, which is convenient to carry and input information, flexible to read chips and adaptive to a variety of construction environments.

Fig. 4.40. Special information chip for segment.

Fig. 4.41. Segment information acquisition equipment.

(3) **Tablet PC:** In order to facilitate the operators in quickly recording all kinds of information in the production process at the construction site, touch-tablet PCs are used to conduct information input, information query, chip information reading and writing and other functions via special applications.

4.4.3.2 *Software functions*

The software system consists of the information input system installed on terminals and the information management system installed on the server. The information management system, mainly functioning to manage the input data, is composed of the raw material information management module, the production process information management module, the product testing information management module, and the segment identification management module. Various third-party testing reports are uploaded via this system. As it is set on the server side, all networked computers can be easily accessed so as to realize information storage, querying, analysis and other functions.

(1) **Raw material information management module:** This module manages the information of various raw materials of segments, digitizing the quality data related to raw materials, including steel bars, concrete raw materials and embedded parts.
(2) **Production process information management module:** This module manages the information in the segment production process, digitalizing the quality data such as steel bar processing, reinforcement framework, placement of reinforcement framework in the fold, concrete pouring and segment steam curing, as shown in the segment reinforcement processing quality data entry interface (Fig. 4.42).
(3) **Product testing information management module:** This module manages the product testing information of finished segments, digitalizing testing data such as segment size, segment test omission and segment horizontal assembly, as shown in the horizontal assembly data entry interface (Fig. 4.43).
(4) **Segment identification management module:** This module manages independent identification for segments and establishes a segment identification system to correspond to relevant quality data. Moreover, the generated label in the form of a QR code is pasted

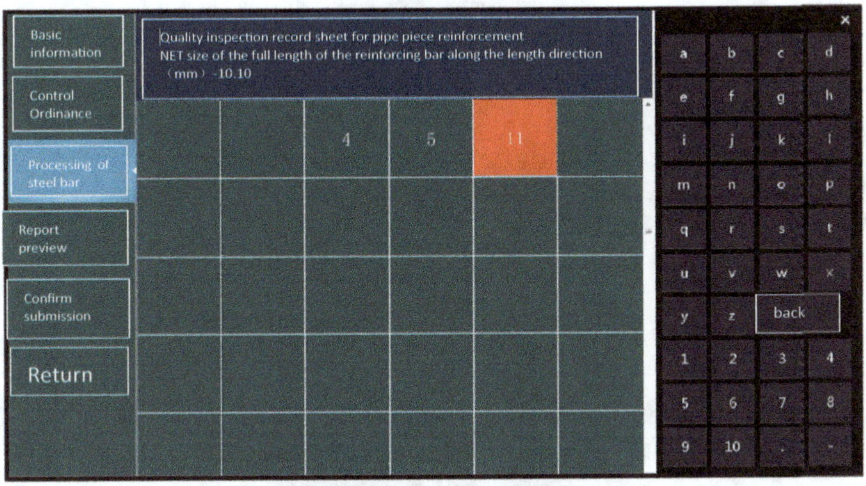

Fig. 4.42. Segment reinforcement processing quality data entry interface.

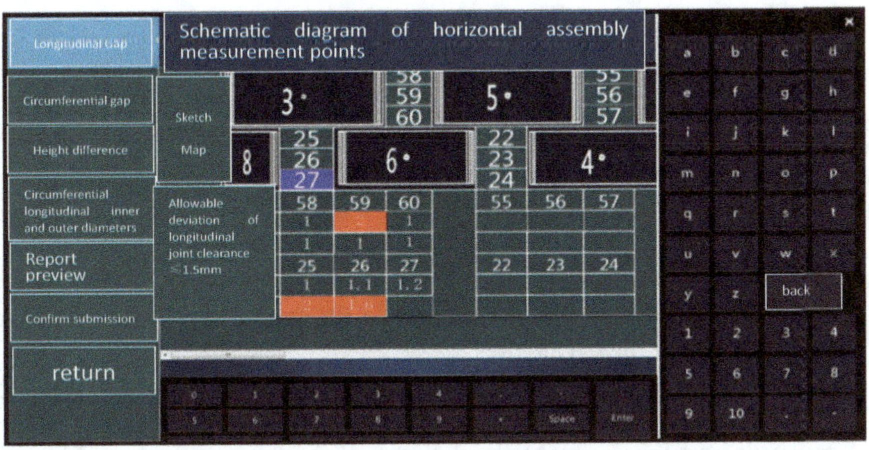

Fig. 4.43. Horizontal assembly data entry interface.

on the segment surface, enabling independent information traceability (Fig. 4.44).

This module manages the testing information of finished products, digitalizing the testing data such as segment size, segment testing omission and segment horizontal assembly.

Fig. 4.44. QR code identification of segments.

4.4.3.4 *Application of segment life-cycle management system*

The segment life-cycle system management was adopted for segment fabrication in "Hongmei South Road — Jinhai Road Passageway", and the management process is shown in Fig. 4.45.

The segment life-cycle management system is adopted at several key process nodes in the construction process to improve information integrity and quality traceability. The application of the segment information system has the following advantages:

(1) **Full-process information coverage:** The whole process, from receiving raw materials to the delivery of finished products and lining construction in tunnels, is recorded. The production process of tunnel segments is shown in Fig. 4.46.
(2) **Dynamic and timely information data:** The data input to the server can be immediately retrieved so that production and quality data can be timely transmitted and adjustments can be made.
(3) **Traceability of quality information:** The data of traditional segment fabrication are divided according to batches, which cannot reflect

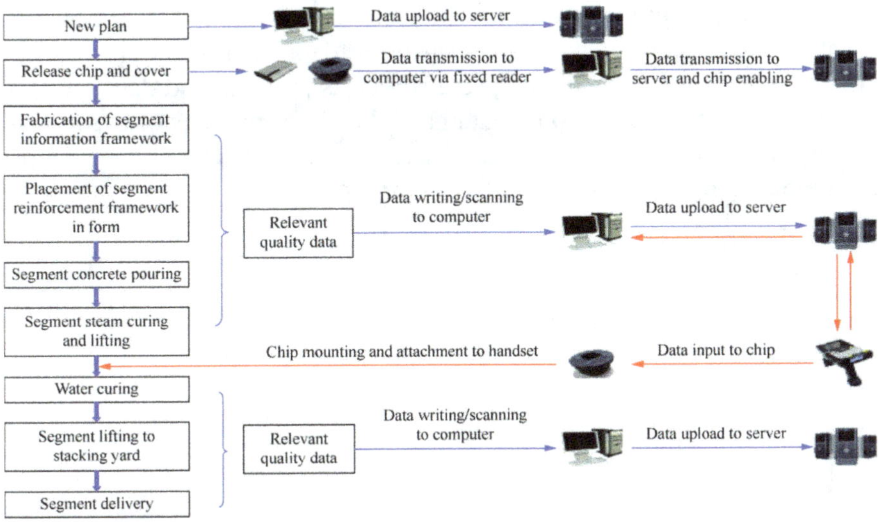

Fig. 4.45. Schematic diagram of management process of segment life-cycle system.

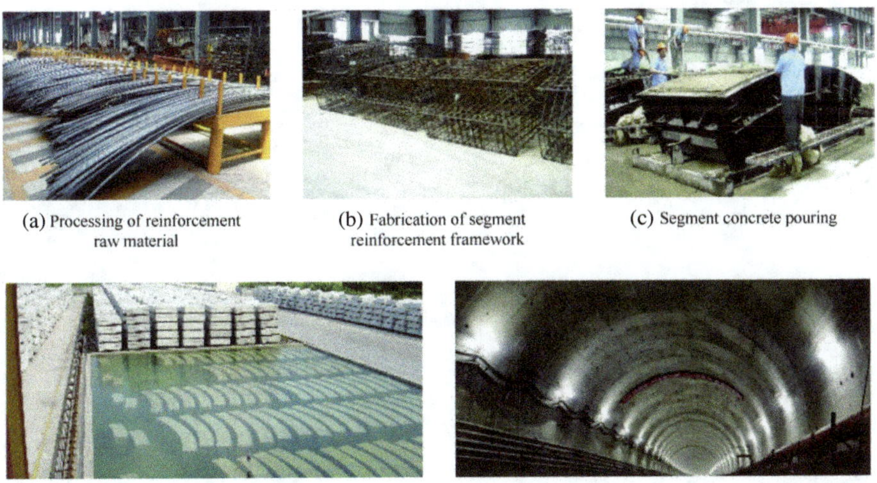

(a) Processing of reinforcement raw material

(b) Fabrication of segment reinforcement framework

(c) Segment concrete pouring

(d) Segment curing and stacking

(e) Segment erection into tunnel lining

Fig. 4.46. Segment production process.

the quality of each segment in the production process. Through independently numbering segments and establishing retrieval rules, each segment is attached with corresponding data to improve traceability.

(4) **Long retention time of data:** The segment information system inputs the process control data and inspection data of production into the computer and stores them on the server as pictures and data for permanent storage. Electronic data have the characteristics of high security, long storage time and easy searchability, as well as a retention time as long as the service life of tunnels.

Chapter 5

Shield Tunneling

Disturbance to soil mass during the excavation of large-diameter tunnels is far greater than that of tunnels in conventional diameters. Therefore, various technologies of shield tunneling need to be improved and innovated to cope with different special conditions and ensure construction stability during the excavation process. This chapter mainly introduces several key construction technologies such as large-diameter shield tunneling, erection and grouting, mainly including large-diameter shield launching and receiving, excavation face stability, shield attitude management, segment erection technology, synchronous grouting technology and tunneling in special sections.

5.1 Shield launching and receiving

5.1.1 Ground treatment

Due to the requirements of the construction process, the shield should not be launched or received until the enclosure at the tunnel portal is completely broken. Therefore, the soil mass within a certain range of the tunnel portal must be treated, mainly aimed to enhance the strength and independence of the ground soil, avoid adverse conditions such as water and soil loss during the launching and receiving of the shield machine, and ensure the safety of the crew, equipment and environment.

If there are weak points in the ground reinforcement quality at the outer side of the tunnel portal, water and soil leakage into the shaft will probably occur during the breaking process of the tunnel portal, which will cause surface subsidence around the tunnel portal and affect the shield machine construction and the safety of the surrounding environment. On the contrary, reinforcing soil with too much strength will bring cutting difficulties to the cutterhead, causing machine failure and affecting tunneling progress.

5.1.1.1 *Theoretical analysis method for stability of reinforced soil mass*

(1) **Analysis method based on the theory of plates:** The idea for this analysis method comes from the theory of calculating the thickness of reinforced soil mass adopted by JSG Group of Japan (Fig. 5.1). In this theory, it is assumed that the reinforced soil mass is an integral plate, and the thickness of the reinforced soil mass can be expressed as

$$h = \left[\frac{K_0 \beta p d^2}{4\sigma_t} \right]^{\frac{1}{2}}, \tag{5.1}$$

where

p — resultant force of water and soil pressure at the center of the portal sealing, $p = p_a + p_w$; P_a — earth pressure; P_w — water pressure;
d — diameter of portal sealing;
σ_t — ultimate tensile strength of reinforced soil mass;
K_0 — safety factor;

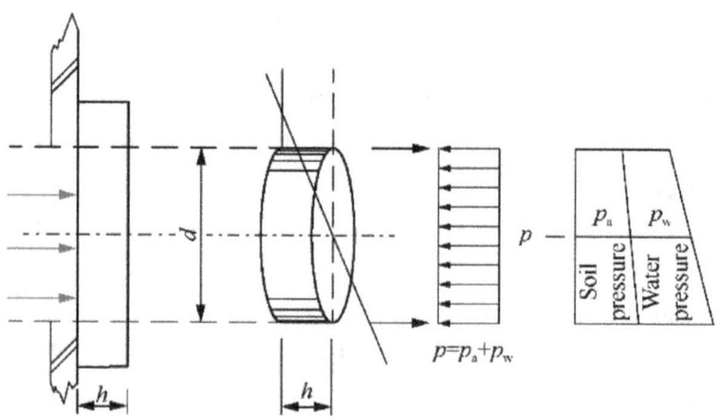

Fig. 5.1. Calculation model of reinforced soil mas based on theory of plates.

β — calculation coefficient;
h — thickness of reinforced soil mass.

Substitute $K_s = K_0\beta$ into Equation (5.1) as follows:

$$K_s = \sigma_t \frac{4h^2}{pd^2}. \tag{5.2}$$

When K_s>1, the reinforced soil mass is stable, otherwise it is unstable. With σ_t predefined, the safety factor K_s can be calculated after measuring and calculating the water and soil pressure p at the center of the portal sealing.

(2) **Analysis method based on the elasto-plastic statics theory:** The analysis method based on the elasto-plastic statics theory is based on the thin plate theory in elasto-plastic mechanics. Figure 5.2 shows a thin plate with equal thickness. In the theory of elastic mechanics, the basic equation of thin plate bending is as follows:

$$\nabla^2\nabla^2 w = \frac{q}{D}, \tag{5.3}$$

whcrc

D — flexural stiffness of plate (kN/m).

The reinforced soil mass is considered a simply supported elastic circular plate with a thickness of h, and the earth pressure and water pressure on the side of the reinforced soil mass are simplified as a uniform load q, as shown in Fig. 5.2. Equation (5.3) can be transformed through coordinate conversion as follows:

$$\left(\frac{\partial^2}{\partial r^2} + \frac{1}{r}\frac{\partial^2}{\partial r}\right)\left(\frac{\partial^2 w}{\partial r^2} + \frac{1}{r}\frac{\partial w}{\partial r}\right) = \frac{q}{D}. \tag{5.4}$$

At $r = 0$, the deflection and internal force are finite. According to the thin plate theory, the maximum deflection and the maximum bending moment occur at the center of the plate. Combined with the boundary conditions of the simply supported thin plate, $\begin{cases}(w)_{r=d/2}=0\\(M_r)_{r=a}=0\end{cases}$, the corresponding maximum bending stress is expressed as follows:

$$(\sigma_r)_{\max} = (\sigma_\theta)_{\max} = \frac{6M_r}{h^2} = \frac{3(3+v)qa^2}{8h^2} = \frac{3(3+v)q\left(\frac{d}{2}\right)^2}{8h^2}, \tag{5.5}$$

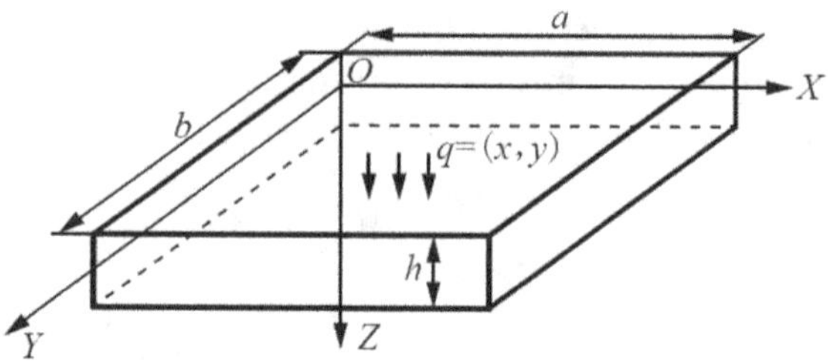

Fig. 5.2. Equal-thickness plate in rectangular coordinate system.

where

v — Poisson's ratio of reinforced soil mass;

q — uniform load from simplified water and soil pressure at outer side.

Then, the maximum shear stress at the boundary is expressed as

$$\tau_{\max} = \frac{q\left(\frac{d}{2}\right)}{2h}. \tag{5.6}$$

K_{s1} and K_{s2} are assumed to be the calculated safety factors of maximum bending stress and maximum shear stress, respectively, and according to the equations $\sigma_{\max} = \sigma_t/K_{s1}$, $\tau_{\max} = \tau_s/K_{s2}$, deduction can be made as

$$K_{s1} = \frac{32\sigma_t h^2}{3qd^2(3+v)}, \tag{5.7}$$

$$K_{s2} = \frac{4h\tau_s}{qd}. \tag{5.8}$$

The overall safety factor of the reinforced soil mass shall not exceed the safety factor corresponding to the maximum bending stress and maximum shear stress, resulting in $K_s = \min(K_{s1}, K_{s2})$.

(3) **Analysis method based on slip line theory:** The first two analysis methods represent stability judgments of the reinforced soil mass from a microscopic perspective, while the analysis method based on the slip line theory gives judgments from an overall perspective (Fig. 5.3). Under the action of ground load p and upper soil mass, the reinforced soil mass in the model may slide along a certain sliding surface as a whole into the tunnel. Assuming that the sliding surface

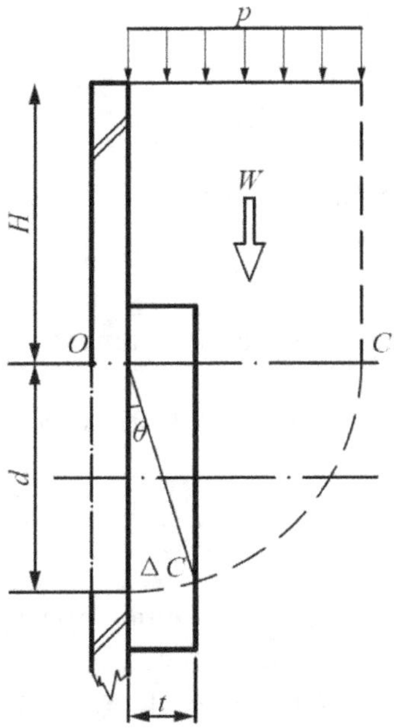

Fig. 5.3. Calculation model of slip instability theory.

is an arc surface with the vertex O as the center and the opening diameter d as the radius, the sliding moment caused is

$$M = M_1 + M_2, \tag{5.9}$$

where

$$M_1 = \frac{1}{2}\gamma H d^2;$$

M_1 — sliding moment caused by dead weight of overlying soil (kN · m),

$$M_2 = \frac{1}{3}\gamma d^3,$$

M_2 — sliding moment of soil mass in sliding ring line (kN · m).

The anti-sliding moment generated by soil mass is expressed as follows:

$$M_d = M_r + \Delta M_r, \tag{5.10}$$

where

$$M_r = \frac{1}{2}c\pi d^2 + Hcd;$$

M_r — resistance moment before soil improvement (kN · m);

$$\Delta M_r = \Delta c\theta d^2,$$

ΔM_r — increased resistance moment after soil improvement (kN · m).

The anti-sliding safety factor of the reinforced soil is expressed as

$$K_s = \frac{M_d}{M}. \tag{5.11}$$

(4) **Analysis method based on disturbance theory:** The size of the reinforced cross-section can be determined by reference to the calculation method of grouting range in the grouting method; that is, according to the plastic range (or loosening range) generated around the section when excavating the tunnel, the plastic range R can be calculated by the method of loosening the upper part of the tunnel. In the process of excavation, the soil mass stress grows out of balance, and additional stress is generated around the end face of the excavation. As shown in Fig. 5.4, in the range of $a < r < R$, the equilibrium equation can be deduced according to Mohr's envelope failure conditions, in consideration of the stress balance and failure conditions of the plastic loose circle:

$$\begin{cases} \sigma_\theta - \sigma_r = 2c \\ \dfrac{\partial \sigma_r}{\partial r} = \dfrac{(\sigma_\theta - \sigma_r)}{r} \end{cases}. \tag{5.12}$$

Substituting the grouting reinforcement boundary conditions, $r = R$, $\sigma_r = \sigma_m$, $r = a$, $\sigma_r = 0$, the expression is deduced as follows:

$$\ln R + \frac{R\gamma_t}{2c} = \frac{H\gamma_t}{2c} + \ln a, \tag{5.13}$$

where

R — distance to the outer side of the plastic range (m);
γ_t — average unit weight of overlying reinforcing soil mass (kN/m^3);
c — cohesion of improved soil mass (kPa);

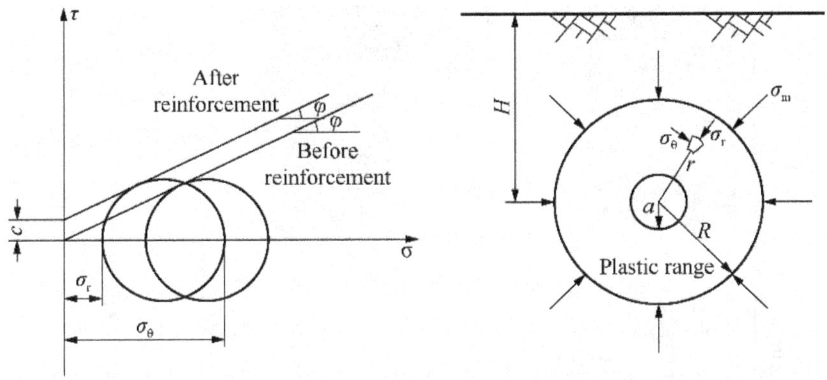

(a) Analysis on stress circle before and after reinforcement

(b) Plastic range generated in tunneling

Fig. 5.4. Calculation model of disturbance theory.

H — converted thickness of overburden to the tunnel center (m);
a — cutting radius or outer diameter of shield machine (m).

According to Formula (5.13), the size of the plastic range R can be obtained to guide the design of reinforced soil mass for shield launching and receiving.

5.1.1.2 *Main methods of ground treatment*

For shield launching and receiving, the main ground treatment methods are shown in Table 5.1, including deep-mixing pile, high-pressure jet grouting, Metro Jet System (MJS), ground freezing and well point dewatering.

For shield launching and receiving, the main ground treatment methods are shown in Table 5.1, including deep-mixing pile, high-pressure jet grouting, MJS, ground freezing and well-point dewatering.

The deep-mixing pile is one of the most common methods for soft soil ground reinforcement. In this method, the mixing pile machine is used to spray cement into the soil mass before full stirring until cement and soil are mixed evenly, thus improving the soil mass strength. The deep mixing machines are classified into single-shaft and multi-shaft deep-mixing pile machines according to the number of stirring shafts. During the construction of large-diameter tunnels, three-axis mixing piles are mostly used for launching and receiving reinforcement, with a depth generally above 18 m. According to the burial depth of tunnels, the largest pile forming depth of deep-mixing piles can reach 60 m with the addition of drill pipes.

Table 5.1. Main ground treatment methods.

Reinforcement method	Features and application scope	Key points in construction	Field example
Deep-mixing pile	(1) Low disturbance to soil mass; (2) Fully mixed cement and soil and pile shaft free of joints along full length, contributing to good water tightness; (3) Small impact to surrounding buildings and underground pipelines; (4) Applicable to soft soil ground	Tri-shaft mixing is commonly adopted for large-diameter tunnels: (1) Pile diameter: 850 mm; Overlap between piles: 250 mm; Strength ≥ 0.8 MPa; (2) Drill rod lowering rate: 0.6–0.8 m/min; Lifting speed: 1–1.2 m/min (in line with the grouting rate); (3) Periphery supporting piling necessary in special formations, such as confined aquifer	
High-pressure jet grouting	(1) Grouting position and range controllable, close overlap and good impermeability; (2) Adjustable grouting parameters to produce solidified mass up to the design standards (3) Portable equipment, easy operation, small space demand during construction; (4) Possible impact to surrounding pipelines and structures during construction; (5) Applicable to sandy soil, clay, silt and artificial fill ground	Triple-pipe and double-high pressure jet grouting pile is commonly used for large-diameter tunnels, and the cement grout pressure ≥25 MPa, compared with the conventional high-pressure jet grouting pile. (1) Pile diameter: 1,200 m; Overlap between piles: 400 mm; Strength ≥1.0 MPa; (2) Air pressure: ≥ 0.7 MPa; Water pressure ≥32 MPa; (3) Lifting rate of grout jetting drill pipe: 10 cm/min	

Omnibearing high-pressure jet grouting	Omnibearing jet grouting; (2) Large pile diameter, good pile shaft quality, contributing to sufficient assurance in ultra-deep construction; (3) Small environmental impact, including surrounding ground settlement; (4) Applicable to rock, sandy-cobbly soil, clay, silt and artificial fill ground	Pile diameter: 2,000 mm ~3,000 mm; (2) Grout pressure: ≥ 40 MPa; (3) Air pressure: 0.7 MPa; (4) Withdrawal rate: 80 m/min (full circle); (5) Ground pressure control factor generally taken as 1.3–1.6	
Freezing	(1) High strength and good water tightness in reinforced soil mass; (2) Long construction period, high cost; (2) Certain impact on ground heave and settlement from frozen soil mass; (4) Applicable to sandy soil ground with a high water content, especially the river-crossing tunnels	(1) Monitoring on saline water flow rate and temperature; (2) Monitoring on freezing temperature; (3) Monitoring on pore water pressure in unfrozen soil and thaw collapse; (4) Thaw collapse to be solved by grouting	
Dewatering	(1) Flexible layout of well points for convenient operation; (2) Fast construction and quick effect; (3) No impact to the entire system in case of any damage of individual wells; (4) Applicable to sandy soil ground with a high water content	(1) Checking the calculation of stability against water inrush; (2) Number and depth of wells; (3) Monitoring on surrounding environment of dewatering well turned on; (4) Standby generator	

The high-pressure jet grouting pile is a method to reinforce stratum by using a high-pressure jet to destroy the soil structure and make the soil and solidifying material slurry mixed and stirred into a high-strength consolidated body. At present, the basic processes of high-pressure jet grouting include the single-pipe, double-pipe, triple-pipe and multiple-pipe methods. In the construction of large-diameter tunnels, triple-pipe double-high pressure jet grouting is often selected as the reinforcement form for launching and receiving.

On the basis of the traditional high-pressure jet grouting process, the MJS adopts a unique perforated pipe and a front-end slurry forming device to realize the forced slurry discharge in the hole and the monitoring of the ground pressure, and controls the ground pressure by adjusting the amount of forced slurry discharge, greatly reducing the impact on the environment. It is a new ground reinforcement method.

The freezing method is used to freeze the soil mass to be excavated by manual drilling and refrigeration with the circulating low-temperature refrigerant to make it a high strength waterproof curtain to achieve the reinforcement effect. According to the method of refrigerating the stratum, it can be divided into direct freezing and indirect freezing. Liquid nitrogen is used for direct freezing, and brine is used for indirect freezing. According to the position in which the refrigeration is performed, it can be divided into horizontal freezing and vertical freezing. When it is difficult to stabilize the soil mass at the excavation face by other methods, the freezing method can achieve better results.

The dewatering method is an effective measure to prevent quicksand during underground construction by draining groundwater to stabilize the soil mass of the excavation face, which is more economical than other drainage methods. However, since wells should be drilled from the ground downward, the application scope and area of the dewatering method are limited. It is generally used in the launching and receiving stages of shield machine construction. Auxiliary dewatering wells are set around the cement system reinforcement area to reduce the infiltration of groundwater into the construction scope and ensure the safety of tunnel portal chiseling during the launching and receiving process.

5.1.1.3 *Scope of ground reinforcement area*

The soil mass should be reinforced to meet the strength and stability requirements, and it is particularly important to determine the scope of ground reinforcement for shield launching and receiving. The reinforcement scope mainly includes the longitudinal length, transverse width and vertical depth of the reinforcement area, as shown in Fig. 5.5.

According to previous experiences in launching and receiving in large-diameter tunnel construction, the longitudinal length of the reinforcement area is generally L = shield machine body length + D, where D is the segment ring width. The transverse width and vertical depth can be determined according to the theoretical analysis of the stability of the reinforced soil mass.

5.1.2 Water stop of tunnel portal

During the launching and receiving of the shield machine, there will be a circular structure void left between the tunnel and the shield machine shell. In order to prevent soil loss from such a void during the shield machine launching, which affects the stability of the soil mass on the excavation face and the construction in the working shaft and the shield machine, the tunnel collar sealing device should be set at the portal.

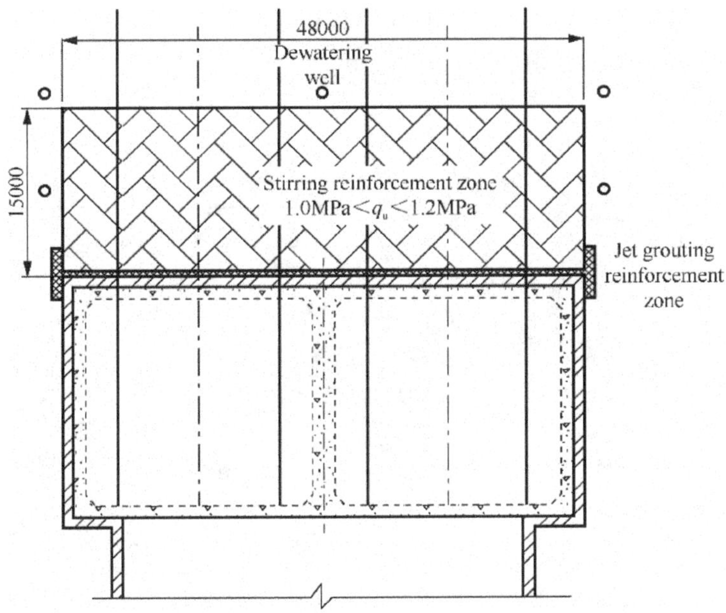

Fig. 5.5. Schematic diagram of the scope of ground reinforcement area (unit: mm).

The tunnel collar sealing device is usually designed as the water stop box, which is divided into the external box type and the built-in box type. The external box is a box structure welded on the embedded steel plate of the tunnel collar, in which a sealing device consisting of rubber curtain tape, ring plate, hinge plate, etc., is installed, and grouting holes are set as the preventive measures for waterproofing and leakage stoppage at the portal, as shown in Fig. 5.6. The built-in box is to provide a groove on the embedded steel plate of the tunnel collar and install curtain tape, ring plate, hinge plate, etc., in the groove, as shown in Fig. 5.7. As the burial depth is different at the time of shield launching and receiving, the stress condition of

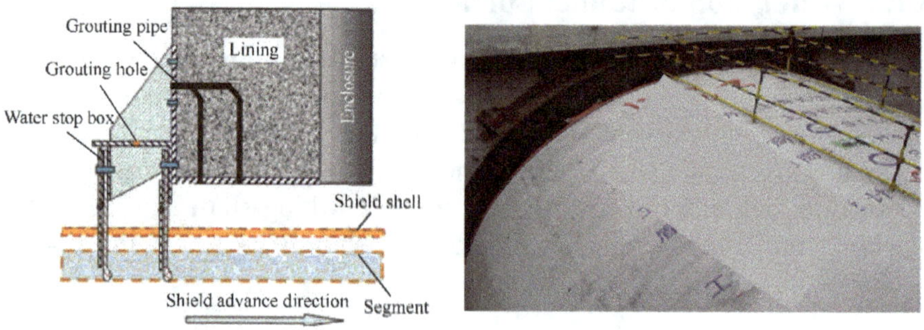

(a) Structural sketch (b) External box for shield launching

Fig. 5.6. Schematic diagram of external water stop box.

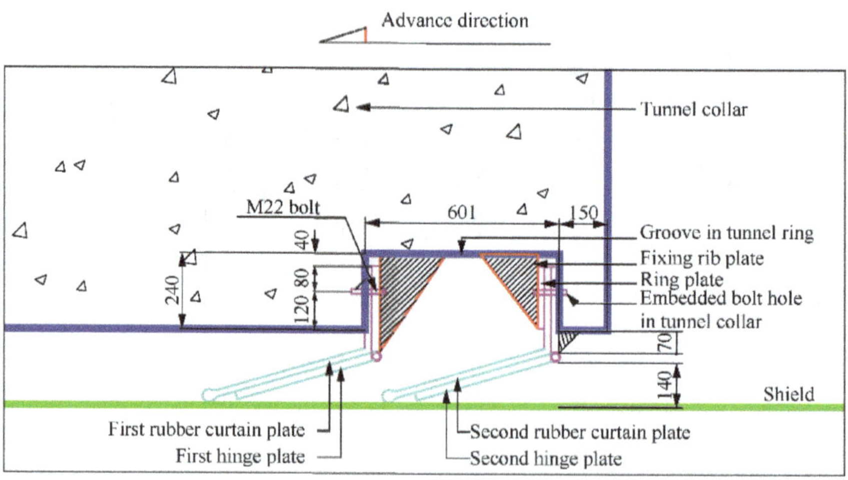

Fig. 5.7. Schematic diagram of built-in water stop box (unit: mm).

the box varies. The external or built-in box structure should be selected as the water-sealing device of the tunnel portal according to different working conditions.

5.1.3 Shield launching

5.1.3.1 *Shield machine base*

As shown in Fig. 5.8, the launching base of large-diameter shield machines is generally built with cast-in-situ concrete, on which the guide rail of the shield machine is laid, and then the guide rail is extended at the interface between the base and the tunnel portal.

During positioning and setting out of the launching base, it is required to ensure that:

(1) The center line of the base is aligned to the measured center of the tunnel portal; the distance between the front end of the base and the portal water stop device is about 1 m, for the sake of plugging the bottom portal.
(2) When launching in a curve section, the base should be located in a secant or tangent direction of the axis by fitting, so as to minimize the deviation between the shield machine and the tunnel axis after launching and keep it within the allowable range of the specification.
(3) The longitudinal gradient is consistent with the longitudinal gradient of the tunnel axis.

Fig. 5.8. Shield machine launching base.

(4) The cast-in-situ concrete base is provided with embedded iron parts before pouring, with horizontal bracings on both sides and manholes in the middle.

It is not appropriate to correct the deviation of the shield machine from launching to the time when the shield tail is separated from the guide rail. The shield machine base is placed accurately according to the secant direction of the axis in combination with the axis deviation of launching on a small-radius curve and the sealing effect of the portal ring sealing device, as shown in Fig. 5.9. Moreover, horizontal bracings are added on

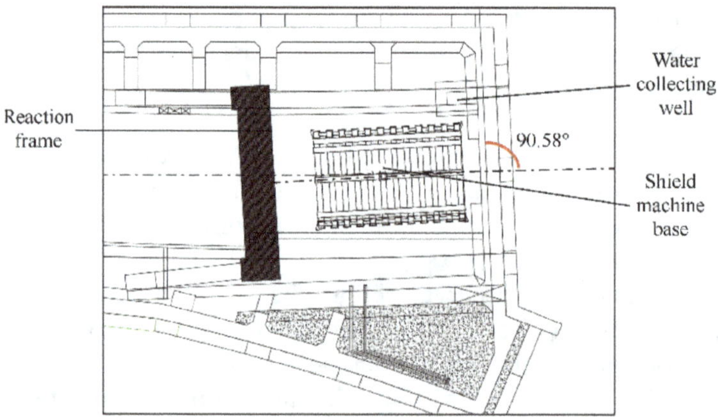

Fig. 5.9. Schematic diagram of base located according to secant of axis.

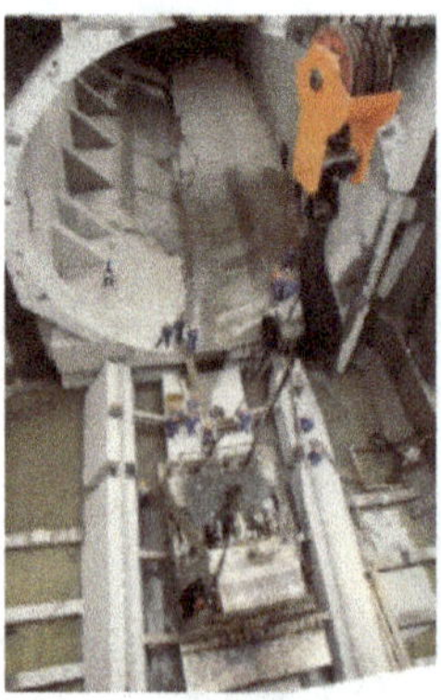

(a) Reinforced concrete backing

(b) Steel backing

Fig. 5.10. Shield machine backing.

both sides of the base to prevent deformation or fracture of the base and rail due to uneven stress.

5.1.3.2 *Backing of shield machine*

As shown in Fig. 5.10, the backing structure of the shield machine is a cast-in-situ octagonal reinforced concrete structure or a steel structure. Reactions occurring when the shield machine is launched are transferred to the corresponding structure in the working shaft and buried section through the backing of the shield machine.

5.1.3.3 *Construction process for shield launching*

The specific construction process for shield launching is shown in Fig. 5.11. If dewatering wells are arranged around the launching reinforcement area, a pumping test should be carried out immediately after the dewatering wells are bored, and corresponding reports are issued. Before the tunnel portal is chiseled, the dewatering well is opened to reduce the water level to meet the construction requirements. The normal operation of the power supply and the presence of a standby water pump should be ensured to ensure uninterrupted dewatering.

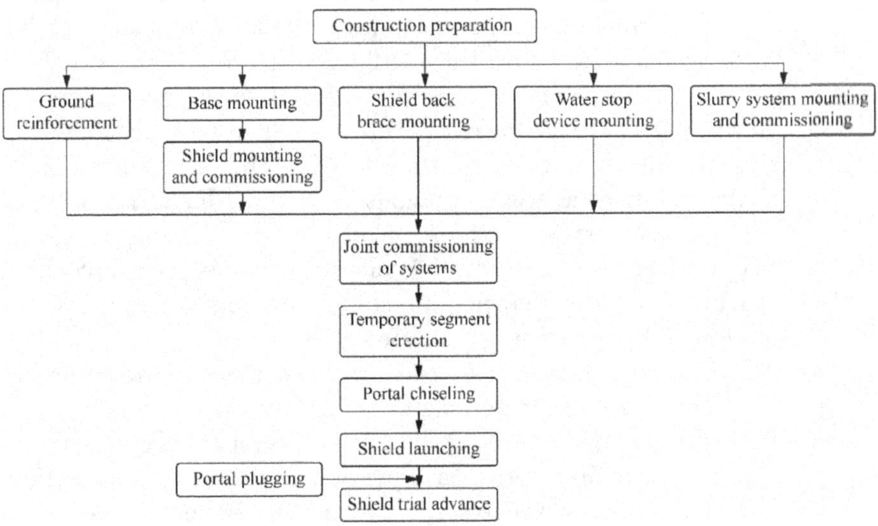

Fig. 5.11. Construction process of shield launching.

(a) Layout of horizontal test boreholes (b) Effect of horizontal test borehole

Fig. 5.12. Horizontal test borehole in portal.

(1) **Acceptance of ground reinforcement effect**

(i) *Acceptance of cement system reinforcement*: The ground reinforcement is subject to acceptance before the tunnel portal is chiseled. After the reinforcement strength meets the design requirements, the vertical coring and inclined coring in the reinforcement area are carried out. Moreover, the "cross"-shaped horizontal test boreholes are set on the tunnel portal, as shown in Fig. 5.12, to check leakage of the trial tunnel section and the soil sample coring and judge the soil mass reinforcement effect in the reinforcement area. The horizontal test boreholes must penetrate the enclosure structure of the working shaft and be drilled into the reinforced soil mass, with a general depth of 3 m. Under special circumstances, inclined coring may be used to judge the overall effect of the reinforcement area.

The launching tunnel portal should not be chiseled until it is confirmed that there is no obvious leakage in the horizontal test boreholes; otherwise, reinforcement measures should be taken. The cement system reinforcement area can be reinforced by grouting, and the freezing method reinforcement area can be reinforced by backfilling and refreezing.

(ii) *Acceptance of freezing reinforcement*: The freezing method reinforcement is subject to acceptance before the tunnel portal is chiseled, and the temperature of the vertical temperature measuring hole and the temperature measuring hole of the trial tunnel section at the portal are checked to see if they meet the design requirements.

(2) **Temporary segment erection:** The number of temporary segments is set according to the length of the working shaft. In order to ensure the

overall rigidity of the temporary segment ring, improve the flatness of the segments erected, and reduce segment fragmentation, the temporary segments should be erected into a closed ring with staggered joints.

(i) *Erect the temporary segments of the first ring*: The positioning of the temporary segments of the first ring is very important, which plays a role as the datum plane for the subsequent segment erection. Therefore, the roundness, gradient and flatness at the ring face of the segments should be controlled as much as possible. Before the erection of the first ring temporary segments, the positioning blocks must be welded at the tail of the shield, evenly arranged at nine places within 180° of the lower part of the shield (Fig. 5.13).

First, assemble the ground blocks, and then assemble the remaining segments in the order of each standard block → adjacent block → block F. Attention should be paid to the opening size when assembling two adjacent blocks to facilitate the smooth insertion of block F.

(ii) *Coat the shield tail brush*: Before the first ring of temporary segments is pushed out, the shield tail brush must be smeared with grease. Before application, check the condition of the wire brush and remove any foreign matter from the wire brush. The initial coating of grease on the shield tail is carried out by hand, and special tools are used for even coating. The clearance of the wire brush is filled with grease to ensure the service life of the shield tail brush.

(iii) *Temporary segment ring pushing and reinforcement for connection*: After each temporary segment ring is erected, the longitudinal and circumferential bolts must be tightened, and the embedded parts at the inner arc face (Fig. 5.14) must be welded firmly before the integral temporary segment ring is pushed out.

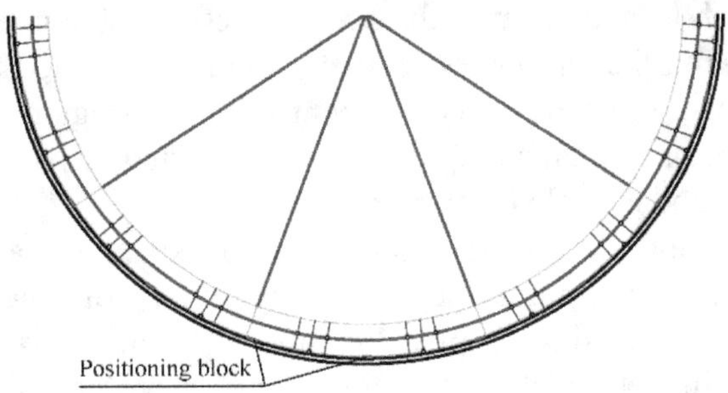

Fig. 5.13. Schematic diagram of positioning block.

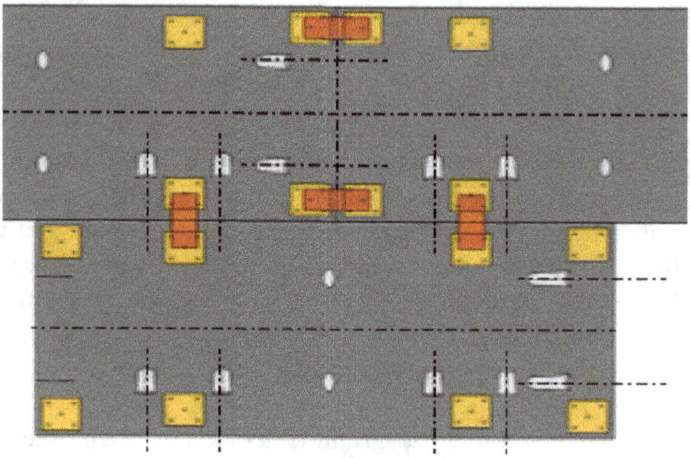

Fig. 5.14. Setting of embedded parts.

Before the temporary segment ring is pushed out, the inner cushion block at the shield tail and the base guide rail at the rear of the shield tail should be greased to reduce friction. Attention should be paid to the consistency of the jack stroke when pushing, through frequent measurement and correction.

During the erection of each ring, the circumferential bolts and longitudinal bolts are tightened in time, and the segments are connected and strengthened transversely and horizontally with embedded steel plates to strengthen the overall stiffness of the temporary segments.

When the segment is exposed out of the shield tail, the horizontal cross brace and 45° diagonal brace are erected in time to maintain the true roundness of the segment ring. Moreover, any void between the segment bottom and the base is also be filled with concrete, so as to connect the temporary segment ring and the base as a whole and enhance the overall stress, as shown in Fig. 5.15.

After the temporary segment ring is pushed to the target position (the last temporary segment ring leans against the concrete backing), any uneven gap with the backing is padded with steel plates and a circle of mortar.

When the shield machine is launched on a small-radius curve, the temporary segments separated from the shield tail is welded with the rib plate in time, and the horizontal brace is strengthened on the other side of the curve direction to prevent the temporary segment ring displacement.

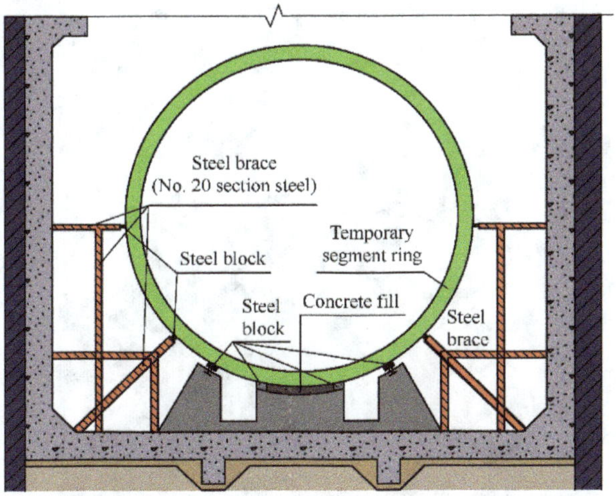

Fig. 5.15. Schematic diagram of temporary segment ring bracing and reinforcement.

(3) **Portal chiseling:** The launching tunnel portal is chiseled by an air pick in layered breaking. The main part of the diaphragm wall is chiseled twice. The first chiseling is carried out during the installation of the shield machine, and the second chiseling is carried out just before the cutterhead of the shield machine reaches the water stop at the tunnel portal. During chiseling, the depth of each chiseling is strictly controlled, and the thickness of each chiseling is about 500 mm. In the process of chiseling, the inner and outer rows of reinforcement in the portal are cut in turn.
(4) **Launching and advancing in reinforcement area:** Launching on a small radius curve will cause a large horizontal angle with the portal ring. Therefore, the processing size of the water stop box needs to be simulated to ensure the portal plugging at once. Moreover, the spacing between the shield machine and the water stop box is considered in the design of the water stop box to prevent deformation of the water stop box, as shown in Fig. 5.16.

According to the known launching attitude of the shield machine, curve fitting of the shield machine tunneling axis is carried out, and the deviation of the shield machine advance axis is simulated in type selection to ensure the quality of the formed tunnel.

When the shield is launched close to the soil mass, the sealing effect of the water stop device at the portal should be checked again to ensure launching safety. A slurry balanced shield machine needs to establish the

Fig. 5.16. Processing of water stop box.

slurry balance before the cutterhead is close to the soil mass, bringing high water stop requirements to the portal, so it is particularly important to check the sealing effect of the water stop box.

When tunneling in the shield machine reinforcement area, the mixer must be started, and the construction parameters must be strictly controlled.

(i) *Strict control of support pressure forward*: Since the shield machine is close to the reinforced soil mass of independence and strength to a certain degree, the cutting face pressure can be set at a low level. Moreover, the cutterhead torque, total thrust, settlement report and other construction parameters are analyzed for adjustment. For a slurry balanced shield machine, in order to prevent slurry from flowing back into the working shaft and other adverse conditions, the cutting face water pressure should be set at the lower limit value, as long as normal slurry circulation can be maintained.

(ii) *Control of advance rate*: As the shield machine is located in the reinforcement area, in order to effectively control the advance axis and protect the cutterhead, the shield machine should advance slowly and steadily instead of at a high advance rate. The shield machine should be controlled at about 10 mm/min for advancing in the reinforcement area.

(iii) *Slurry management*: During the construction of a slurry balanced shield machine, the stability of the excavation face mainly depends on a high-quality slurry membrane. The slurry requirements during shield machine advance in the reinforcement area: a density of 1.10–1.18 g/cm^3 and a viscosity of 16–18 s (funnel viscosity). During the launching construction, it is necessary to monitor leakage at the water stop device, and inject HS materials to stop leakage if necessary. In addition, while ensuring that sufficient slurry meeting the standards can be fed, adequate preparation must be made for slurry discharge to ensure stable and smooth operation of the slurry system. When the shield machine passes through the reinforcement area and enters the undisturbed soil mass, the slurry indexes should be appropriately increased on the premise of increasing the cutting face pressure, i.e. density of 1.10–1.30 g/cm^3 and viscosity of 18–20 s (funnel viscosity).

(5) **Portal plugging:** After the shield tail shell enters the box, the structure void at this part suddenly increases, so the box type water stop sealing device should be strictly monitored in real time. The tunnel portal can be plugged when the segment ring backed with the embedded steel plate is 150–200 mm out of the shield tail (the water curtain plate still acts on the shield tail shell), as shown in Fig. 5.17, so as to prevent soil

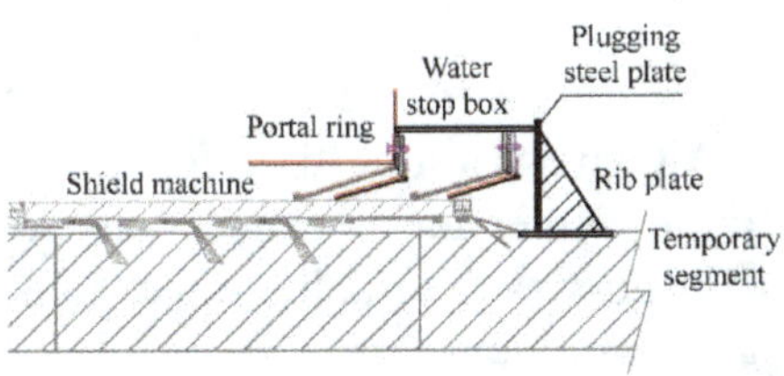

(a) Structural sketch

(b) Site construction

Fig. 5.17. Portal plugging.

loss from such a void, causing ground collapse. Moreover, after the shield tail leaves the tunnel collar for about 10 rings, grouting is carried out through the provided grouting hole at the tunnel portal under the premise of controlling the grouting pressure until full filling.

(6) **Monitoring control:** During advance in the base section of the shield machine, the control of the shield machine attitude is strengthened, and the construction is carried out in strict accordance with the requirements of straight line advance, that is, the stroke difference between the left and right jacks should be consistent.

   (i) After the tunnel portal is plugged, the water stop box and the tunnel collar are grouted. After the shield machine is out of this range, the tunnel collar is grouted with a single-liquid slurry to prevent temporary segment displacement within the range of the tunnel collar.

   (ii) The increased ground loss and correction frequency during advance in the curve section increase disturbance to the soil mass. The synchronous grouting amount should be strictly controlled during advance in the curve section. The grouting amount is adjusted during advance ring by ring according to the deformation monitoring results. The quality of grout, grouting amount and grouting pressure must be strictly controlled. When grouting fails to meet the requirements, the shield machine should be suspended.

   (iii) During the construction at small-radius curves, specially designated personnel should be appointed to correct the deviation of the shield machine frame wheels to ensure that the frame is basically kept level.

5.1.3.4 *Frame conversion*

(1) The shield machine needs to be shut down for a long time for frame conversion. Therefore, before the frame conversion, it is necessary to select a proper parking position with good stratigraphic conditions and little impact on the surrounding environment.

(2) When the frame moves from the non-drop area of the frame section structure to the step of the drop area, a steel structure platform should be arranged in the base area of the inner arc face of the square tunnel under the structural floor step. After the entire frame moves onto the

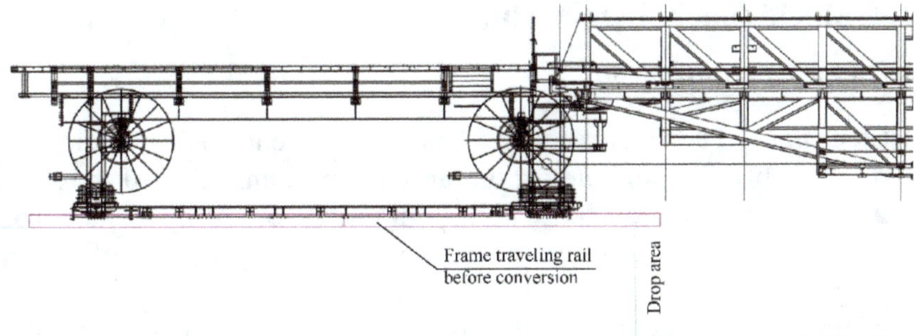

(a) Before frame conversion

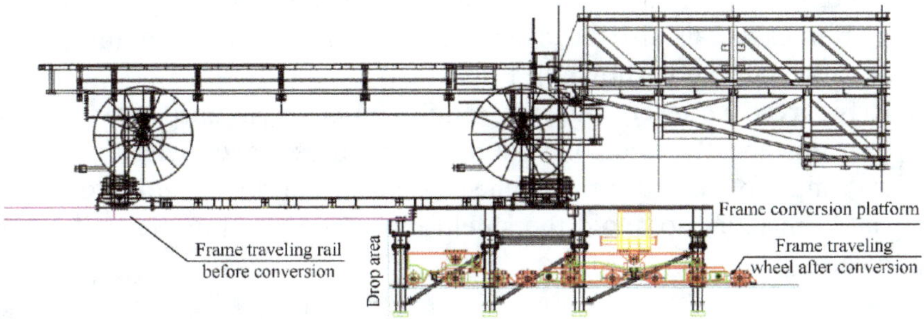

(b) Frame conversion

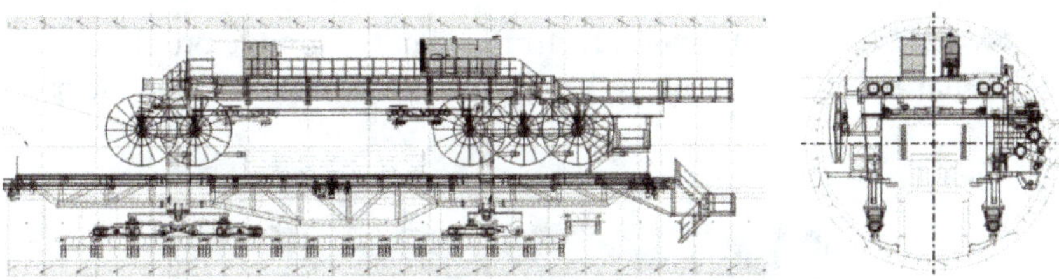

(c) After frame conversion

Fig. 5.18. Schematic diagram of frame traveling mechanism conversion.

platform, the traveling mechanism should be converted to install the traveling components of the standard section and removing the temporary traveling wheel and steel structure platform, as shown in Fig. 5.18.

(3) After the frame is converted, the vertical transport and horizontal transport in the tunnel are switched to the normal mode.

5.1.4 Shield machine receiving

5.1.4.1 *Receiving base*

(1) **Steel base:** The steel receiving base, convenient for installation and disassembly and suitable for turnaround and translation of the shield machine, can be recycled with a high utilization rate. The steel base must be firmly supported to ensure its stability during placement (Fig. 5.19).
(2) **Mortar base:** The mortar base is low in strength, usually M5–M10, which is easy to be cut by the shield machine and chiseled in the later stage, and is suitable for entry in water. The bearing capacity of the mortar base can also be calculated. The mortar base of the shield machine cutting is shown in Fig. 5.20.
(3) **Reinforced concrete base:** A reinforced concrete base has a strong bearing capacity. The concrete base should be placed with the height and angle of the base rail rechecked according to the attitude of the shield machine to avoid the shield machine shoveling the rail when receiving.

5.1.4.2 *Portal sealing device*

As shown in Fig. 5.21, air bags are arranged in the groove provided in the portal collar before receiving. When the shield is received, the air bag is

Fig. 5.19. Placement of steel base.

Fig. 5.20. Mortar base of shield machine cutting.

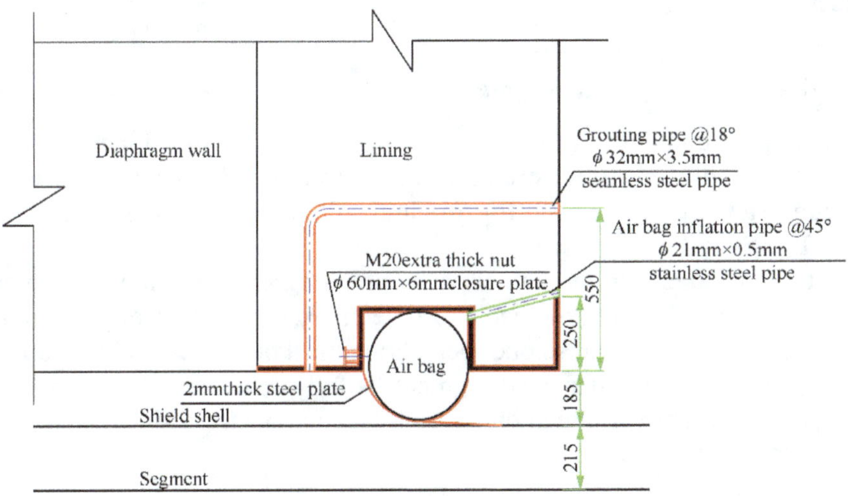

Fig. 5.21. Schematic diagram of air bag opening (unit: mm).

opened for inflation to prevent the sediment from flowing into the tunnel collar. Before air bag installation, an on-site inflation and deflation test is conducted to ensure the air tightness of the airbag.

During the tunnel portal chiseling, attention should be paid to protecting the air bag against any damage affecting the sealing device.

5.1.4.3 *Directional survey*

The directional survey before holing through by the shield machine is an important basis for checking the orientation of the shield machine, confirming the attitude of the shield machine, evaluating the attitude of the shield machine when entering the tunnel, formulating the construction axis of the shield machine in the portal section, the control value of the advance gradient, and preparing the method statement so that the shield machine construction can always be implemented according to the developed plan during the construction of this stage and entering the tunnel with a good attitude.

When the shield machine advances from about 200 m away from the tunnel portal, a directional survey is carried out, which include a resurvey of the ground control network, determination of the central position of the receiving shaft portal, a shaft connection survey and an underground traverse survey.

5.1.4.4 *Sing of special segment*

Generally, the integrity of the longitudinal connection of segments in the portal section needs to be strengthened. Commonly, long screws or connecting steel plates are used for longitudinal connection, with a distance of 20 rings in the portal section.

In order to ensure the tunnel portal plugging after the shield machine enters the tunnel, a ring of special segments backed with steel plate must be set at the portal collar, so one special ring must be processed in advance. The back of this ring of special segments is fully covered with steel plates, and steel plates are arranged on the ring face in contact with the jack.

5.1.4.5 *Construction process of shield machine receiving*

The construction process of shield machine receiving is shown in Fig. 5.22.

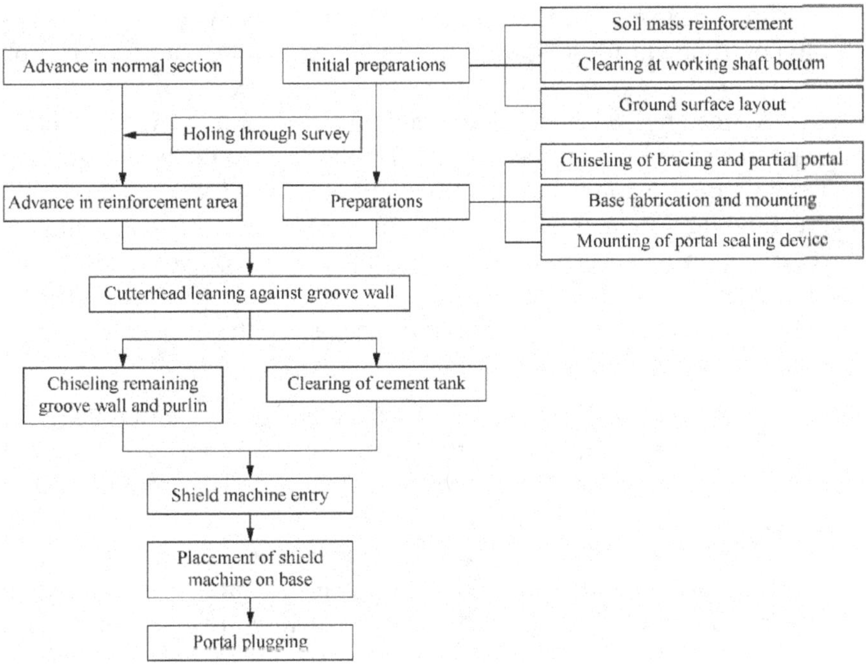

Fig. 5.22. Construction process of shield machine receiving.

(1) **Acceptance of ground reinforcement effect:** The ground reinforcement is subject to reinforcement before the tunnel portal is chiseled, and the acceptance method is the same as that of the shield machine launching. The "cross"-shaped horizontal test boreholes are set on the tunnel portal to check the leakage of the trial tunnel section and the soil sample coring and to judge the soil mass reinforcement effect in the reinforcement area. Receiving and tunnel portal chiseling should not be carried out until no obvious leakage is confirmed; otherwise, reinforcement measures should be taken.

(2) **Dewatering or freezing construction treatment:** For any dewatering wells around the reinforcement area at the tunnel portal, the pumping test should be carried out immediately after the dewatering wells are drilled, and the corresponding report is issued. Before the tunnel portal is chiseled, the dewatering wells are opened to ensure that the water level drops to the required level. During dewatering, it is necessary to ensure power supply, and a standby water pump is

provided to ensure uninterrupted dewatering. Moreover, the surrounding environment must be closely monitored, and dewatering must be stopped in time after the shield machine is received.

If freezing method is adopted for reinforcement, the shield machine advancing adjacent to the frozen area should be stopped, the freezing pipes should be pulled out and placed at 1 m over the shield machine, and then freezing is resumed. After the shield machine is received, unfreezing, pipe pulling, backfilling and other processes can be carried out.

(3) **Advance in receiving reinforcement area:** Control of the main construction parameters:

(i) *Axis control*: Taking the portal center as the reference, the deviation should be controlled within ±20 mm, and the void between segments and the shield shell should be maintained as uniformly as possible.

(ii) *Advance rate*: It should be controlled at about 10 mm/min.

(iii) *Support pressure forward*: The support pressure forward is gradually adjusted according to the volume of excavated earth and total thrust. At this stage, the slurry tank of a slurry balanced shield machine is gradually replaced with clean water so as to achieve the purpose of tank clearing.

(iv) *Synchronous grouting*: In the receiving reinforcement area, the synchronous grouting should be conducted with early-strength grout, and the synchronous grouting amount is the same as that of the normal excavation section. In order to ensure that the grouting pipeline is unblocked, the grout barrel is cleaned in time. At this stage, pressure compensation grouting is also be done to ensure that the void around segments can be completely filled.

(v) The bolts of segments in rings should be re-tightened within 20 m of the receiving section, and the connection and reinforcement between the annular joints of segments should be strengthened to ensure that the segments in rings can meet the strength and stiffness requirements during the receiving construction.

(vi) The speed of information feedback should be accelerated, contact and communication should be strengthened between operators, and construction accuracy and quality should be improved by means of real-time and informatization.

(4) **Support pressure release:** When the shield machine advances close to the diaphragm wall at the tunnel portal, the support pressure forward is gradually released to 0, and the earth chamber is emptied.

(5) **Portal chiseling:** When the shield machine is close to the enclosure structure of the working shaft at the portal and the support pressure forward is fully released, the tunnel portal chiseling can be started. As shown in Fig. 5.23, the receiving tunnel portal is chiseled manually in combination with a pick breaker in layers on the working face. The tunnel portal is chiseled continuously to shorten the operation time as much as possible. Once the shield machine leans against the groove wall, the tunnel portal chiseling should be started.

(6) **Shield machine receiving in water:** After entering the working shaft, the shield machine maintains stable construction, mainly relying on its dead weight, the buoyancy of water in the shaft, the support force of the base and the support reaction of the segments. Therefore, the construction control in the process of receiving water is mainly to ensure balance under these four forces. After the portal is chiseled, water is added to the working shaft. As buoyancy is a function of the volume of the shield machine entering the working shaft and the slurry level, the slurry level in the shaft needs to be adjusted at different stages of the shield machine entering the working shaft to ensure that the buoyancy of slurry to the shield machine is greater than the sum of the shield machine dead weight and the friction between segments.

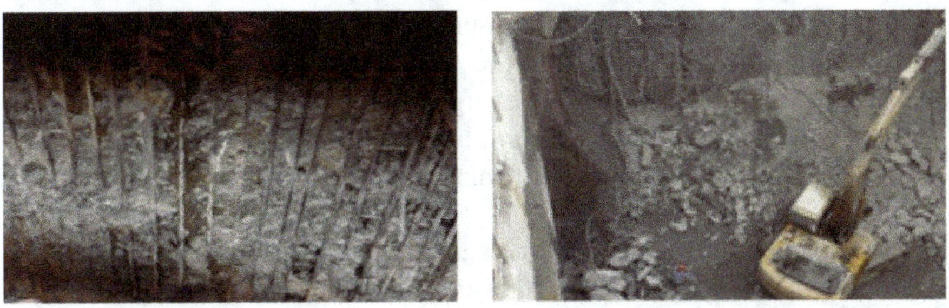

(a) Manual stripping of concrete (b) Chiseling by pick breaker

Fig. 5.23. Receiving tunnel portal chiseling.

Fig. 5.24. Shield machine receiving in tunnel under water.

(i) *Advance in the working shaft*

(a) In the process of receiving water, a large amount of slurry is discharged into the working shaft from the slurry tank after the tunnel portal is chiseled open, as shown in Fig. 5.24. The shield machine should be stopped, and the slurry circulating system should be switched to bypass mode. After stabilization, the bubble chamber pressure is set according to the pressure sensor in the center of the cutterhead, and the slurry circulating system is restarted for advancement. During the circulation, the slurry inflow and slurry discharge flow will be close to each other. Moreover, a liquid level observer in the shaft is arranged to be on duty and communicate with the shield machine driver at any time to adjust the slurry inflow and discharge to ensure that the slurry level in the working shaft is unchanged.
(b) After the cutterhead of the shield machine enters the working shaft, close monitoring should be conducted on the relative position between the shield machine and the water stop device to prevent scratching or other damage to the water stop device.
(c) Synchronous grouting, pressure compensation grouting and segment re-tightening must also be done well in the process of advancing the working shaft.
(d) When receiving water, the receiving base is generally built with mortar.

(ii) *Advance with shield tail leaving portal*

(a) Compaction grouting is carried out through the grouting hole on segments to fill the void between segments and the tunnel portal.
(b) After grouting, water is pumped and soil is cleared out of the working shaft.

(iii) *The portal is plugged*
When receiving water, the liquid level in the working shaft should be gradually reduced, and the structure void in the tunnel collar should be grouted after the steel plate is welded between the shield shell and the tunnel portal. After the grouting is completed, the shield machine moves forward until the shield tail is completely separated from the portal ring, and the tunnel portal is plugged immediately, and then the tunnel portal is grouted for blockage.

5.1.5 Shield turnaround

The Programmable Logic Controller (PLC)-based integral synchronous lifting technology is adopted for the turnaround of the shield machine. Lowering, leveling, translating, rotating, lifting and gradient adjusting of the shield machine can be achieved by adjusting the movable shield machine base stand (connected with the shield machine as a movable unit).

5.1.5.1 *Turnaround process*

The shield turnaround process is shown in Fig. 5.25.

5.1.5.2 *Installation and commissioning of turnaround device*

(1) **Leveling and laying of steel plates:** After the construction of the working shaft is completed, the working shaft floor is leveled and fully paved with 16 mm thick steel plates, which are connected by groove welding to ensure that the joints are flat and firm.
(2) **Steel plate fixation:** In order to prevent the dislocation of the steel plate caused by the integral translation and rotation of the shield machine, after the steel plate is laid, the steel plate should be drilled, and so should the concrete floor at the corresponding positions, and the steel bars should be embedded into the concrete floor and fixed, with the embedment depth not less than 400 mm.

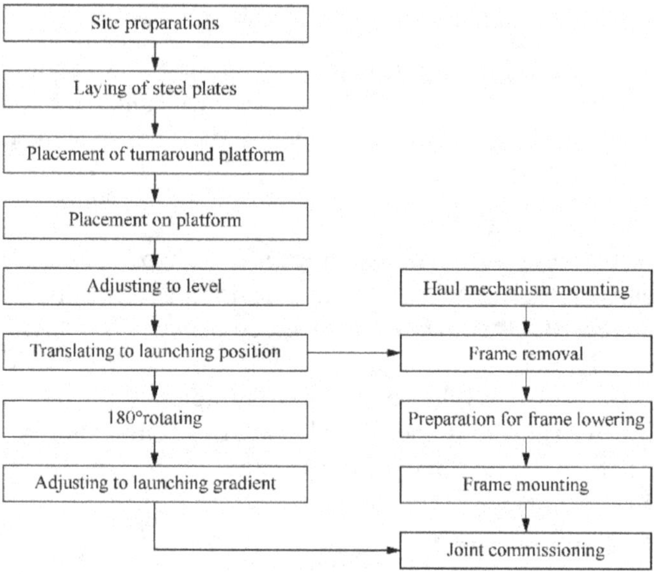

Fig. 5.25. Shield turnaround process.

(3) **Laying of steel structure base:** According to the design center line of the tunnel, the baseline should be positioned, and the steel structure base stand, jack and controllable suspended sliding device should be mounted in the receiving shaft. The installed base stand should be fixed with longitudinal and transverse braces, as shown in Figs. 5.26 and 5.27.
(4) **Gradient adjustment and jacking test:** After the hydraulic equipment is mobilized, according to the actual jack layout and grouping, the whole base stand should be commissioned in steps such as pre-jacking and simulated jacking for decreasing gradient adjustment to ensure that the equipment can be jacked successfully at one shot after entering the foundation pit.

The formal jacking lifting should be carried out after the trial jacking. When the shield machine is adjusted to decrease the gradient in the receiving shaft, the method of “first adjusting to level, then descending as a whole” is adopted.

5.1.5.3 *Key points of operation*

The schematic diagram of translation and rotation construction is shown in Fig. 5.28.

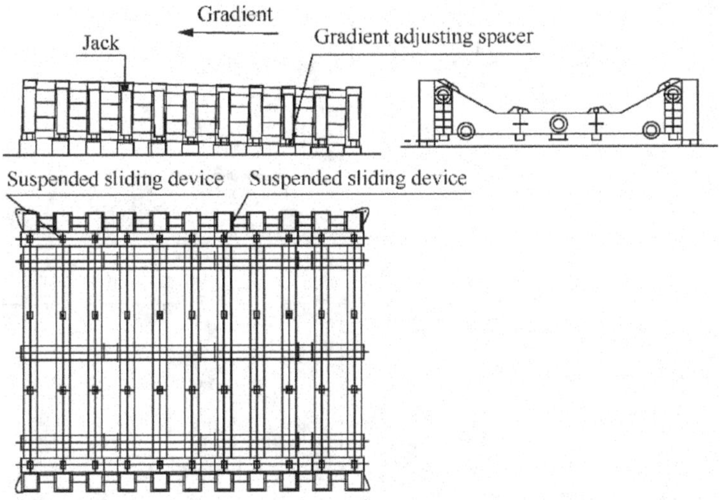

Fig. 5.26. Schematic diagram of base stand mounting.

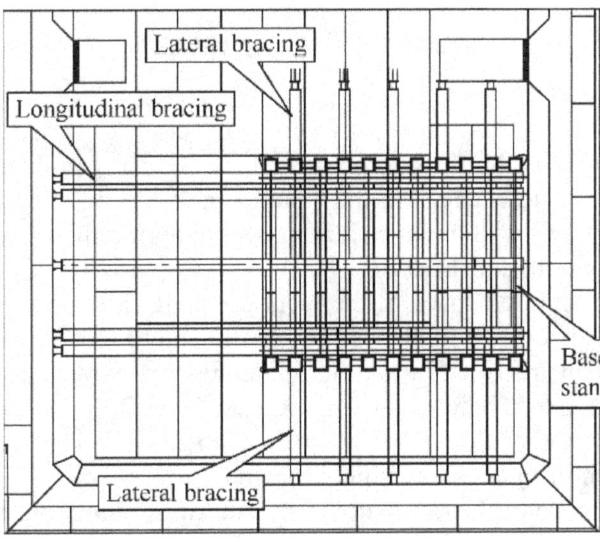

Fig. 5.27. Schematic diagram of bracing of base stand.

During translation and rotation construction, the following key points are considered:

(1) The translation distance is the theoretically calculated value. Before the on-site translation construction, the translation trajectory is set out

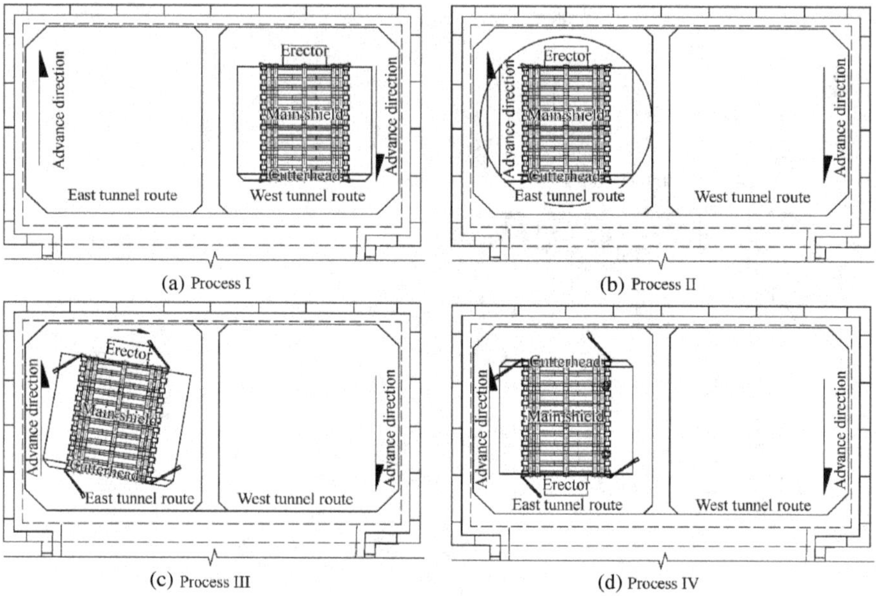

Fig. 5.28. Schematic diagram of shield machine translation and rotation.

according to the base point provided by the civil construction contractor through on-site measurement, so as to clear any problems in time and avoid large changes in the actual conditions on site, which may cause the shaft wall and the shell to collide during the translation and prevent smooth construction.

(2) Due to the limited space of the working shaft, it is required to accurately control the central position of the shield machine during the rotating construction. In addition to computer monitoring, marking lines are made on the stand and the shield machine shell, and a theodolite is used to aid measurement so as to ensure the accuracy of the translation distance of the shield machine.

(3) In order to control the accuracy of translation direction, a stopping device is set for translation.

5.2 Stability of excavation face

Shield machines advance by the action of jacks while cutting soil mass with the front rotating cutterhead. The soil mass in the area ahead of the cutterhead is called the excavation face (or working face), which is subject to the support pressure from the shield machine. When the force on the excavation face is too large, passive failure will occur, and the soil mass ahead will be heaved; when the force on the excavation face is too small, active failure will occur, and the soil mass ahead will collapse. In engineering practices, the active and passive failures of excavation face are often referred to as instability, which will cause huge environmental and construction quality problems. How to maintain the support pressure in a reasonable range and ensure the continuous stability of the excavation face is the key point of shield tunneling technology. According to the different balanced methods, shield machines are classified into two types: earth pressure balanced type and slurry balanced type, and their corresponding stability mechanisms of the excavation face are also different.

5.2.1 Excavation face stability of earth pressure balanced shield machine

5.2.1.1 *Stability mechanism of excavation face*

As shown in Fig. 5.29, the earth pressure balanced shield machine uses the full-section cutting cutterhead installed in the front of the shield machine to cut the soil mass ahead and convey it into the storage and sealing chamber behind the cutterhead and maintain the chamber under an appropriate pressure to balance with the water and soil pressures on the excavation face so as to reduce the disturbance of the shield machine to

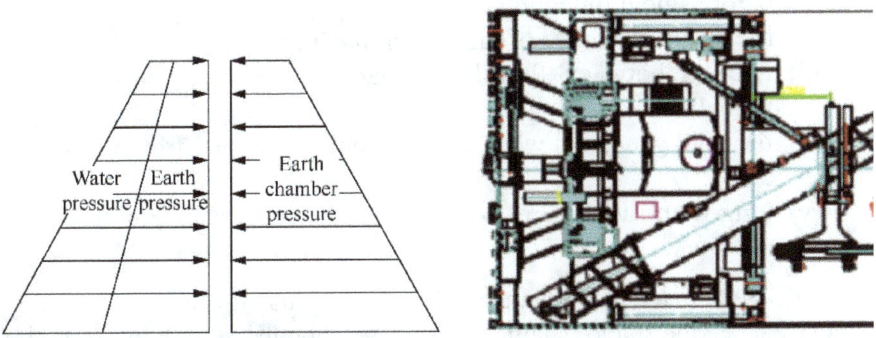

(a) Pressure balance on excavation face (b) Shield machine excavation face

Fig. 5.29. Schematic diagram of stability mechanism of excavation face of EPB shield machine.

the stratum and control the surface settlement. A too large earth pressure in the shield machine pressure chamber will cause heave of the upper stratum at the working face; a too small earth chamber pressure will make the pressure exerted on the excavation face insufficient to support its stability, so that the soil mass is loosened and then collapses, causing the surface subsidence. Therefore, the key to the stability of the excavation face lies in the determination of a reasonable support force range on the excavation face.

5.2.1.2 *Factors influencing the stability of excavation face*

(1) **Disturbance of shield machine construction to soil mass:** The influence of shield tunneling on surrounding strata mainly includes the stress relief of soil mass and the change in water content, volume or pore water pressure. The deformation of soil mass is attributed to the position change of soil particles, which also affects the mechanical parameters of soil mass, instead of the deformation of soil particles due to external forces. In the originally stable stratum, the shield tunneling will lead to a change in the original stress state around the stratum, which will cause the surrounding soil mass to present complex mechanical behaviors such as loading or unloading and destroy the original equilibrium state of the soil mass, thus causing disturbance to the soil mass.

Stress changes in soil mass during shield tunneling are mainly caused by the following:

(i) before shield tunneling,
(ii) the excavation face of the shield machine,
(iii) friction between the shield machine and soil mass during tunneling,
(iv) the void generated when the segment is supplied from the shield tail,
(v) long-term rheology of soil mass in the process of ground stress recovery.

Shield machines advance relying on thrust from the rear jacks. Only when the shield machine jacks have enough strength to overcome all kinds of resistance encountered in the process of moving forward can the shield machine move forward; moreover, these resistances react with the soil mass and generate additional stress of the soil. The resistance causing soil mass disturbance mainly includes the friction F_1

between the shield machine shell and the surrounding soil mass, the resistance F_2 from the soil mass cut, the friction F_3 between the shield machine and the supporting frame, the friction F_4 between segments and the shield machine, and the active earth pressure F_5 from the soil mass on the excavation face. The specific conditions are as follows.

When the total thrust of the jacks $T \geq F_1+F_2+F_3+F_4+F_5$, the soil mass cut by the shield machine is loaded and elastic-plastic deformation occurs, resulting in a squeezing disturbance area. The soil mass is compressed by the squeezing action on the excavation face, causing the soil mass to move forward and rise, under which the shield machine works normally; when the total thrust of the jacks $T < F_1+F_2+F_3+F_4+F_5$, the shield machine is in a standstill state. This state corresponds to the situation where the soil mass ahead of the excavation face is not supported in time or the soil mass is seriously overexcavated. The soil mass with stress released slides toward the free face in the shield machine pressure chamber, causing the surface to sink.

Friction resistance is generated between the shield shell and the surrounding soil mass during shield tunneling. This resistance creates a shear disturbance in the soil mass containing the shield shell, which is in a small range. Outside the shear disturbance area, due to the existence of a structure void at the shield tail, the soil mass moves into the void, causing the soil mass to loosen and the surface to sink; the soil mass over the shield machine moves downward due to its dead weight and ground overload, forming an unloading disturbance area; the soil mass under the shield machine forms an unloading disturbance area due to heave.

The disturbance of shield machine construction to the surrounding soil mass can be divided into two forms:

(i) The squeezing action on the excavation face or lateral soil mass causes the soil mass to compress and make the soil mass move forward and rise.

(ii) The over-excavation of the excavation face and the existence of a void at the shield tail cause the soil mass to be loosened and collapse, resulting in the surface settlement. The disturbance range and degree of the soil mass depend on the construction state of the shield machine. When the earth pressure is large, the advance rate is high, and the amount of soil discharged is small, the soil mass is squeezed into a plastic zone, and the disturbance range depends on the under-excavation amount and soil properties; when the earth pressure is small, the advance rate is low, and the amount of soil discharge is large, the soil mass tends to create a relaxation zone, and its scope depends on the amount of over-excavation and soil properties.

(2) **Support pressure of excavation face:** The stability of the excavation face of the earth pressure balanced shield machine is mainly maintained by the support pressure forward, that is, the earth pressure in the front earth chamber of the shield machine. Therefore, the stability of the excavation face directly depends on the rationality of the support pressure forward. During the shield machine construction, the stability of the excavation face is achieved by the accumulation action of muck in the excavation chamber. In the actual excavation process, the support pressure of the excavation face cannot always be maintained in a constant ideal state but is always in a constantly changing state. The main reasons for the change in support pressure are as follows:

(i) *Stratum diversity*: During shield tunneling, the soil conditions ahead of the excavation face change constantly. Under different stratum conditions, the same excavation face support pressure produces different effects. The determination of support pressure is closely related to soil layer conditions, groundwater conditions, excavation methods, tunnel diameter and burial depth. Such facts require that the support pressure of the excavation face change constantly to meet the requirements of different excavation methods in the construction process.

(ii) *Fluidity of soil mass in the chamber*: The flow state of muck entering the excavation chamber is different, and the occurrence of problems such as "blocking" and "mud caking" during the excavation process leads to great changes in the support pressure. Moreover, due to the immature technologies related to the application of additives, the soil in the excavation chamber cannot reach a more stable flow state, which also affects the control of the support pressure on the excavation face.

(iii) *Influence of tunnel diameter and burial depth*: The additional stress on the soil caused by shield tunneling is extremely complex. When the shield machine advances, it causes additional stress in the soil ahead and lateral soil, making the soil mass around the shield machine in an active or passive state, while elastic recovery occurs after the shield machine moves forward.

The additional stress in soil caused by tunneling is not only related to the physical parameters of the soil but also closely related to the construction parameters of the shield machine. The larger the diameter of the shield machine, the more difficult it is to support the whole cross section; at a small burial depth, the support pressure is generally calculated according to the full overburden, but at a large burial depth, a certain soil arching effect should be considered when calculating the support pressure.

5.2.1.3 *Stability measures for excavation face*

The main measure to stabilize the excavation face is to control and maintain the water and soil pressure in the excavation chamber within the target range.

The control of water and soil pressure is realized by adjusting the tunneling rate of the shield machine and controlling the speed of the screw conveyor and the amount of soil discharged. The automatic speed control of the screw conveyor can be divided into volume control and earth pressure control. The volume control means that the speed of the screw conveyor is controlled by the excavation volume in direct proportion to the tunneling rate of the shield machine, and the earth chamber pressure control means that the speed of the screw conveyor is controlled by maintaining a certain level of earth pressure in the pressure chamber. When controlling and managing the earth chamber pressure, the control range of the earth chamber pressure is generally determined according to the stratum characteristics and surface environment.

(1) **Targeted design of shield machine**

(i) *Mixing device in the earth chamber*: The large-diameter shield machine has many disadvantages, such as the large size of the excavation face, the obvious difference in earth chamber pressure distribution and the wide range of complex soil layers to be cut, so the excavation face tends to be unstable. In order to mitigate the adverse effects of the nonlinear distribution of earth pressure and differences in the physical and mechanical properties of soil layers, a mixing device can be added to the earth chamber, as shown in Fig. 5.30.

(ii) *Improvement of muck on excavation face*: Due to the influence of overburden and water and soil pressure (confined aquifer), there is a risk of water surge on the excavation face during the shield

Fig. 5.30. Mixing device in earth chamber.

machine tunneling, which increases the difficulty of the earth chamber pressure control. If the pressure fluctuation of the earth chamber is too large, disturbance to the soil mass ahead increases, causing the loss of the soil mass ahead. Therefore, the fluctuation of the soil pressure on the cutting face should be reduced as much as possible. Injecting a certain amount of muck improver (foam or bentonite) into the excavation face can reduce the permeability of the soil mass ahead, improve its flow plasticity, and better control the earth chamber pressure.

(a) **Foam improvement:** The foam improvement technology is to inject the foam produced by the action of foaming agent and compressed air into the earth chamber, cutterhead and screw conveyer of the shield machine to mix with the .oil for improvement. Due to foam injection, soil particles are wrapped in the elastic cover, forming a three-phase system. The consistency and friction force among soil particles are reduced, and the soil mass has an ideal plastic state within the deformation range, which makes it possible to control the support pressure of the excavation face. With the pressure decrease in the earth chamber, gas in the particle structure will expand, and the soil mass in the earth chamber will deform under the additional pressure; when the pressure in the earth chamber increases, the void volume of the soil mass will decrease, thus preventing cutterhead torque from rising.

The micro-schematic diagram of foam is shown in Fig. 5.31.

In terms of foam application, the following parameters are mainly used: foam concentration (volume ratio of foam

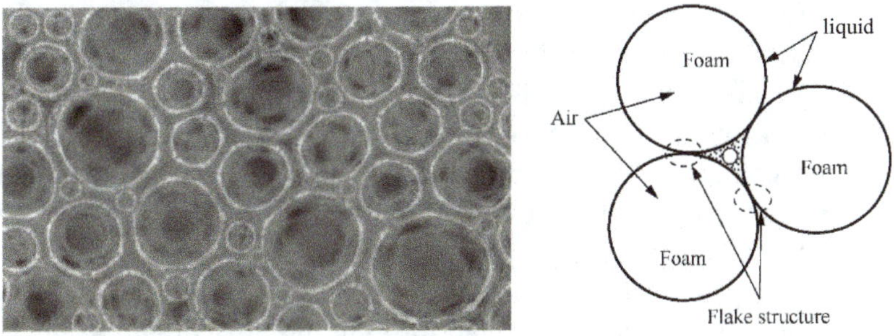

(a) Structure of foam liquid (b) Structural sketch

Fig. 5.31. Micro schematic diagram of foam.

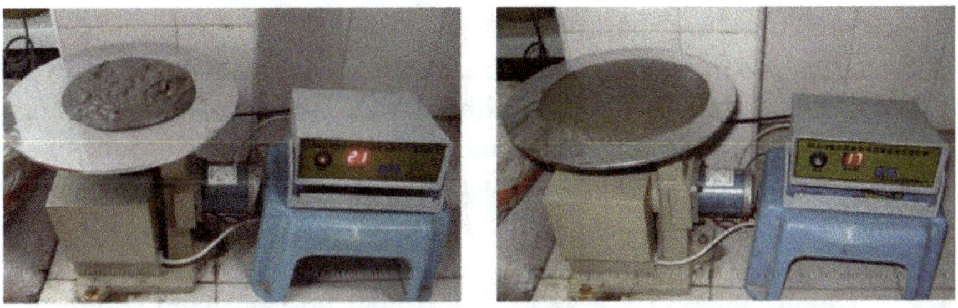

(a) Fluidity of original soilt (b) Fluidity of improved soil

Fig. 5.32. Test of foam injection into soil.

liquid to water), foaming capacity (expansion volume factor of foam mixture under the action of compressed air), and injection rate (volume ratio of injected foam to improved soil). The test of foam injection into soil is shown in Fig. 5.32.

(b) **Bentonite improvement:** Bentonite features hygroscopic expansion, low permeability, high adsorption and good self-sealing performance. The sodium-based bentonite solution injected into the earth chamber can increase the soil fluidity in the earth chamber, play a role as a retaining wall during the process of cutting soil by the rotating cutterhead, and help to maintain the earth pressure balance in the earth chamber, thus avoiding soil collapse on the excavation face, and maintaining the continuous and smooth progress of the excavation.

(c) **Polymer improvement:** The polymers used in early shield tunneling are generally compounded by adding soda ash, sodium carboxymethyl cellulose (CMC) and other polymer materials to ordinary bentonite.

(iii) *Blowout prevention measures for screw conveyer*: Under the action of deeply buried, confined water, gushing may occur on the screw conveyer. Therefore, in the process of advance, the muck on the excavation face should be improved to reduce its water permeability and eliminate the potential risk of gushing; the opening of the screw conveyer gate can be set within 50% to prevent large-scale gushing; in case of gushing, the screw conveyer gate should be closed immediately; emergency water stop materials should be injected through the breast plate of the earth chamber and the injection inlet of the screw conveyer to cut off the water flow channel and restore the soil plug effect of the screw conveyer.

(2) **Support pressure setting:** The stability of the excavation face is ensured by maintaining pressure in the excavation chamber. In order to improve the characteristics of the muck in the excavation chamber, lubricating materials such as bentonite are usually added to the excavation face. The pressure chamber and the screw conveyor are filled with improved soil to keep the earth pressure in balance with the excavation face.

The principle of setting earth chamber pressure is to control the set value between active earth pressure and passive earth pressure. If the actual earth chamber pressure is less than the active earth pressure, ground loss will occur in the soil mass ahead of the shield machine due to stress relaxation, resulting in ground subsidence; if the set value is greater than the passive earth pressure, ground heave will occur in the soil mass ahead of the shield machine due to stress concentration. The earth chamber pressure should be kept appropriate to ensure that the ground above the construction axis of the shield machine neither rises nor collapses.

In addition, if the difference between the earth chamber pressure and the static earth pressure is large, the soil will be disturbed. Therefore, a value as close as possible to the static earth pressure should be taken in combination with the actual situation.

The balance pressure ahead is

$$P = K_0(\gamma_s - \gamma_w)h_s + \gamma_w h_w, \tag{5.14}$$

where

P — equilibrium pressure (including groundwater) (kPa);
γ_s — average unit weight of soil mass (kN/m^3);

γ_w — unit weight of water (kN/m^3), taken as 10 kN/m^3;
h_s — burial depth of tunnel center (m);
h_w — depth from groundwater level to tunnel center (m);
K_0 — coefficient of static earth pressure.

The set value of the balance pressure can be derived from referring to the above methods during the tunneling construction of the shield machine. The specific construction set value is subject to constant adjustment according to the settlement monitoring data.

(3) The control on volume of excavated earth is

$$V = \frac{\pi}{4} \times D^2 \times L, \tag{5.15}$$

where
D — tunnel excavation diameter (m);
L — width of each ring (m).

The optimum volume of excavated earth by the shield machine advance is set by summarizing and analyzing the parameters in the actual construction process.

The excavation process should be observed to follow the curve changes of the mucking volume and the statistical curve of the ring under cutting. When the volume of excavated earth is found to be too large, the surface subsidence and support pressure should be checked immediately. In addition, a detection device can also be used to investigate the soil collapse, and the relevant parameters should be adjusted in time after the cause is found to ensure the stability of the excavation face.

A horizontal belt conveyor of synchronous extension is erected in the tunnel. The muck cut by the shield machine is transported to the sump of the working shaft via this conveyor to complete the horizontal transport of muck in the tunnel.

(4) **Surface subsidence monitoring and control:** The surface deformation generated during shield tunneling is roughly divided into the following three types:

(i) *Surface deformation caused by the shield machine before arrival*: Before the shield machine reaches the measuring point, the disturbance of excavation to the soil mass will lead to a change in the soil stress state and then cause the ground to heave or settle.

(ii) *Surface deformation when the shield machine arrives*: The horizontal support stress of the soil mass on the excavation face may be greater or less than the original static earth pressure, which may cause the uplift or settlement of the soil mass ahead of the excavation face.

(iii) *Surface deformation caused by segments separated from the shield tail*: After the segment is separated from the shield tail, an annular void is formed on the tunnel excavation wall and around the segment lining, which may cause surface deformation due to the setting of synchronous grouting pressure and grouting amount.

During shield tunneling, the monitoring results should be timely fed back, construction parameters should be adjusted reasonably, and corresponding technical measures should be taken to minimize ground movement, and the ground loss rate should be generally controlled within 5‰. For special parts, the settlement control standard should be appropriately improved according to the protection requirements to ensure construction safety and stability in the surrounding environment.

5.2.2 Excavation face stability of slurry balanced shield machine

5.2.2.1 *Stability mechanism of excavation face*

During the construction using a slurry balanced shield machine, the slurry pressure is used to balance the earth pressure and water pressure of the excavation face so as to maintain the stability of the excavation face. The newly injected slurry forms a weakly permeable slurry membrane on the excavation face to keep the slurry pressure effectively acting on the excavation face. At the excavation face, as the pressurized slurry continuously seeps into the soil mass, the sandy soil particles in the slurry fill the soil pores, forming a slurry membrane with a very low permeability coefficient. Since the slurry membrane reduces the pressure loss of the excavation face, the slurry pressure can effectively act on the excavation face, thus preventing the deformation and collapse of the excavation face and ensuring the stability of the excavation face.

A slurry pressurized shield machine maintains the stability of the excavation face, taking advantage of the pressure and characteristics of the slurry in the pressure chamber. The slurry injected onto the excavation face at a certain pressure will form an impermeable slurry membrane or seepage wall on the excavation face. The water and soil pressure in the surrounding rock on both sides of the slurry membrane will be balanced by the action of the slurry pressure to effectively maintain the stability of the excavation face. The infiltration of the fine slurry particles into the excavation face ahead can increase the cohesion of the soil, which is conducive to maintaining the stability of the excavation face. Therefore, it is very important to select the best slurry pressure for the stability of the excavation face. If the slurry pressure is too small, the excavation face will be damaged under the action of the surrounding rock, soil pressure and water pressure; when the pressure is too high, the slurry will permeate and escape in the sandy soil with good water permeability. When the slurry pressure exceeds the shear strength of the soil, split fractures occur in the ground, causing serious consequences of slurry pumping.

The specific weight, viscosity, stability and filtration characteristics of slurry determine the speed of slurry membrane formation and the quality of the slurry membrane. Generally, the greater the specific weight of the slurry, the better the stability of the excavation face. However, in long-distance tunnel excavations, an excessive specific weight will not only cause overload of the slurry pump but also increase the difficulty of slurry treatment. If the specific weight is too small, a large amount of slurry will escape, which will increase the difficulty of forming a slurry membrane and seepage wall. In saturated sandy soil with good water permeability, the filtration characteristics of the slurry are particularly important. The relative relationship between the particle size in slurry and the pore size in soil directly determines the degree of difficulty of slurry membrane formation. If the particle size is too small, the slurry will only permeate the ground without forming a slurry membrane. If the particle size is too large, slurry will deposit and cause blockage in the slurry delivery pipe and pressure chamber.

When the slurry pressure is greater than the groundwater pressure, a certain proportion of suspended particles will be formed in the soil void. After the slurry seeps into the soil mass of the excavation face, according to Darcy's law, the suspended particles infiltrate into the pores among the soil particles along with the slurry. Under the effects of "blocking" and "bridging", these suspended particles are captured under the intermolecular van der Waals force and accumulate on the contact surface of the soil and slurry to form a slurry membrane, as shown in Fig. 5.33. As time goes on, the thickness of the slurry membrane increases, and the penetration resistance increases gradually. When the resistance of the slurry membrane is far greater than the difference between the water and soil pressure

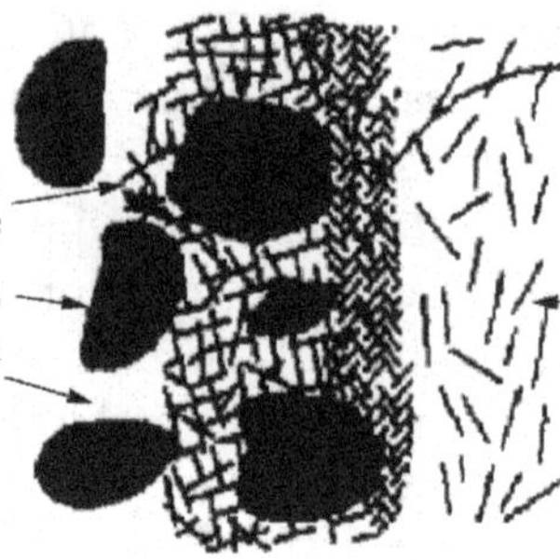

Fig. 5.33. Schematic diagram of slurry membrane formation.

ahead and the slurry pressure, the slurry balance effect will be produced; when the slurry pressure is less than the water and soil pressure before the slurry membrane is formed, it is difficult to form the slurry membrane, causing soil collapse and loosening.

As the pressurized slurry continuously infiltrates into the soil mass, the sandy soil particles in the slurry fill the soil pores, forming a slurry membrane with a very low permeability coefficient (a filter cake layer is formed when the bentonite suspension is supported). Moreover, since the slurry membrane formed reduces the pressure loss of the excavation face, the slurry pressure can effectively act on the excavation face, thus preventing the deformation and collapse of the excavation face and ensuring the stability of the excavation face. Therefore, in the construction process of a slurry balanced shield machine, setting slurry pressure and controlling slurry quality are two key steps.

In order to effectively apply slurry pressure on the cutting face of the cutterhead, it is required that, under the effect of the difference between slurry pressure and groundwater pressure, the bentonite particles suspended in the slurry can quickly form a layer of impermeable slurry membrane on the excavation face to prevent the loss of slurry pressure and prevent the soil particles ahead of the cutterhead from flowing into the shield machine. The effective formation of slurry membrane on the cutterhead excavation face plays an extremely important role in improving the stability of the excavation face, especially in cohesionless sandy soil with poor self-stability and a small uniformity coefficient.

For cohesive soil with a low permeability coefficient and large cohesion, slurry rarely penetrates into the soil. There are two reasons for this. First, the cohesive soil itself has poor water permeability and can effectively prevent the penetration of slurry; second, for the cohesive soil with a high content of montmorillonite and other mineral components, the montmorillonite clay particles have a significant hydraulic effect when encountering water. Under the slurry pressure, a layer of gelatinous slurry membrane is naturally formed on the excavation face. Therefore, along with tunneling by a slurry balanced shield machine in such soil layers, the slurry pressure can effectively act on the excavation face, and the suspended particles retained after the separation treatment of the soil cut by the cutterhead can also be fully used as the slurry constituent, thus reducing the addition of bentonite and other slurry materials. For cohesionless soil with a high permeability coefficient, in order to prevent the loss of slurry pressure, a layer of slurry membrane needs to be formed on the excavation face to prevent slurry from penetrating through the pores among soil particles and penetrating into the deeper range of the soil layer, resulting in a large consumption of slurry and unstable slurry pressure. Therefore, it is necessary to add a large amount of bentonite to the slurry. The slurry mixed with bentonite can effectively form a slurry membrane in silty soil or sandy soil. However, for the soil mass with coarser particles, it is sometimes necessary to add other materials to increase the density and viscosity to ensure the formation of the slurry membrane and further prevent slurry from entering the soil layer and causing slurry pressure loss. Therefore, for a cohesionless sandy soil layer with a high permeability coefficient, the formation of a slurry membrane is very important to maintain the stability of slurry pressure on the excavation face and the stability of the soil layer ahead of the cutterhead.

In addition to the above actions of slurry pressure and slurry membrane, the stability of the soil mass on the cutterhead excavation face during shield tunneling is related to the supporting action of the cutterhead on the soil mass. Therefore, strictly, the stability of soil mass on the cutterhead excavation face is the result of the combined action of shield machine thrust (mainly transmitted by the cutterhead) and slurry pressure. The slurry pressure mainly contributes to the stability of the soil mass within the opening range between the cutterheads, and the soil mass within the action range of the cutterheads is mainly borne by the cutterheads of shield machines.

5.2.2.2 *Factors influencing the stability of excavation face*

(1) **Formation of slurry membrane:** Slurry is a kind of polymer (mixed metal hydroxide slurry system) produced from polymers and active soils mixed with various inactive soils. Under the effect of pressure

difference, a dense and strong slurry membrane is formed on the surface of the permeable layer, which plays a role in reducing slurry filtration and stabilizing the excavation face. From the microscopic mechanism analysis, the formation of the slurry membrane is the result of interaction among active particles. The positively charged active particles are easily dispersed and suspended in water. The particle surface carries a positive charge of not less than 35 mV, which will have a directional effect on the polar water molecules, polarize the water molecules, and form a stable hydration film around them. The outer edge of the hydration film is positively charged, while the clay particles are negatively charged, which will have a similar effect in water. Therefore, when two particles with a strong hydration film are close to each other, a uniform stable body stabilized by water molecules will be formed in the whole space, which is also called a "complex". Countless "complexes" are combined into a stable system.

As shown in Fig. 5.34, the positively charged active particles in the slurry form countless complexes with the negatively charged clay particles on the excavated face. When water loss occurs, more positively charged active particles will cling to the surface of the excavation face, repel the inorganic cations on the surface, and form a potential energy layer with positive potential, which will be completed instantly on the excavation face and continuously build protective films along with excavation to ensure the stability of the excavation face. The positively charged active particle solution can inhibit the clay. The active particle will keep away the cations on the clay surface, and the interlayer cations will leave together with the

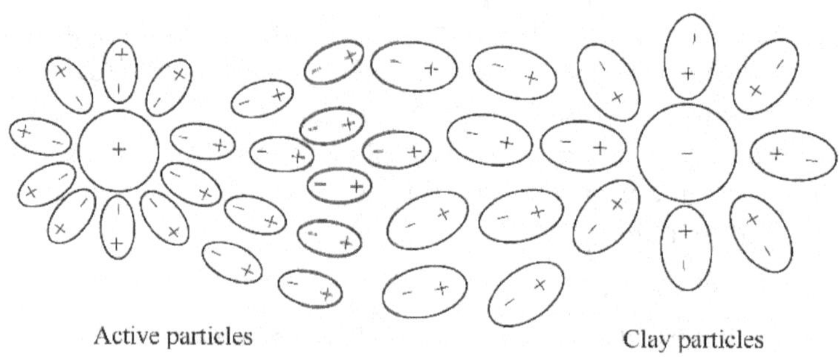

Fig. 5.34. Schematic diagram of combination of active particle hydration membrane and clay particle hydration membrane.

hydration membrane, which will disperse and weaken the clay mineral layers. Moreover, the water molecules are arranged orderly under the action of two kinds of charges, reducing the movement range of the water molecules so that the water absorption of clay can be inhibited.

From the view of macroscopic analysis, in order to effectively apply slurry pressure to the cutterhead working face, it is required that the soil particles suspended in slurry quickly form an impermeable slurry membrane on the excavation face under the effect of the difference between the slurry pressure and groundwater pressure to prevent the slurry pressure loss on the one hand and prevent the soil particles ahead of the cutterhead from flowing into the shield machine on the other hand. Therefore, the effective formation of slurry membrane on the cutterhead excavation face plays an extremely important role in improving the stability of the excavation face.

When the cutterhead cuts the slurry membrane, a new slurry membrane is formed quickly, maintaining the stability of the excavation face. In order to form the slurry membrane reliably and rapidly, so that the pressure can effectively act on the excavation face, the slurry should have the following characteristics:

(i) *Density*: In order to maintain the stability of the excavation face, the deformation of the excavation face should be controlled to the minimum, which brings certain density requirements on the slurry. Theoretically, the slurry density should reach the density of the excavated soil. However, the slurry at a high density will cause overload in the operation of the slurry pump and difficulty in slurry treatment; the slurry in a low density can reduce the load of slurry pump, but the increase in slurry particle leakage causes slow formation of the slurry membrane, which is unfavorable to the stability of the excavation face. Therefore, when selecting the slurry density, the stratum structure of the soil mass must be fully considered, and the capacity of the equipment must be considered while ensuring the stability of the excavation face.

(ii) *Sand content*: In highly permeable soil masses, the speed of slurry membrane formation is closely related to the maximum particle size and sand content of sand particles mixed in the slurry due to the role of sand particles in filling soil pores. In order to give full play to this role, the particle size of sand should be larger than the soil pores, and the content should be moderate.

(iii) *Viscosity of slurry*: Slurry must have a proper viscosity to achieve the following effects:

(a) Prevent the clay and sand particles in the slurry from depositing in the slurry tank, and keep the excavation face stable.
(b) Increase viscosity and resistance to prevent slurry escape.
(c) Transport the excavated spoil by fluid and use the post-treatment equipment to remove waste residue for slurry separation.

The functions of slurry are as follows:

(a) Fill the excavated space, immerse the soil mass by the action of pressure difference, and block the water passage.
(b) Balance the lateral pressure of the formation, form a slurry membrane on the excavation face, maintain the stability of the excavation face, and prevent the collapse of the formation.
(c) Restrain hydration dispersion of the stratum clay and prevent the expansion and collapse of the clay layer caused by hydration dispersion and subsequent failure if the shield machine is in normal advance.
(d) Suspend the muck and carry it out of the ground.

(2) **Slurry membrane quality:** From the slurry balance theory, it can be seen that in the tunneling process of a slurry balance shield machine, it is a key link to form a slurry membrane with a very low permeability coefficient as soon as possible. The analysis of the formation mechanism of slurry membrane shows that the formation of slurry membrane is related to both the quality of the slurry and the nature of the ground. The quality of slurry includes the maximum particle size of slurry, the mix proportion of slurry, the concentration of slurry and the pressure of slurry, while the soil properties include the soil type, the particle diameter of soil and the permeability of soil. Therefore, the formation of a slurry membrane is the result of an interaction between slurry quality and ground properties.

(i) *Maximum particle size of slurry*: As the slurry membrane is formed by blocking and bridging effects, the maximum diameter of particles contained in the slurry has a great impact on this effect.

Only when the maximum particle size of the slurry matches the permeability coefficient and the particle size of the soil mass can it be conducive to the formation of the slurry membrane. The correlation among soil layer, permeability coefficient and maximum particle size of slurry is shown in Table 5.2.

(ii) *Particle grading*: The particle grading of the slurry also has a great influence on the formation of the slurry membrane. The optimum particle grading of slurry needs to be determined through a large number of experiments. Similar to grouting, slurry membrane formation and bleeding are related to viscosity and the hole plugging effect. When the particle size of slurry is larger than one-tenth of the effective particle size of undisturbed soil, slurry cannot penetrate undisturbed soil; when the particle size of slurry is larger than one-third of the undisturbed soil pores, slurry cannot pass through the undisturbed soil pores. During construction, slurry suspension is prepared by mixing soil particles cut by the cutterhead, and fresh slurry is used to form the slurry membrane and balance the excavation face.

(iii) *Slurry density*: Slurry density is closely related to viscosity, bleeding rate and the mix proportion of the slurry. The increase in slurry density can increase the yield value of slurry and enhance the stability of the slurry membrane. High density slurry can produce a high-quality slurry membrane. The density of slurry varies with different soil layers, and the adjustment range is 1.2–1.25 g/cm^3. The slurry with a low density can be used in cohesive soil, and the slurry with a high density is used in sandy soil (sand or sand-gravel layer). The slurry density should not be too high or too low during the excavation process. The former will affect the slurry conveying capacity, while the latter will damage the stability of the excavation face. The range of slurry density shall be 1.15–1.25 g/cm^3, and the upper limit shall be

Table 5.2. Correlation among soil layer, permeability coefficient and maximum particle size of slurry.

Soil layer	Coarse sand	Medium sand	Fine sand	Silt
Permeability coefficient (cm·s^{-1})	1–90	1–9	$(1–9) \times 10^{-1\sim-2}$	$(1–9) \times 10^{-3}$
Maximum particle size of slurry (mm)	0.84–2.00	0.42–0.84	0.074–0.42	<0.074

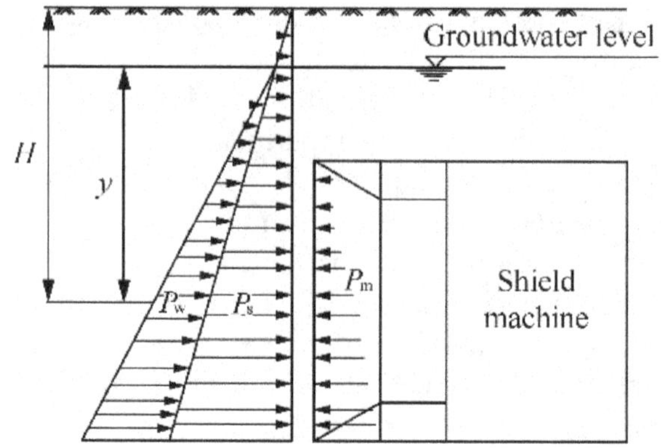

Fig. 5.35. Schematic diagram of stability mechanism on excavation face of slurry balanced shield machine.

determined according to the special requirements of construction. For construction in sandy soil, shield tunneling through shallow overburdens, protection of ground buildings, etc., the slurry density can reach 1.30 g/cm^3, or even 1.35 g/cm^3. The increase in slurry density and viscosity makes the coarse and fine particles in the slurry membrane closely bound, with small pores and high compactness.

(3) **Setting of slurry pressure:** The soil mass excavated by a shield machine is subject to the release of original stress and deformation toward the stress relief surface. In order to control the ground settlement and maintain the stability of the excavation face, a force equivalent to the released stress must be applied to the excavation face, as shown in Fig. 5.35.

The formula for the slurry balance of the excavation face is as follows:

$$P_s + P_w \leq P_m, \tag{5.16}$$

where

P_s — lateral pressure of soil mass at the center of shield machine (kPa);

P_w — Water pressure, $P_w = \gamma_W y$ (y refers to the distance from the groundwater level to the center of the shield machine) (kPa);

P_m — slurry pressure, taken as the average value of slurry pressure at the top and bottom of the shield machine (kPa);

According to the Rankine earth pressure formula, the active earth pressure is expressed as follows:

$$P_s = \sum_{i=1}^{n} H_i \gamma_i \tan^2\left(45° - \frac{\varphi_i}{2}\right) + 2c_i \tan\left(45° - \frac{\varphi_i}{2}\right), \tag{5.17}$$

where

H_i — thickness of soil at ith layer (m);

γ_i — unit weight of soil at the ith layer, taken as the natural absolute density of soil above the groundwater level, and the effective unit weight of soil below the groundwater level (kN/m^3);

c_i — cohesion of soil at the ith layer, taken as $c_i = 0$, in the case of sandy soil;

φ_i — internal friction angle of soil at the ith layer (°).

During construction, different slurry pressures are set according to different soil layers. The reasonable setting of slurry pressure is very important for stabilizing the ground ahead of the cutterhead. However, due to the influence of slurry quality, slurry density, mix proportion of slurry and others, the slurry pressure ahead of the cutterhead often fluctuates in the actual tunneling process, which requires that the slurry pressure be controlled within a certain range so as to facilitate the construction crew to set the slurry pressure.

5.2.2.3 *Stability measures for excavation face*

During the tunneling process of a slurry balanced shield machine, slurry is continuously circulated, and the slurry membrane on the excavation face is in the process of "formation — destruction — formation" due to the cutterhead cutting. Due to the change in stratum and other factors, the balance of the excavation face is relative. In slurry shield tunneling, the stability of the excavation face is one of the most important management items, which directly affects the tunneling quality. Reasonable slurry management, water pressure management at the cutting face, synchronous grouting management and control of cutting volume per ring are necessary to ensure the stability of the excavation face. Considering the difference between Japanese and European shield machines, the following methods are usually used to judge the stability of the excavation face according to their types:

(1) **Japanese slurry balanced shield machine**

(i) *Control the volume of soil and sand excavated, W*: The volume of soil and sand excavated per ring, W, is calculated according to the geological conditions, and the theoretical volume of excavation obtained is taken as the approximate target to control the actual volume of soil and sand excavated per ring.

(a) **The actual volume of soil and sand excavated, W':** The actual volume of soil and sand excavated is directly displayed on the operation interface of the tunneling management system in the central control room to reflect the actual volume of soil and sand cut in the actual excavation process.

(b) **Relationship between the actual volume of soil and sand cut, W', (dry sand volume) and deviation flow Δq:** The deviation flow index is to reflect the cutting state of the excavation face, and its instantaneous calculation equation is shown as follows:

$$\Delta q = \Delta q_1 - (\Delta q_0 + Av), \tag{5.18}$$

where

Δq — deviation flow per unit time (m^3/min);
Δq_1 — slurry discharge per unit time (m^3/min);
Δq_0 — slurry delivery per unit time (m^3/min);
A — area of cutterhead of shield machine (m^2);
v — advance rate (m/min).

It can be seen from the above equation that when the deviation flow Δq is positive, the shield machine is in an "overbreak" state, and the volume of soil and sand cut is greater than the standard value; when the deviation flow Δq is negative, the shield machine is in an "under-break" state or slurry leakage occurs, and the volume of soil and sand cut is smaller than the standard value.

(c) **Judgment method of cutting volume:** Statistics are made on the cutting volume of 10–30 rings under the same working conditions and similar geological conditions, with good slurry quality and small surface subsidence over the cutting face of the shield machine behind each ring of tunneling, and the statistical results are entered into the computer. The excavation process should be observed to follow the curve changes of the volume of excavated earth and the statistical curve of the ring under cutting. When the cutting volume is found to be too large, the slurry density and viscosity, as well as the water pressure of the cutting face, should be checked immediately. In addition, a detection device may also be used to investigate the soil collapse, and the relevant parameters should be adjusted in time after the cause is found to ensure the stability of the excavation face.

(ii) *Bleeding rate inspection*: The bleeding rate is a good indicator to verify the quality of the slurry. According to the construction experience, the bleeding rate of slurry with high quality considered during the construction of a slurry balanced shield machine is 6.2A ($L/m^2 \cdot h$), where A is the cutting section area (m^2). When the excavation is stopped, the central control room observes the accumulated value in unit time. If the slurry bleeding rate is greater than 6.2 A ($L/m^2 \cdot h$), the slurry quality and the slurry condition in the pipeline system should be checked.

(iii) *Surface subsidence and information feedback*: The surface settlement is also a viewpoint to judge the stability of the shield machine excavation face. Therefore, settlement measuring points are arranged along the shield tunneling route to track and measure the surface settlement caused by shield tunneling. Generally, the settlement points within the influence range of 10–20 m in front of the shield and 30–50 m behind the shield along the axis are monitored every day. Moreover, the points beyond 30–50 m are measured regularly until the settlement is stable. The surface settlement caused by the instability of the excavation face often occurs over the cutting face of the shield machine, so the quality of the slurry and the water pressure on the cutting face should be checked. Large settlements occurring behind the shield machine are mostly caused by insufficient synchronous grouting, so the synchronous grouting rate should be increased to improve the grouting effect.

(iv) *Inspection of water pressure signal on excavation face*: When checking the water pressure on the excavation face, attention should be paid to checking the water pressure signal sensor on the excavation face. Sometimes, normal data collection may be affected due to the blockage of the pipeline.

(2) **European slurry pressure balanced shield machine**
Main differences in the aspect of the control measures from the Japanese slurry balanced shield machine:

(i) Due to the existence of a bubble chamber, the fluctuation range of liquid level in the bubble chamber can be accurately controlled to reduce disturbance to the excavation face and achieve the purpose of stabilizing the excavation face.
(ii) When the slurry flow deviation is large, the liquid level of the bubble chamber may be adjusted proactively by air supplementation or discharge in time to achieve the pressure stability of the excavation face.

5.3 Shield attitude management

The attitude management of the shield machine is a construction method to control the spatial state of the shield machine in real time through the hydraulic thrust system, taking the tunnel design axis as the reference, under the assistance of the measurement system (Fig. 5.36), in consideration of the segment erection situation, in an attempt to control the deviation of the formed tunnel within a reasonable range.

In this process, the spatial state of the shield machine is deemed the shield machine attitude, which is usually described by parameters such as horizontal deviation, elevation deviation, pitch angle, azimuth angle, roll

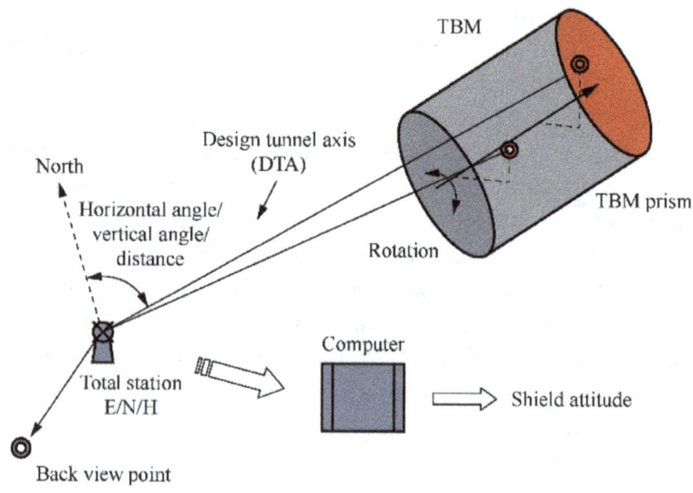

Fig. 5.36. Schematic diagram of shield machine attitude control.

angle and cutting mileage. Therefore, shield machine attitude management is an important link in the shield tunneling process, which directly affects the deviation of the formed tunnel from the axis and the segment erection condition. In the actual construction process, the above parameters are measured and calculated mainly through tunneling measurement technology. See Table 5.3 for data accuracy requirements.

The shield machine attitude management mainly depends on the prediction and control of shield tunneling. See Table 5.4 for the specification requirements for the plane and elevation deviation of highway tunnels. Any wrong tunneling trend that forms needs a long-term process to correct the deviation, which not only affects the construction progress, but also affects the tunnel quality and environmental safety. For example, in the case of a large axis deviation, excessive correction will cause excessive disturbance to the soil mass ahead and aggravate the impact of surface settlement. Uneven jack thrust may also cause quality problems

Table 5.3. Precision requirements of shield machine attitude parameters.

Item	Unit	Precision
Horizontal deviation	mm	1
Elevation deviation	mm	1
Pitch angle	'	1
Azimuth angle	'	1
Roll angle	'	1
Cutting mileage	m	0.01

Table 5.4. Specification requirements on plane and elevation deviation of highway tunnels.

Item	Allowable deviation (mm)	Inspection method	Inspection frequency
Plane deviation of erected segments	±75	Measurement by total station along center line	1 point/ring
Elevation deviation of erected segments	±75	Elevation measurement by level gauge	1 point/ring
Plane deviation of bored tunnel	±150	Measurement by total station along center line	1 point/10 rings
Elevation deviation of bored tunnel	±150	Elevation measurement by level gauge	1 point/10 rings

such as faulting of segment ends, fragmentation and water seepage, which may even affect the overall structure of tunnels under serious circumstances.

5.3.1 Factors affecting shield attitude

During the construction of a large-diameter shield machine, the attitude deviation may lead to over-excavation or even wrong tunneling, which may lead to a series of engineering problems such as segment dislocation. The main influencing factors of shield machine attitude deviation include shield machine performance, segment layout, geological conditions and construction technology.

5.3.1.1 *Performance of shield machine*

The performance of a shield machine is directly related to attitude adjustment, mainly reflected in the operability of shield machine equipment, the sensitivity of shield machine steering, and the configuration level of shield machine equipment (measurement system, grouting system, over-cutter, etc.).

The shield machines are equipped with similar thrust systems, which are generally driven by multiple groups of jacks consisting of two or three cylinders each. According to the steering requirements, those jacks are divided into zones. At present, the large-diameter shield machine is mainly designed with six zones. Each zone can control the oil pressure independently to achieve accurate control of the shield attitude.

The sensitivity of the shield machine is mainly reflected in steering and turning. In order to meet the needs of steering and segment erection for shield construction on small-radius curves, there are two different types of articulation mechanisms in the shield machine design, i.e. the active shield machine articulation and passive shield machine articulation. The articulation cylinder is located in front of the thrust cylinder, and the stroke of the articulation cylinder can be actively adjusted to change the tunneling direction of the front shield in the former case, as shown in Fig. 5.37(a); the articulation cylinder is located at the outer ring of the thrust cylinder in the latter case, and the shield tail can only be changed to a certain extent through changing traction on the shield tail, and the heading direction of the front shield is still controlled by the thrust cylinder, as shown in Fig. 5.37(b).

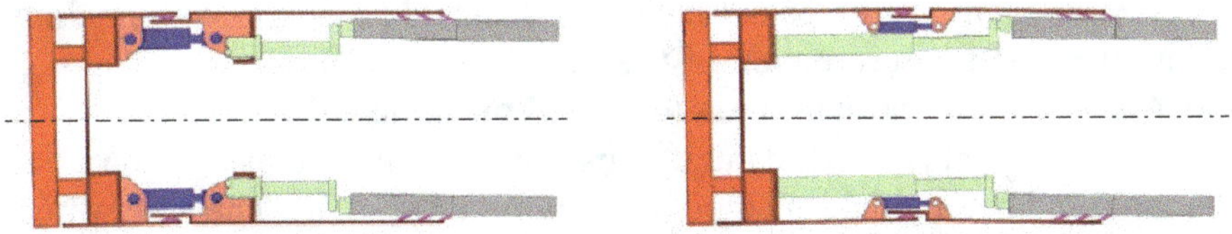

(a) Active mode (b) Passive mode

Fig. 5.37. Schematic diagram of articulation system of shield machine.

The measurement system of the shield machine is very important to the attitude change. The automatic measurement system can collect and visually display all the data necessary for the shield machine on the dashboard, which is convenient for use. Moreover, the measurement system should be checked manually on a regular basis.

The main function of the shield machine grouting system is to fill the structure void at the shield tail in time to prevent settlement. Generally, there are twelve grouting holes for large-diameter shield machines, six for normal service and six for standby. The grouting process can also assist in adjusting the shield machine's attitude. It should be noted that the voids at the shield tail are also critical to steering. In the absence of voids or if there is a super-sized void, steering will be difficult, and the segment may be damaged. Therefore, the shield machine attitude should be kept consistent with the segment attitude as far as possible, and the void at the shield tail should be reasonable and evenly distributed.

In addition, in the construction on a small-radius curve along the axis, the overcutter or profiling cutter is generally set to assist steering. The opening amount of the overcutter represents average overbreak at a certain value; the profiling cutter can be used for overbreak as needed, which is more intelligent.

5.3.1.2 *Segment layout*

In terms of design, attention is mainly paid to the requirements of structural clearance, construction error range, longitudinal slope setting, segment taper design and some special instructions and requirements. Among them, structural clearance and construction error are the standards for controlling the attitude of the shield machine, and targeted corrective measures are taken accordingly.

The segment, as the back bracing of the shield machine in tunneling, determines the main direction of the shield machine's advance. The segment taper is an important basis for shield machine attitude control.

The universal tapered segment is mainly used in the construction using the large-diameter shield method, as shown in Fig. 5.38. The taper direction is changed by rotating the closure block, so as to adjust the lead at the top, bottom, left and right and control the axis of the formed tunnel.

5.3.1.3 *Geological conditions*

In the case of soils varying in hardness, even soft soil on the working face, if the jack thrust on the soft soil side is not adjusted in time, the shield machine will deviate from the design axis toward the soft soil during the advance process, resulting in a large attitude deviation. In order to prevent the above situations, on the premise of ensuring the normal tunneling of the shield machine, the pressure of the oil cylinders in zones should be adjusted, and the stroke display of the oil cylinders in zones is observed at the same time, so as to reduce the tendency of the shield machine

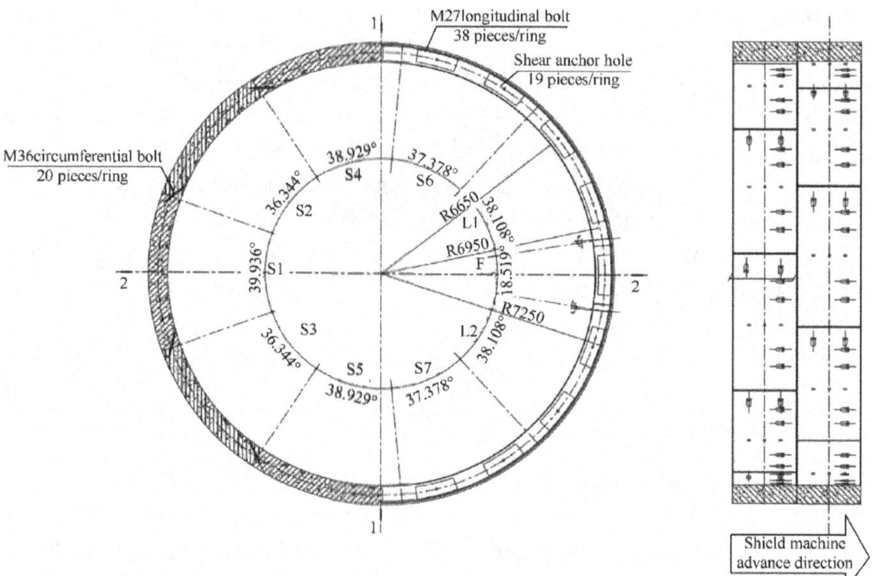

(a) Structure of lining circle (b) Profile

Fig. 5.38. Schematic diagram of tapered segment (unit: mm).

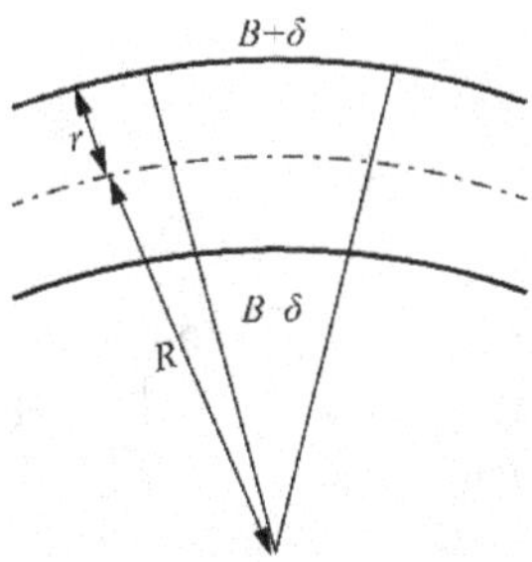

Fig. 5.39. Schematic diagram for calculation of theoretical deviation correction amount in curve section.

shifting to one side and control the attitude of the shield machine during the tunneling process.

5.3.1.4 *Construction technology*

In the process of shield tunneling and segment erection, the control of construction parameters such as pressure on the cutting face, advance rate, mucking volume, and technical measures including segment model selection, overcutters are decisive for the change in shield machine attitude. During construction, the selection of various technical measures is subject to strict calculations, and real-time monitoring should be conducted.

For example, before the shield machine enters a curve section, the deviation correction amount and segment taper of each ring should be adjusted in advance to determine a good trend and avoid sudden and sharp correction after entering the curve section.

5.3.2 Shield attitude correction management

5.3.2.1 *Calculation of deviation correction amount*

As shown in Fig. 5.39, the theoretical correction value of the shield machine in a curve section is 2δ, which is controlled by the stroke difference of the left and right jacks in each ring of tunneling:

$$\frac{B-\delta}{B+\delta}=\frac{R-r}{R+r}, \tag{5.19}$$

where

B — segment width (m);
R — radius of tunnel route (m);
r — tunnel radius (m).

In cases of great deviation of the shield attitude from the axis, it is not allowed to immediately return to a normal attitude through a sudden and sharp correction of the cutting face. Instead, it is necessary to change the unfavorable attitude trend, adjust the shield attitude to a controllable trend parallel to the design axis, and then conduct conventional correction to correct the cutting face deviation.

5.3.2.2 *Management measures*

(1) **Setting of reasonable construction parameters:** The pressure on the cutting face of the shield machine should be set correctly, especially in hard rock or sandy soil ground. A polymer, foam agent or bentonite can be used to improve the soil mass of the working face, so as to ensure good flow plasticity of the muck, avoid the sudden change of the shield attitude caused by the abnormal fluctuation of the working face pressure, strictly control the volume of excavated earth, avoid uneven resistance in all directions caused by over- or under-excavation, and ensure the stability of the shield attitude.
(2) **Control of deviation correction amount:** During tunneling, the starting and ending strokes of the jacks are recorded, and the jack stroke difference of each ring should be reasonably adjusted according to the above correction calculation formula. The oil pressure in zones should be strictly monitored, and the deviation should be corrected in steady advance instead of sudden and sharp correction. Moreover, the segment taper of the current ring can be determined through the calculation of the deviation correction amount to ensure uniform stress behind the lining ring.
(3) **Proper use of the over cutter and articulated jack:** In the composite stratum, the over cutter can be used to conduct local over-excavation on the working face to improve the adverse trend of the shield attitude toward soft ground and avoid hard ground. Under a small curve radius, when the shield has an unfavorable attitude trend, the articulated jack can be used to quickly correct the deviation.
(4) **Construction survey review:** The downhole observation station is displaced frequently for measurement, especially in the curve section.

It should be prepared in advance to avoid blind advance under poor intervisibility conditions. A construction survey review is performed at the specified mileage.

(5) **Roll angle control:** The roll angle of the shield machine refers to the included angle of the shield swinging relative to the preset horizontal line, usually in mm/m or °. The circumferential displacement at the outer diameter of the shield can be simply calculated, and then the allowable roll angle of the shield machine can be obtained. The roll angle of the shield machine can be adjusted by changing the rotation direction of the cutterhead. Generally, the cutterhead rotates counter-clockwise to produce a negative roll angle and clockwise to produce a positive roll angle.

5.3.3 Automatic measurement of shield attitude

The shield attitude measurement includes azimuth angle, pitch angle, roll angle, spatial position (horizontal deviation, vertical deviation) and mileage. Large-diameter shield machines are generally divided into a front shield, a middle shield and a shield tail. The shield attitude is obtained by measuring and calculating the coordinates of the center of the front shield and the center of the shield tail.

In the actual construction measurement, the commonly used shield attitude measurement methods include the manual and automatic measurement methods. Among them, manual measurement methods include the ruler and three-point methods; automatic measurement methods include the laser and prism methods. The automatic measurement methods are described in detail as follows.

5.3.3.1 *Laser method*

With the development of science and technology, laser alignment technology has been applied in tunnel excavation. Its principle is to use a laser to project a good linear beam inside the shield machine so that the operator can know the deviation and deflection of the shield machine in time and correct the direction of advance at any time to ensure the construction quality and improve the construction rate.

At present, there are three kinds of alignment systems: VMT, Poltinger Precision System (PPS) and The Guidance System for Japanese Enterprise Development (ENZAN).

(1) **Main functions of laser alignment system:** In the process of tunneling, due to the difference between the ground resistance, the cutterhead cutting reaction and the thrust jack force, the shield machine may deviate from the established center, bringing a negative impact on the construction quality and progress. The laser alignment system for shield machine construction can point out the advance direction of the shield machine at any time, so that the driver can control the machine to advance on the design route.

(i) The theoretical tunnel axis can be calculated according to the geometric elements of the tunnel design.
(ii) The roll and pitch angles of the shield machine are measured and displayed by the inclinometer.
(iii) On the display screen, the accurate position of the shield machine axis relative to the tunnel design axis is displayed graphically at any time, so that the operator can adjust the attitude and position of the shield machine tunneling at any time according to the deviation, and the tunneling axis of the shield machine approaches the tunnel design axis.
(iv) After the first ring tunneling, the elongation value of the thrust cylinder is obtained from the PLC automatic control system, and the plane position of the last segment ring is calculated accordingly. After entering the data on the void at the shield tail, the segment type suitable for the erection of this ring is calculated.
(v) The heading attitude of each ring and other relevant data can be searched in the database.
(vi) The communication with the computer in the ground office is established via modems and telephone cables, and the shield tunneling data is transmitted to the ground so that the engineering management personnel can monitor the shield tunneling in real time.

(2) **Working principle of laser alignment:** The laser alignment device is shown in Fig. 5.40. The laser generator is mounted on the formed inner wall of the tunnel. The linear beam emitted by the laser alignment technology is projected onto the target plate in the shield machine, and then a supporting system is used to point out the jacking direction of the shield machine in a simple and visible form. The laser measuring line is an eye-catching bright red light beam, and a red

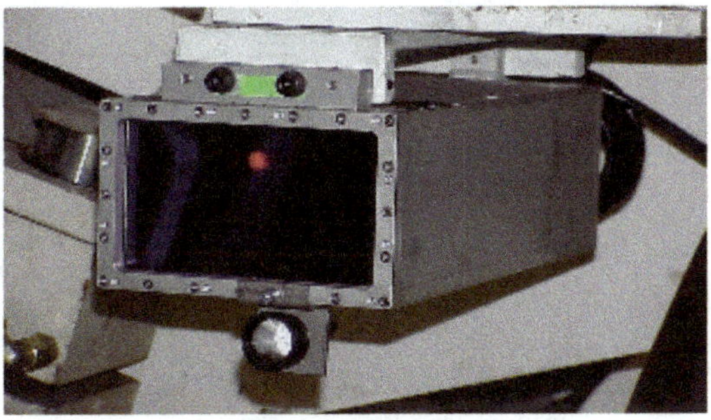

Fig. 5.40. Laser alignment device.

light spot is projected on the plastic target plate in the shield machine. The driver adjusts the top, bottom, left and right positions of the shield machine according to whether the light spot projected by the laser is consistent with the design tunnel centerline position on the target. The equipment calculates the position of the shield machine based on the beam from the laser at the fixed reference point. The deviation from the design axis can be calculated after knowing the position of the shield machine. In order to measure the position of the shield machine, two devices containing sensors are required, namely a target and an inclinometer. These two devices are connected to the control unit through cables and power distribution boxes, which supply power to the sensors. The target measures the hit position of the laser beam and its incidence angle, and the inclinometer measures the deflection angle of the shield machine in both directions.

(3) **Composition of laser alignment system:** Figure 5.41 shows the SLS-TAPD automatic measurement system developed by VMT. The system mainly consists of a laser theodolite, electronic laser target, control box, computer and other supporting hardware and software.

(i) *Laser theodolite*: The laser theodolite (Leica TCA1103/ART/GUS64) is a measuring instrument that simultaneously measures angle (horizontal and vertical) and distance and can emit a beam of visible red laser. The laser theodolite is temporarily fixed on the erected segment. As the shield machine continues to advance, the laser theodolite will also keep moving forward, which is called station displacement.

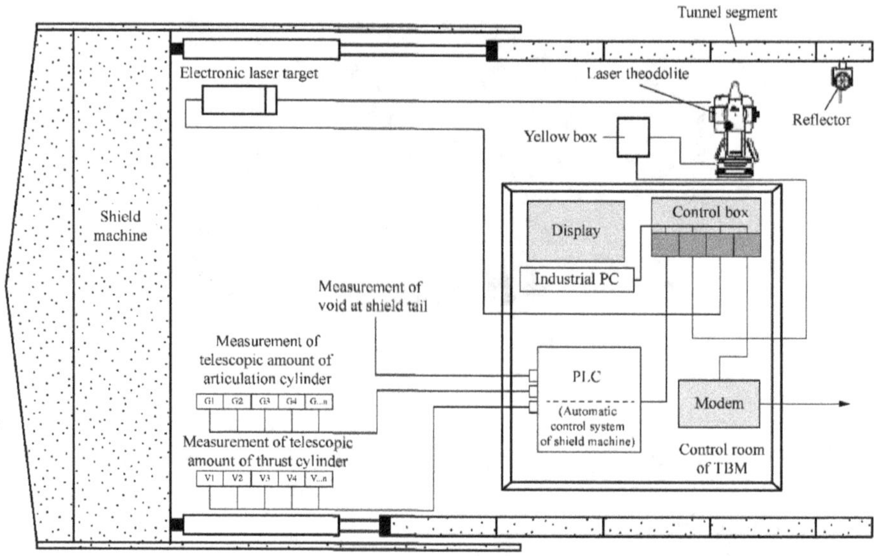

Fig. 5.41. Overall schematic diagram of laser alignment system.

(ii) *Yellow box*: It is mainly used to supply power to the total station and laser, and it also serves for communication data transmission of the PC connecting the total station and the main control room.

(iii) *Cable drum*: When the shield machine advances, the distance between the laser total station and other equipment installed on the shield machine increases, so the cable drum with rollers is required.

(iv) *ELS target*: The laser target is fixed in the middle shield to receive the laser beam. The reference plane of the ELS target is covered with sensing elements, which can transfer the up and down inclinations, the left and right inclinations of the incidence angle and the rotation angle of the incidence point to the centerline of the ELS target. The laser theodolite emits a laser beam on the laser target. The laser target judges the incidence angle and refraction angle of the laser beam. In addition, the inclinometer in the laser target is used to measure the roll and tilt angles of the shield machine.

(v) *Automatic measurement controller unit of shield tail void*: The void between the shield tail of the shield machine and the segment under erection is measured by the instrument installed in the segment erector area.

5.3.3.2 *Prism method*

The prism method of shield tunneling alignment is to place markers inside the shield machine and establish the relative position relationship between them and the cutting face and shield tail of the shield machine. By automatically measuring the three-dimensional urban coordinates of these marker points, the three-dimensional urban coordinates of the cutting face and shield tail are inversely calculated according to the relative relationship. Finally, the three-dimensional urban coordinates of the cutting face and shield tail center are compared with Designed Tunnel Axis (DTA) to obtain the specific degree of the shield machine deviation from the design axis.

The key components of shield tunneling attitude real-time measurement are a measuring instrument and a dual-axis inclinometer.

The measuring instrument can be a TCA1800 total station or a Leica total station of 1,200 series, with Automatic Target Recognition (ATR) function, an automatic target searching function and an automatic target tracking function. When the visibility is low and the human eye cannot find the target, the instrument can still work for measurement.

The inclinometer can be the TSR series inclinometer, which can not only achieve an accuracy of 0.0005° (1.8s) under vibration, impact and different temperature environments but can also measure angles up to ±5° (TSR) and ±10° (TSR10).

The system comprises:

(i) the aiming target installed on TBM, the prism group;
(ii) Leica TCA 1203/1800 total station with ATR;
(iii) a dual-axis electronic inclinometer;
(iv) a rear view prism;
(v) an industrial PC (with at least two COM ports);
(vi) connecting cables (computer to electronic inclinometer, total station to computer).

The wiring is shown in Fig. 5.42.

5.4 Segment erection technology

The segment erection technology, complementary with shield machine attitude management, is the most important quality control point of the formed tunnel. As shown in Fig. 5.43, since the segment erection is located

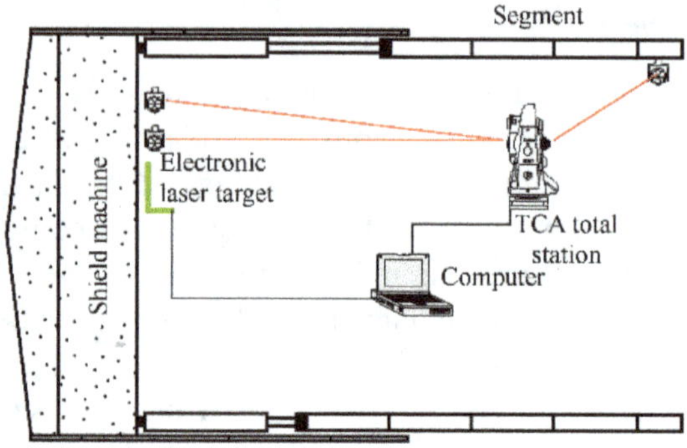

Fig. 5.42. Overall schematic diagram of prism alignment system.

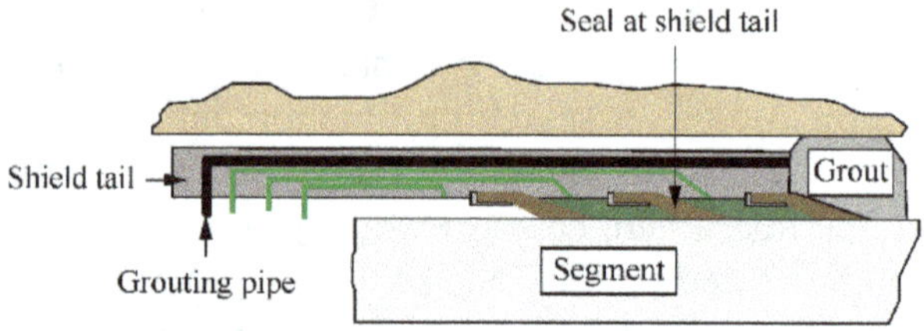

Fig. 5.43. Schematic diagram of the relationship between segments and shield tail.

at the shield tail, it will inevitably be restricted by the shield machine attitude.

The end face of segments are perpendicular to the axis of the shield machine as far as possible, so that the thrust cylinder of the shield machine can be vertically pushed on the segments and the segments can be evenly stressed to avoid breaking during tunneling. Moreover, the void between the segments and shield tail is also be taken into account to avoid damage to segments caused by sealing the wire brush at the shield tail when the shield machine is advancing.

The shield machine is advanced as far as possible according to the design route to avoid unnecessary deviation. In the actual tunneling process, due to varying geological conditions, uneven thrust and other reasons, the attitude of the shield machine often deviates from the tunnel design route. For correction of such shield machine deviation, special attention should be paid to the segment types selected to avoid accidents such as segment damage caused by a too small void at the shield tail. Sudden and sharp corrections must be avoided in order to ensure that the articulation seal and brush seal at the shield tail work well and to avoid any damage to the segment. The deviation correction value of the shield tail void in a tunneling cycle should be controlled within ±8 mm; otherwise, the deviation of segments in the turning ring cannot keep up with the deviation correction amplitude of the shield machine, resulting in segment crushing by the shield tail.

5.4.1 Basis for segment model selection

Large-diameter shield machines usually adopt the erection method with staggered joints, and the segment type selection is to select the appropriate "erection point". The so-called "erection point" refers to the location of the closure block during the segment erection. As shown in Fig. 5.44, during the actual erection of the turning ring, the offset in different directions and the curve alignment of a shield tunnel can be controlled according to different erection points so as to achieve the axis control of the tunnel.

The main factors affecting segment model selection include:

(1) unncl dcsign alignment;
(2) attitude of the shield machine (position and trend of deviation from the centerline);
(3) shield tail void;
(4) cylinder stroke, articulation cylinder stroke;
(5) position of available closure block (meeting the number of through joints and other design and safety requirements);
(6) the number of through joints shall, which should meet the design requirements (generally, the number in two rings shall not be greater than that in three rings, and the three rings shall not contain any continuous through seams).

In the actual construction, the calculation table of erection points is generally prepared according to the design taper. At the end of advance for each ring, the calculation is conducted for model selection according to the above factors, and the optimum point is selected for erection.

5.4.2 Segment model selection software

The segment model selection software can quickly judge the position of the closure block according to the actual advance status and accurately predict

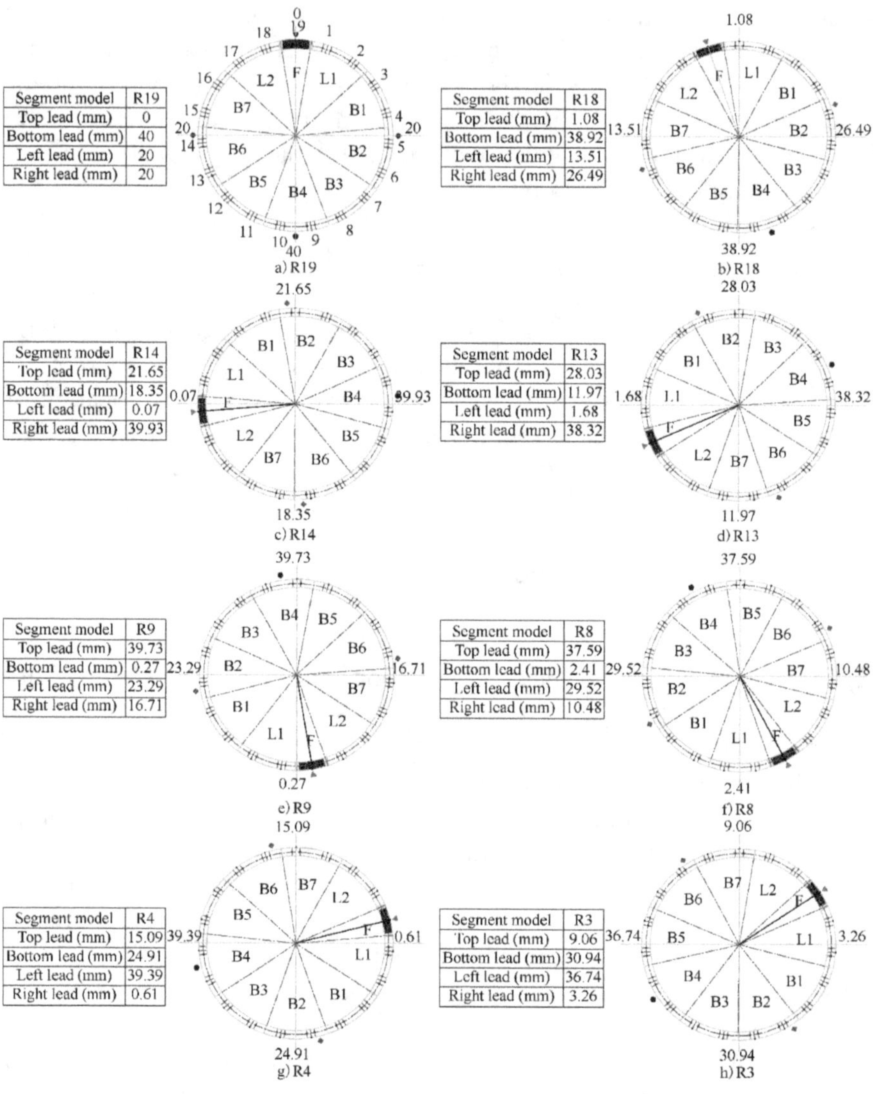

Fig. 5.44. Schematic diagram for calculation of erection points.

the tapered segment selection of the next three rings based on the segment information of the current ring, so as to improve the axis control accuracy.

The segment selection software can display the advance trend and through-seam situation after erection of the segments under current selection through a chart based on some relevant data in the acquisition system,

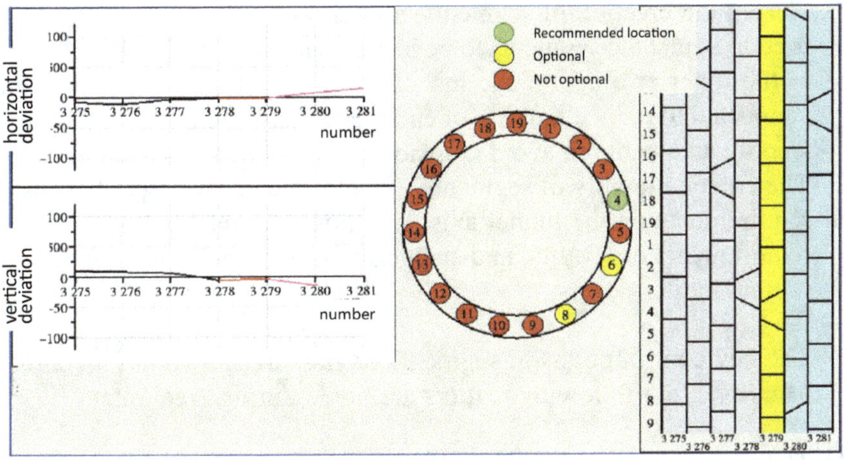

Fig. 5.45. Segment model selection software interface.

such as the jack stroke difference, shield machine attitude, the closure block position of the previous ring and the through-joint condition, combined with the shield tail void. The model selection software interface is shown in Fig. 5.45.

5.4.3 Segment erection process

The quality control measures for segment erection include the following:

(1) Strengthen the inspection of incoming segments, and do not use segments with damage or cracks.
(2) Before pasting the water stop strip and cork gasket, the segment should be thoroughly cleaned to ensure stable and firm pasting. The segment stacking area on the construction site is equipped with rainproof facilities.
(3) Pay attention to protecting segments and water stop strips when lifting and transporting segments down in the shaft to avoid damage.
(4) Before segment erection, the segment erection area should be cleaned, and the sludge and sewage should be removed to ensure that the erection area and the segment interface are clean.
(5) The thrust cylinder at the segment erection position and other positions should not retract at the same time.

(6) During the erection of segments, a micro-adjustment device must be used to adjust the segments to be installed and the longitudinal surface at the inner arc face of the installed segment block to be smoothly connected to reduce faulting of ends. The adjustment action should be smooth and stable to avoid collisions and damage to segments.
(7) The erection quality of segments is controlled according to the design requirements on the tunnel axis deviation, and relevant specification requirements on ovality, and staggering of circumferential and longitudinal joints.

The key control points of segment erection are shown in Fig. 5.46. Moreover, the following matters are noted during erection:

(1) The segment model is selected on the premise of meeting the tunnel alignment, and the shield tail void after segment erection shall meet the void value required for the next excavation, so as to ensure that there is enough shield tail void to prevent the shield tail from directly contacting the segment.
(2) When the segment is rotated to approach the previous segment, control the rotation and movement speed to prevent damage to the positioning rod and shear pin.
(3) For the segments with pre-tensioned bolts, the first ring of segments shall be tensioned on the ground, and the remaining segments shall be connected with bolts during erection, and prestressed tensioning shall be carried out after erection.

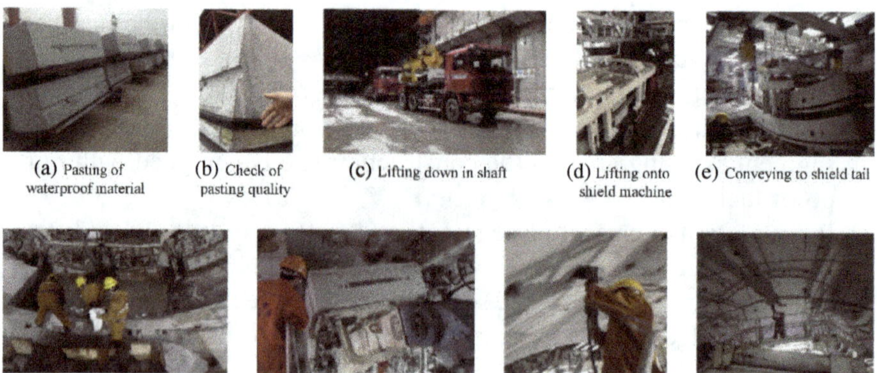

(a) Pasting of waterproof material (b) Check of pasting quality (c) Lifting down in shaft (d) Lifting onto shield machine (e) Conveying to shield tail

(f) Clearing at shield tail (g) Telescopic cylinder action to place segment (h) Bolting and segment separation from shield tail (i) Secondary fastening of bolts

Fig. 5.46. Key control points for segment erection.

(4) Before the erection of the closure block, the water stop strip shall be lubricated. The segment block is generally inserted radially first and then slowly jacked longitudinally after the position is adjusted.
(5) After the segment block is installed in place, the thrust cylinder at the corresponding position shall be extended in time to jack the segment until its jacking force is greater than the thrust required to stabilize the segment, and then the segment erector can be removed.
(6) After the segment ring is separated from the shield tail, the segment connecting bolts shall be tightened again.

5.4.4 Quality problems of segment erection and corresponding solutions

(1) **Segment fragmentation:** Segment fragmentation is a common quality problem in shield tunneling. Broken corners, edges and cracks in the erected segments will affect the structural strength and cause leakage, as shown in Fig. 5.47. The main causes of segment fragmentation are as follows:

   (i) During erection, due to the excessive eccentricity of the segment in the shield tail, the segment collides with the shield tail, and

Fig. 5.47. Segment fragmentation.

the shield shell causes the segment to become stuck and damaged when the shield machine is advancing.

(ii) The segment for locating mortise-and-tenons is not correctly placed during erection, so that the mortise-and-tenons are not aligned. When the jack approaches, the segment will be damaged due to the radial force component of the tenon to the mortise when the jacks approach.

(iii) During segment erection, the dislocation of segments causes no surface contact between segments. When the shield machine is advancing, stress concentration occurs at the contact point, which causes the segment angle to break.

(iv) The uneven ring face of segments in the previous ring makes the next ring segments in unilateral contact, like a seesaw under jacking, and the segment is broken due to additional bending moment. The opening amount of the erected adjacent block is not enough, and the gap for inserting the closure block is too small. Forcible insertion will lead to the corner collapse of the closure block segment or the adjacent block segment.

(v) The speed of the erector is too high during operation, and the segments collide with each other, causing corners to collapse during erection.

(vi) The whole ring of segments is deflected, and the segment suffers from local stress imbalance, resulting in a bending moment and damage to the segment corner.

(vii) The segments are excessively faulted at the ends, and the inner arc face of the segments is damaged under the thrust of the oil cylinder.

The measures to prevent segment fragmentation mainly include:

(i) Before erection, inspect the ring face of segments in the previous ring, and then determine the deviation correction amount and measures during the erection of segments in the current ring.

(ii) Remove various foreign matters from the ring face and shield tail.

(iii) Control the jacking force to be even.

(iv) Improve the pasting quality of the deviation correction taper.

(v) Check the pasting of the water stop strip to ensure its reliability.

Solution:
For the segments erected with an uneven ring face, add tapers to the next ring to correct the ring face in time and to make the ring face flat.

(2) **Excessive segment opening and faulting of ends:** The problems such as front and rear flares, internal and external flare angles along longitudinal joints and circumferential joints, and excessive steps between segments will cause serious harm to the waterproofing of tunnels and the stress on segments. The main causes are as follows:

   (i) The erected segment is not placed correctly, and there are foreign matters in the shield shell, which makes the bottom segment block not rightly placed, curled up or turned down. The foreign matters are included in the circumferential seam of the ring face, which will also make the longitudinal seam in front and rear flares.
   (ii) The segment erected fails to form a perfect circle, resulting in excessive internal and external flare angles and steps.
   (iii) The datum of the previous ring segment is inaccurate, resulting in an inaccurate position of the newly erected segment.
   (iv) The tunnel axis is inconsistent with the actual center line of the shield machine, causing the segment to collide with the shield shell.

Preventive measures:
(i) Before erection, clean all faces of the shield shell and segment to prevent foreign matter from getting caught between segments.
(ii) Conduct deviation correction frequently to minimize the deviation between the axis of the shield machine and the design axis, so as to ensure that the erected segment is centered and there is enough building void around the segment to make the segment erected into a perfect circle.
(iii) The deviation on the ring face shall be corrected in time to reduce the error between the centerline of the segment erected and the design axis, so that the segment erected from the shield tail can always be centered.
(iv) The segments shall be correctly placed, and the approaching jacks shall uniformly exert force. Except for the closure block, each segment shall be allocated at least two jacks.

Solution:

(i) The deformation of the longitudinal joint can be corrected by re-tightening the circumferential bolt after the segment is fed from the shield tail. After the segments are wrapped by the surrounding soil, accordingly resulting in a decrease in ovality, the longitudinal joints are compressed, and the bolts can be re-tightened to achieve a better effect.

(ii) The method of adding tapers locally shall be adopted to correct the quality of flare angles and steps.

(3) **Segment rotation:** Compared with the erection position required by the design, the segments erected into a ring have been rotated at a certain angle, which makes the erection of the closure block more difficult, mainly due to the following:

(i) The jacks are unreasonably grouped, which causes uneven stress on the segments and relative rotation of the segments.

(ii) The skew ring face of segments causes the jacking direction to not be perpendicular to the ring face, and the advancing shield machine will produce a torque to cause the segments to rotate.

(iii) The first segment erected is placed inaccurately, resulting in rotation during erection.

(iv) A gap of 5–8 mm is generally provided between the bolt hole and the bolt on the segment to facilitate erection, which provides conditions for dislocation between two rings; segment placement without strictly following requirements may cause rotation deviation or failure to tighten the segment bolts, impeding effective transmission of torque.

(v) The segment under erection collides with the segments in place, causing displacement of the erected segments. If the segments are erected in the same sequence for a long time, rotation deviation will occur on the erected segments in this direction and to a large degree.

(vi) The cutterhead rotates in one direction all the time, instead of even distribution in both the clockwise and anticlockwise directions.

Preventive measures:

(i) The advance attitude of the shield machine shall be well controlled. The jacks shall be grouped to make the change of thrust uniform.

The ring face angle of the segment shall be adjusted to reduce the rotating torque generated during the advance process.

(ii) The segments shall be placed correctly before erection, and the approaching jacks shall exert enough jacking force to prevent the segments from relative sliding.

(iii) The erector shall be operated smoothly and slowly since slow rotation is conducive to the accuracy of erection.

(iv) The rotation of segments in rings shall be measured frequently and corrected in time.

(v) The rotation time in a single direction shall be shortened as far as possible so that the forward and reverse rotation times and torque tend to be balanced.

(vi) Each segment is placed and erected after the bolts are initially tightened, and the bolts are tightened again after segments are erected into a complete ring. In addition, the advance pressure during the advance process is far greater than the jacking pressure for segment erection. Therefore, the segment bolts shall be tightened again to achieve a better tightening effect and keep the segment in good integrity to resist twisting.

(vii) During tunneling, the pressure difference between the upper and lower parts of the thrust jacks shall be as small as possible.

Solution:

By making use of the margin of dislocation between segments, after the longitudinal bolt of the first segment is inserted, the erector is used to take the segment to rotate for an appropriate angle in the correction direction, and then the jacks approach and the longitudinal bolts are tightened. With the first segment as the benchmark, the remaining segments shall be erected correctly, and the rotation error can be gradually corrected by using this method for the erection of segments in consecutive rings.

(4) **The segment ovality is too large:** The difference between the horizontal diameter and the vertical diameter of the erected segment is excessive, leading to an ovality exceeding the limit. The main reasons are:

(i) The center of the segment erection position does not coincide with the center of the shield tail.

(ii) The angle included between the segment ring face and the axis of the shield machine is too large, so that the segment is stressed to produce oblique components.
(iii) The grouting pressure is excessive, which causes uneven stress on the segment.

Preventive measures:
(i) Frequently correct the axis of the shield machine to keep the shield machine moving along the design axis and the erected segments centered.
(ii) Frequently correct the segment ring face to make the ring face perpendicular to the axis of the shield machine, and the segment always follows the axis of the shield machine, so that the structure void between the segment and the shield tail is uniform.
(iii) The grouting principle of double control of pressure and grouting amount shall be followed to make the segments uniformly stressed.

Control method:
(i) The location of the closure block shall be reasonably selected so that the segment erection position is kept in the center of the shield tail.
(ii) The deviation correction of the shield machine shall be controlled so that segments erected from the shield tail are centered.
(iii) After the segment is separated from the shield tail, the circumferential bolts of the segment shall be retightened to ensure reliable connection of the segments.

5.5 Synchronous grouting technology

5.5.1 Purpose and method of synchronous grouting

Synchronous grouting technology is critical to the stability control of segment structure and the safety of the surrounding environment. The synchronous grouting process is to inject the filling material into the outer space of the segment ring at the shield tail by the pumping action of the

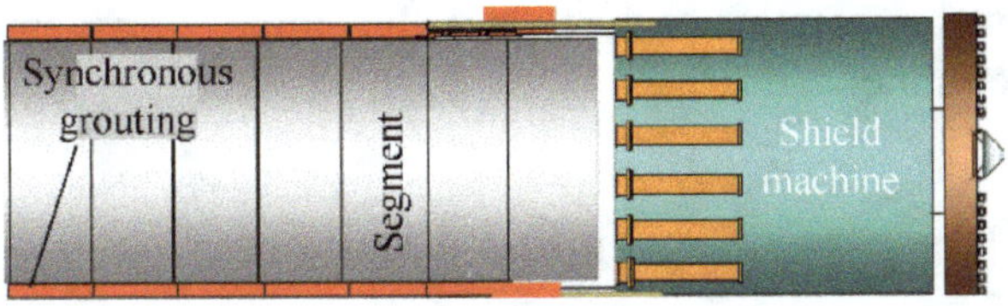

Fig. 5.48. Schematic diagram of shield machine synchronous grouting technology.

grouting pump while the shield machine is tunneling. Its technical principle is shown in Fig. 5.48.

The purpose of synchronous grouting mainly includes the following aspects:

(1) **Control ground deformation:** Since the excavation diameter of the cutterhead of the shield machine is larger than the outer diameter of the segments, the segments will produce an annular void with the soil mass after being erected and separated from the shield tail, which is called the shield tail void. If the shield tail void is not filled in time, it is bound to cause formation deformation, settlement of adjacent surface buildings or deviation of the tunnel itself. The main purpose of synchronous grouting is to fill the shield tail void in time to prevent large deformation of the stratum due to the existence of the shield tail void. After the shield tail is separated from the segment, there is a gap between the soil mass and the segment. If grouting is carried out synchronously, the grout can quickly fill the gap to reduce the movement of the soil layer, thus reducing surface deformation.
(2) **Ensure the stability of tunnel structure:** The shield tail void should be filled timely with grout with a certain early strength to ensure the stability of the segment lining during the construction and beyond. Shield tunnels are in a structurally stable structure jointly acted by the segment lining and surrounding soil mass. Filling the void at the back of the segment with synchronous grouting is a prerequisite to ensuring the uniform action of soil mass, ensuring the stability of the tunnel structure and preventing floating or sinking.
(3) **Improve the impermeability of the tunnel:** After the grout is synchronously injected into the back of the segment wall, it is solidified and hardened to build a certain impermeability, which can be used as

the first water stop line of the tunnel, thus improving the impermeability of the tunnel.

5.5.2 New large specific weight shear mortar

5.5.2.1 *Grout performance*

The requirements for synchronous grouting technology in large-diameter shield tunnel construction are mainly reflected in the following aspects:

(1) Shield tunneling has a large impact on the surrounding environment. How to effectively control the disturbance of shield tunneling to the surrounding environment, especially the long-term settlement control of the ground after construction, is a core research subject.
(2) In terms of construction technology, grouting materials, grouting equipment and grouting parameters should meet the requirements of synchronous full and uniform filling of super-large-diameter shield machines.
(3) During synchronous grouting construction, the grouting point stress borne by the segment at the grouting position is relatively large, which has a great impact on the stability of the ring structure. How to properly control the grouting parameters is key to ensuring the grouting effect and tunnel quality.
(4) The construction quality of synchronous grouting determines the stability of the ring tunnel structure, and the structure stability has an important impact on the control of the shield attitude and other related thrust parameters.

With sand as aggregate and lime and fly ash as main binders, a new type of large-specific-weight shear mortar material is prepared by adding an appropriate dosage of bentonite, large-specific-weight curing materials and additives. Featuring good fluidity, pressure thixotropy and early strength, this kind of grout can meet the construction requirements of synchronous grouting for large-diameter shield tunnels. It has good fluidity and pressure thixotropy at the initial stage after mixing and can evenly fill the shield tail void under the action of grouting pressure so as to avoid slurry accumulation near the grouting port, which will lead to the accumulation of grouting pressure and damage to the segment; it also has good pressure transmissibility. After the segment is separated from the shield tail, the strength built by the locking force and internal friction between the aggregate and sand particles can resist the tunnel floating and transmit the uplift and grouting pressure to the surrounding strata, thus ensuring the stability of the tunnel.

Different from the requirements of conventional synchronous grouting materials for initial setting time, early compressive strength and other indicators, taking the internal friction force in the grout materials and the strength built by the shear force as the main indicators to control the performance of synchronous grouting, the "double control" mode of grouting amount and grouting pressure is highlighted in the construction with this new type of grout to manage the synchronous grouting construction during the shield tunneling.

According to the construction characteristics of synchronous grouting in large-diameter shield tunneling, combined with the construction theory of new large-specific-weight shear mortar, the grouting material must have the following properties:

(1) **Filling performance:** The grout flow distance behind the segment wall is long, so the grout must have good fluidity and filling performance to ensure that the grout can fill the gap between the segment and the stratum after being pumped out of the shield tail via the pipeline, so that the grout can completely wrap the segment and form a dense layer of protection.
(2) **Internal friction characteristics:** The grout should have good particle grading to provide effective internal mechanical locking force, form good internal friction, and prevent the tunnel from floating up and surrounding soil deformation, taking advantage of the shear yield strength.
(3) **Post-construction stability:** The grouting material should have certain post-construction hardenability characteristics, provide strength growth and make the grouting layer reach or exceed the strength of the surrounding undisturbed soil in a short time, and the grout consolidation body should have a small shrinkage rate, have good long-term stability to control the stability of the ring structure of tunnels and the post-construction surface settlement.
(4) **Workability:** The grout must have a certain plastic consistency, including a low bleeding rate and good anti-segregation performance, to meet the pumping and storage requirements of grout during transportation.

(5) **Impermeability:** As the first waterproof wall of the tunnel, the grout should have a low permeability coefficient after consolidation.

5.5.2.2 *Grouting material*

The mix proportion of grouting materials mainly involves two aspects of testing the selection of raw materials and the adjustment of the mix proportion.

(1) **Selection of raw materials**

(i) *Sand*: As an important constituent of grouting materials, sand plays an important role in improving grout density, increasing grout yield strength and post-construction consolidation strength, and reducing bleeding and layered sedimentation. Only by selecting medium sand (river sand) with good grain grading as the grout material can the performance index requirements for grout use be met and the grout be used better in construction.

(ii) *Fly ash*: The grout mixed with high-calcium fly ash features a rough surface, large bleeding and fast slump loss. High bleeding rates and slump loss are the main reasons for pipeline blockage, which is not conducive to grout pumping. The low-calcium fly ash has the function of reducing bleeding and slump loss. The grout mixed with low calcium fly ash is lubricated on the surface and features high glossiness, a low bleeding rate, slow slump loss and strong water retention capacity to effectively resist slump during grout storage and transportation, which is conducive to grout conveying.

(iii) *Lime*: Lime is used to stimulate the activity of fly ash by $Ca(OH)_2$, promote its pozzolanic reaction, and improve the later consolidation strength of grout. Hydration reaction is required for quick lime to generate $Ca(OH)_2$ during use, so slaked lime with a $Ca(OH)_2$ content greater than 95% has a good effect during use. In addition, slaked lime also has a certain water retention property, which can significantly increase the viscosity and improve the workability of grout.

(iv) *Bentonite*: The grout mixed with calcium bentonite is of poor quality, prone to precipitation and segregation, and poor in water retention. If it is directly mixed with water, the expansion effect is poor. Comparatively, the sodium bentonite grout presents a better quality.

(v) *Curing materials*: Cement, flue gas desulfuration gypsum and other binders are used to accelerate the early hydration rate of grouting materials so as to improve early strength performance.

(vi) *Admixture*: The special admixture has the functions of water reduction and water retention, which can obviously promote the water reduction and retarding of the grouting material and improve the water retention performance.

(2) **Mix proportion**

(i) *Binder–aggregate ratio*: With the increase in binder–aggregate ratio, the water consumption of grout increases, the water–cement ratio increases, the density decreases, the fluidity falls, and the yield value decreases. The bleeding property of grout is improved with the increase in binder–aggregate ratio, which indicates that fly ash can obviously improve the workability of grout, but such an effect is no longer obvious when the dosage of binders reaches a certain threshold; with the increase in binder–aggregate ratio, the early strength of grout increases rapidly, but the later strength is obviously insufficient; the increase also brings some consolidation and shrinkage effects to grout. The grout with a binder–aggregate ratio between 0.6 and 1 has the best performance in construction.

(ii) *Lime*: Lime has an obvious consolidation effect on the grout, which plays an important role in ensuring a high later strength of the grout. With the increase in lime dosage, the compressive strength and shear strength of the later consolidation body of the grout are significantly improved. However, an appropriate dosage of lime plays an important role in the strength-building of the grout, while the later strength of the grout is poor in cases of excessive dosage of lime.

(iii) *Bentonite*: With the increase in bentonite dosage, the water retention effect of grout is significantly improved. The addition of bentonite not only increases the sliding effect of mortar but also increases its stability. To a certain extent, it can increase the pumpability of synchronous grouting, and prevent pump plugging and pipe plugging. However, due to the low consistency of the grout itself, the increase in the bentonite dosage also makes the liquidity of the grout significantly reduced and viscous. This is because the bentonite itself is a kind of water-absorbing swelling material, which reduces free water in grout, leading to a decrease in liquidity and an increase in viscosity.

(iv) *Curing materials*: Cement, flue gas desulfuration gypsum and other binders can produce strength due to their own hydration reactions to improve the early strength performance of the mixture, but the too-early strength building is not conducive to the conveying and long-term storage of grout in shield tunneling. Therefore, in order to balance the workable time of grout and the early strength performance index, the dosage of curing materials should be strictly controlled. According to the requirements of working conditions, the dosage of general curing materials should be controlled between 2% and 5%.

(v) *Admixture dosage*: The admixture is produced by mixing a water-reducing agent and a water retaining agent, and the selection of compound dosage is obtained from the application effect of grouting materials.

(3) **Grout performance index:** See Table 5.5 for performance control indexes of the new large-specific-weight shear mortar.

5.5.3 Grouting process

According to the characteristics of grouting materials, the "multi-point grouting method" and the construction control method of "double control of grouting amount and grouting pressure" are proposed.

Table 5.5. Performance indexes of new large specific weight shear mortar.

Slump (cm)		Density	Bleeding rate	Shear yield strength (Pa)		Workable time	Fluidity	7d strength
0 h	8 h	(g/cm^3)	(%)	0 h	8 h	(h)	(mm)	(MPa)
9~14	>5	>1.90	<1	>300	>800	0~24	>200	>0.10

5.5.3.1 *Construction technology and supporting system equipment*

The construction equipment of the synchronous grouting system consists of material mixing equipment, grout transportation equipment and grouting equipment. The adaptability of supporting equipment and grouting materials is related to the construction quality and effect of synchronous grouting.

(1) **Grout mixing system:** Because the new large-specific-weight shear mortar has the characteristics of low consistency, high sand content and high density, higher requirements are put forward for the grout mixing equipment, which is a key component of the whole shield machine synchronous grouting system, and has a significant impact on the mixing quality and efficiency of grout.

Instead of the conventional simple mixing equipment on the ground, an automatic mixing station should be adopted, which is composed of a sand screening system, sand feeding system, independent powder storage and metering system, liquid additive storage and metering system, mixer and control room. The mixing system is equipped with a continuous metering device to realize continuous production; the PLC control system with high reliability is adopted to realize automatic and manual functions and has the functions of automatic data collection and storage; the grout production capacity of a single system can reach 24–30 m^3/h. The grout mixing system is shown in Fig. 5.49.

(2) **Slurry conveying system:** In the synchronous grouting transportation system for super-large-diameter shield tunnels, there are two grout conveying methods:

(i) *Direct hoisting mode of grout storage barrel*: Directly hoist the grout storage barrel on the shield machine frame through the ground grout mixing system → vertically hoist it onto the underground transport vehicle → horizontally transport it by the underground transport vehicle in the tunnel → reach the shield machine frame and hoist it over the fixed grout barrel of the shield machine.

(ii) *Underground transport by a truck mixer*: The grout mixed on the ground is directly transported to the truck mixer underground by a delivery pump or by gravity, and the truck mixer is used for horizontal transport in the tunnel to the shield machine frame position and then pumped to the grout storage barrel on the frame by a delivery pump.

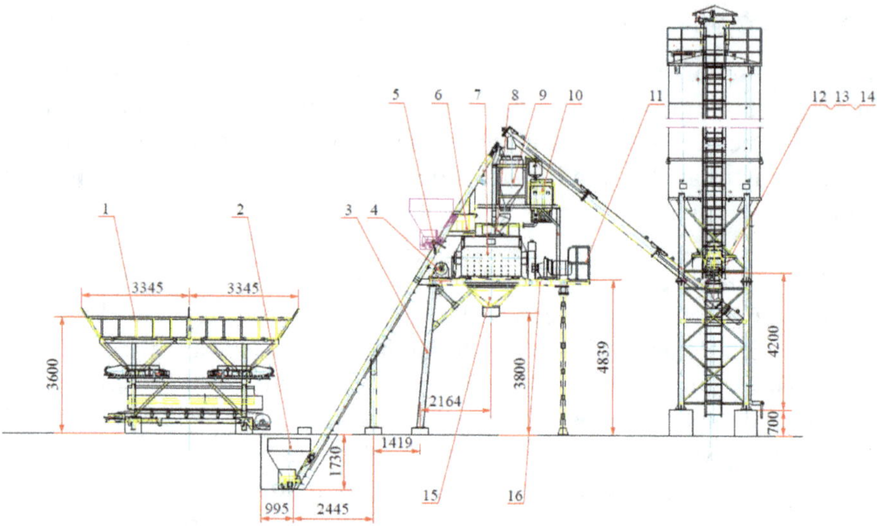

Fig. 5.49. Schematic diagram of grout mixing system (unit: mm).

Notes: 1 — Two-bucket batching machine (heightening plate); 2 — Hopper; 3 — Outrigger; 4 — Winch system; 5 — Loading rack; 6 — Funnel; 7 — Mixing console; 8 — Console cover; 9 — Fly ash and mineral powder scale; 10 — Fluid piping system; 11 — Walkway; 12 — φ 219 screw conveyor; 13 — φ 273 screw conveyor; 14 — Cement cabin; 15 — Collecting hopper; 16 — Chassis.

(3) **Grouting system:** The grouting system is equipped with a full hydraulic double-cylinder double-outlet piston grouting pump, and the commonly used grouting pump is the SchwingKSP20. In order to adapt to the advance rate, the grouting device can control the amount of grouting through pressure regulation. A grouting pressure can be selected in advance in the range from the minimum to the maximum, which can not only ensure that the shield tail seal will not be damaged and the segment will not be imposed with excessive pressure but also minimize the disturbance to the surrounding soil ground.

Grouting may be conducted in either the manual mode or the automatic mode. Any one of the grouting pipes can be selected under the manual mode, and the operator can operate the button on the site console to start the grouting system. The automatic mode is preset on the operation console of the grouting site, and the grouting system will be started immediately after the shield tunneling is started to advance, as shown in Fig. 5.50.

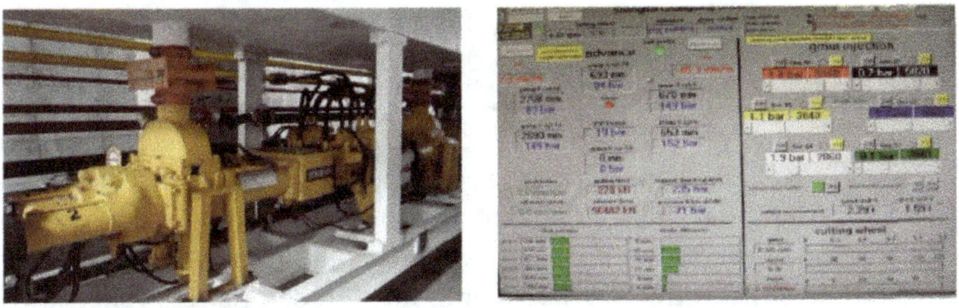

(a) Physical pump (b) Operation interface

Fig. 5.50. Synchronous grouting pump and operation interface.

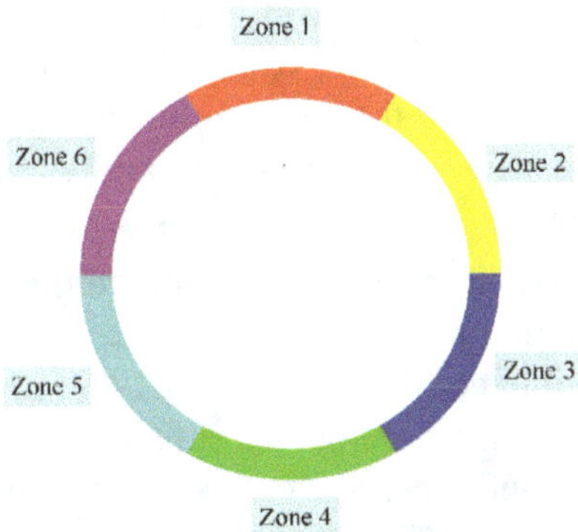

Fig. 5.51. Schematic diagram of multi-point grouting method.

5.5.3.2 *Construction control method for synchronous grouting*

(1) **"Multi-point grouting" construction method:** Shear mortar has good fluidity and pressure thixotropy. Under the multi-point grouting (6–8 grouting points) method, the structure void outside the shield tail segment is grouted synchronously, and the shield void can be filled timely and effectively, as shown in Fig. 5.51.
(2) **Construction method of "double control of grouting pressure and grouting amount":** Shield tunneling will inevitably disturb the surrounding strata, resulting in surface subsidence and affecting the stability of surrounding buildings and structures. After the tunnel segment is separated from the shield tail, due to the shield tail void, the surrounding soil mass experiences stress relief, and the segment displacement occurs. During synchronous grouting of large-diameter

shield tunnel construction, the grouting pressure and grouting amount should be mainly controlled so that the structure void can be effectively filled without affecting the quality of segments in rings.

As the new shear mortar is prepared with sand as aggregate, it can be automatically adjusted to the required compactness depending on the locking force and friction among sand particles under the effect of grouting pressure and can transfer the grouting pressure to the surrounding strata, effectively reducing the degree of stress relief in the stratum. Therefore, the grouting pressure can be controlled around the initial ground stress level of the surrounding strata to reduce the stress disturbance to the strata. For the soft soil stratum with a very high water content in Shanghai, controlling the grouting pressure has significant effects on reducing the change of pore water pressure, controlling the disturbance to the stratum during the construction period and long-term consolidation settlement, and can also avoid the damage to tunnel segments caused by excessive local grouting pressure.

(i) *Grouting pressure control*: According to the purpose and requirements of grouting (that is, fully fill the stratum void generated by shield machine construction to prevent the resulting surface subsidence from affecting the safety of surface buildings and underground pipelines; prevent excessive grouting pressure from causing surface uplift or damaging segment lining and prevent grouting from damaging the shield tail seal), the grouting pressure calculation is shown in Fig. 5.52, according to the following formula:

$$P_G = \gamma h K_0 + P_h, \tag{5.20}$$

where

P_G — grouting pressure of grouting hole (MPa);
γ — unit weight of grout (kN/m^3);
h — depth from grouting hole to overburden (m);
K_0 — horizontal pressure coefficient or lateral pressure coefficient;
P_h — pipe resistance of grouting hole (MPa).

When the grouting pressure is too small, the soil mass will rapidly deform toward the segment direction, the stratum void generated by shield machine construction will be narrowed, and the amount of grouting will be reduced, which will cause surface settlement and affect the safety of surface buildings and

$$P_G = \gamma h K_0 + P_{h_i} \quad (5.20)$$

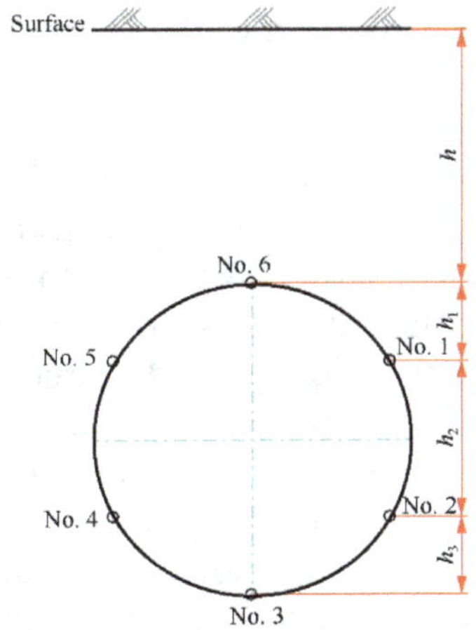

Fig. 5.52. Schematic diagram of grouting pressure calculation.

underground pipelines. When the overburden is too shallow, the grouting pressure should not be too high, since excessive pressure increases the disturbance to the stratum and the deformation of the structure, affecting the surface and surrounding buildings. The grouting rate should match the shield machine advance rate to ensure uniform grouting and timely filling of the shield tail void. In soft strata, due to the collapse-proneness of soil mass after excavation, the shield tail void is naturally narrowed, resulting in unsmooth mortar flow. In order to obtain effective filling, the grouting pressure required is large; on the contrary, in the stratum with good conditions, the self-stability of soil mass is strong, and the mortar flows smoothly in the void, so the grouting pressure required is small. According to the burial depth of the tunnel and the surrounding water and soil pressure, the synchronous grouting pressure needs to be properly adjusted to meet the uniform filling of grout and construction safety requirements.

(ii) *Grouting amount control*: Theoretically, it is only required to 100% fill the structure void with motor, but the following factors need to be considered:

(a) The effective injection amount is less than the actual injection amount due to grout shrinkage and consolidation after water loss.

(b) Some grout will spill into the surrounding stratum.
(c) The actual excavation section is elliptical in the case of advance on curves, deviation correction or shield machine pitching.

Therefore, an appropriate grouting amount should be larger than the theoretical one. Grouting with the new high-density shear mortar will not cause the surrounding strata to split. Therefore, the actual grouting amount should be controlled at about 120% of the theoretical structure void to meet the requirements for filling the structure void. The grouting amount and pressure depend on the pressure value during grouting and the monitoring data of formation deformation. As advance in curve sections increases the frequency of deviation correction and also increases the disturbance to the soil mass, the amount of grouting should be strictly controlled during the advance in curve sections, and the grouting amount and pressure should be adjusted at any time according to the monitored settlement value during advance in each ring.

5.5.4 Synchronous dry-mixed mortar

Dry-mixed mortar solves the difficulty of controlling the mix proportion in the traditional process. With the advantages of accurate metering, reliable quality, convenient use, etc., it is conducive to the application of automatic construction machines and tools and gradual improvement from the backward traditional construction process. In addition, the on-site mixing of mortar requires a large-area raw material storage site, and dry-mixed mortar has solved the dilemma that it is impossible to mix mortar on site in some projects located in the city center due to the small construction site (Fig. 5.53).

(a) Dry power material (b) Mixing test (c) Mixing plant

Fig. 5.53. Synchronous dry-mixed mortar.

Table 5.6. Grout performance requirements.

Index	Slump (mm)	Density (g/cm³)	Pressure bleeding (mL)	20 h yield shear strength (Pa)
Reference value	100~180	Greater than 1.90	Less than 35	Not less than 800

5.5.4.1 *Grout performance*

(1) The mortar has good fluidity and pumping performance.
(2) The construction operation is simple, thanks to the good filling performance, water retention and a certain workability and uniformity.
(3) It has good corrosion resistance to acid, alkali and brine.
(4) The hardened grout has good shear resistance.
(5) It is pollution-free for the environment and harmless to the human body.

5.5.4.2 *Grout indexes*

See Table 5.6 for grout performance requirements.

5.6 Key technologies of tunneling in special sections

5.6.1 Tunneling in sections with shallow overburden

Shallow overburden construction refers to tunneling with a burial depth at the tunnel roof less than the tunnel outer diameter; ultra-shallow overburden construction is deemed if the burial depth is less than half of the tunnel outer diameter.

5.6.1.1 *Risk analysis*

In the case of a shallow or ultra-shallow overburden, it is difficult to control construction parameters such as cutting face pressure, grouting parameters and slurry quality. It is more difficult to control the stability of the excavation face during construction due to its high proneness to ground mud spillover, settlement, uplift and others. Moreover, it is also difficult to control the stability of the tunnel, especially when serious ground mud spillover, settlement or uplift occurs. After the erected tunnel is separated from the shield tail, it is difficult to wrap the tunnel with the synchronous grout, and the constraining force on the tunnel structure is weak, resulting in segments floating under the buoyancy of the surrounding groundwater or the shield tail grout, and subsequent large shield machine attitude deviation, segment bolt shearing, large faulting of ends, and an uneven shield tail void.

5.6.1.2 *Targeted measures*

(1) **Design measures**

(i) *Stiffness of tunnel structure is enhanced*: Shear pins (Fig. 5.54) are set circumferentially in the segments in the shallow overburden section to increase the longitudinal stiffness of the tunnel and prevent the tunnel ring separated from the shield tail from floating, especially to prevent faulting between rings.

Moreover, pre-tensioned bolts and welded connecting plates can be used to strengthen the circumferential and longitudinal connection strength of segments and improve the integrity of the lining (Fig. 5.55).

(ii) *Segments are provided with additional grouting holes*: In order to strengthen the secondary grouting and emergency handling capacity of the shield machine construction in the shallow overburden section, additional grouting holes can be provided in the segment design for the shallow overburden section. As shown in Fig. 5.56, two additional grouting holes are provided for each segment, except for the closure block.

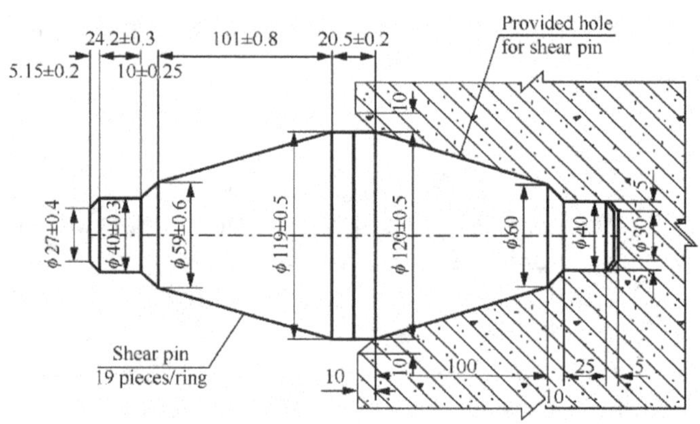

Fig. 5.54. Shear pin (unit: mm).

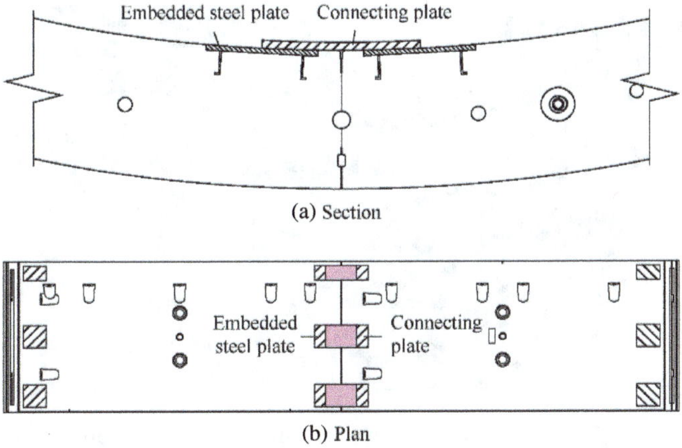

Fig. 5.55. Schematic diagram of embedded steel plate of segment.

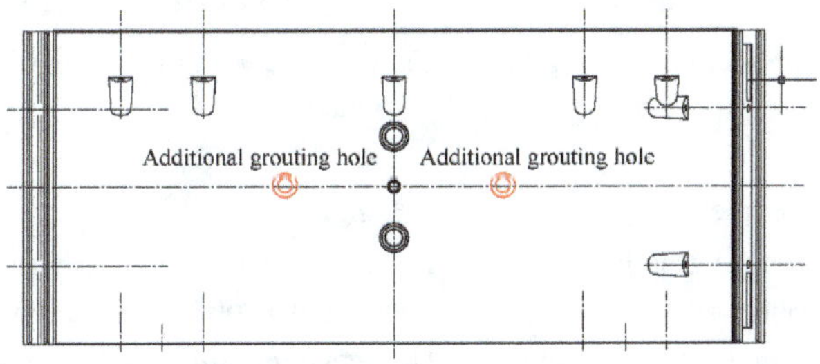

Fig. 5.56. Schematic diagram of additional grouting holes.

(iii) *Anti-floating slab and uplift pile*: For the construction of the ultra-shallow overburden tunnel section, the anti-floating slab and uplift pile can be set for pretreatment before construction, as shown in Fig. 5.57.

(2) **Construction measures**

(i) *Optimization of construction parameters*: The construction parameters are optimized to stabilize the excavation face and reduce the disturbance to the stratum and the impact on the environment. The construction parameters are shown in Table 5.7.
(ii) *Synchronous grouting management*: Strict control should be exerted on the quality of synchronous grout to ensure symmetrical distribution on the left and right and synchronous pressure

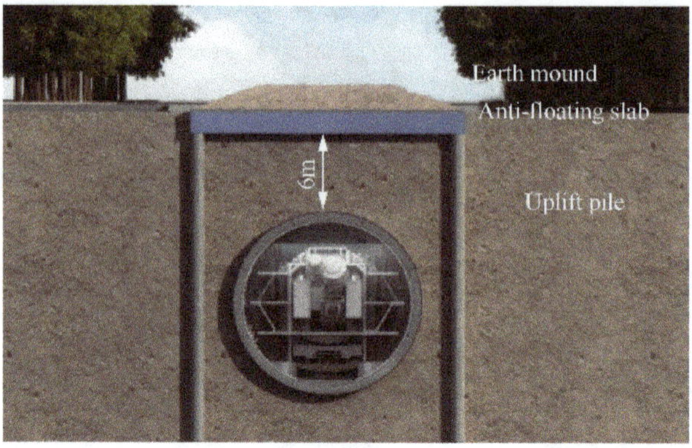

Fig. 5.57. Schematic diagram of anti-floating slab and uplift pile.

Table 5.7. Construction parameters.

S. No.	Item	Parameter requirements
1	Water pressure on cutting face	To be set according to overburden depth and geological conditions
2	Pressure fluctuation	−0.005–0.005 MPa
3	Advance rate	2–3 cm/min
4	Synchronous grouting pressure	To be set according to overburden depth and geological conditions
5	Synchronous grouting amount	The actual grouting amount should be 110–130% of the theoretical structure void

injection. When it is found that the tunnel floats largely, the grouting amount should be adjusted for float control. It should be ensured that the injected grout has a good waterproof effect and that the grout right ahead of the shield machine is not channeled back to the periphery of the segment. Therefore, the grout should be tested on site before use, regarding slump, shear strength and other indicators, as shown in Fig. 5.58.

The early strength of synchronous grout should be improved to quickly stabilize the segment at the rear of the shield tail.

Under the circumstance that a large-diameter shield machine advances in the shallow overburden section, additional embedded grouting holes can be used when necessary to take timely, repeated and sufficient grouting measures to reinforce the part behind the segment wall. The construction process is shown in Fig. 5.59, and the grouting range is shown in Fig. 5.60.

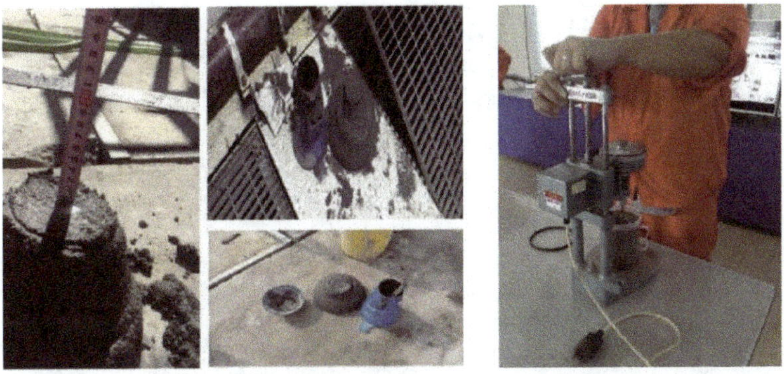

(a) Slump test

(b) Examples and shear test

Fig. 5.58. Field grout test.

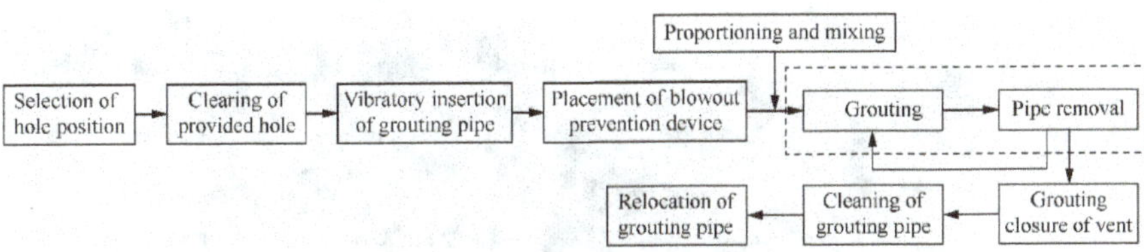

Fig. 5.59. Construction process of secondary grouting.

(iii) *Surcharge loading on the ground or in the tunnel*: According to the actual overburden thickness of the tunnel, the ground surcharge loading or tunnel internal ballasting is adopted to limit the tunnel floating, as shown in Figs. 5.61 and 5.62. For the soft and shallow overburden section, cement slurry pre-reinforcement can also be used to improve the soil mass and increase the anti-disturbance capacity.

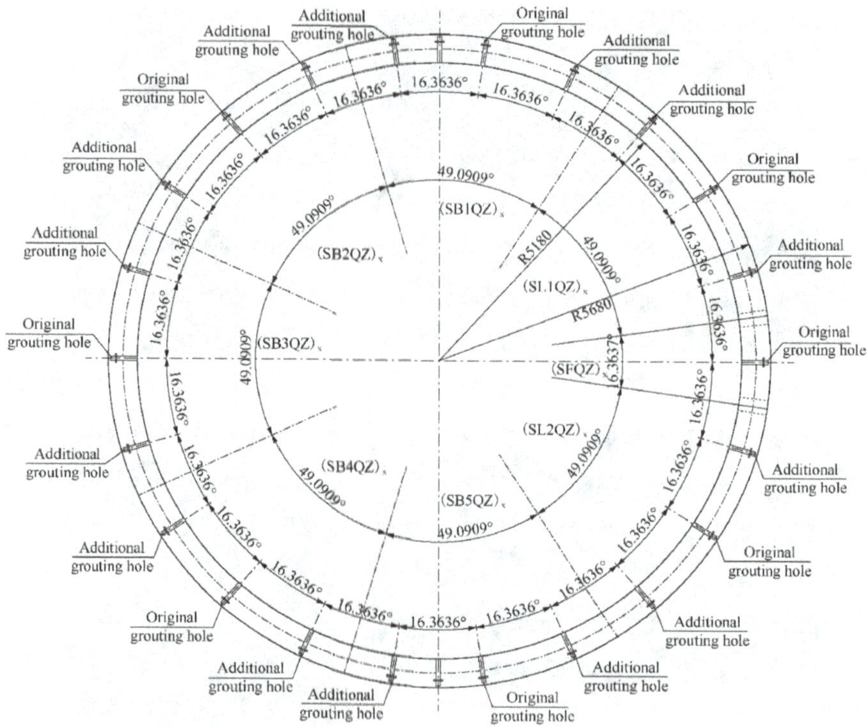

Fig. 5.60. Schematic diagram for layout of additional embedded grouting holes.

Fig. 5.61. Ground surcharge loading.

(iv) *Construction monitoring*: Under the condition of shallow overburden, shield machine construction is prone to causing ground heave and tunnel floating to varying degrees, which will affect

Fig. 5.62. Ballasting in tunnel.

the normal use of surrounding ground buildings, underground pipelines and other facilities. Therefore, the following monitoring is mainly conducted:

(a) *Tunnel stability monitoring*: Tunnel uplift, tunnel displacement, segment faulting at ends, opening, etc. Generally, the tunnel uplift should be monitored per team, and if necessary, should be monitored for each ring until the segments separated from the frame are stable.
(b) Ground surface and pipeline settlement monitoring over the tunnel axis.
(c) Settlement and crack monitoring of general buildings (structures) and automatic monitoring of sensitive buildings.

5.6.2 Tunneling in small-radius curve section

The tunnel curve less than 40 times the outer diameter of the shield machine is regarded as a small-radius curve.

Numerous large-diameter shield tunnels with small-radius curve sections have been constructed in Shanghai, among which the second tube of the Dapu Road Tunnel has a ratio of the minimum horizontal curve turning radius to the outer diameter of 35 and the smallest R/D value among the shield tunnels of the same specifications in China at present. The model fitting curve of a combination of conventional segments in 1.5 m ring width and segments in 0.75 m ring width is used in the small-radius curve section. The tunnel was built without any abnormal conditions in the construction process, and the tunnel structure is now in good service condition (Fig. 5.63).

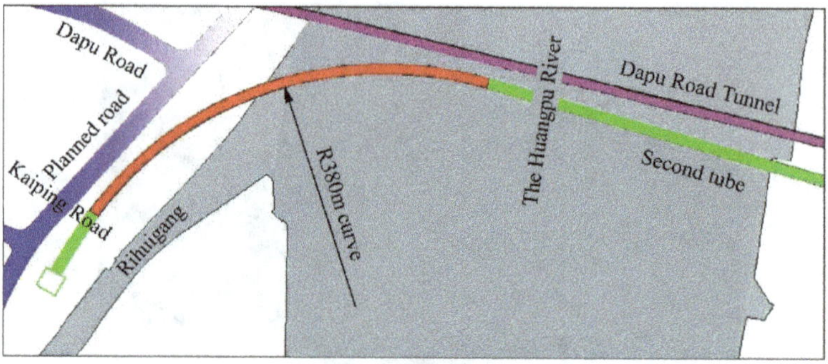

Fig. 5.63. Shield machine advance on small-radius curve in second tube for Dapu Road Tunnel.

5.6.2.1 *Risk analysis*

The main risk faced by the construction of a small-radius curve section with a large-diameter shield machine lies in the difficulty of controlling the quality of the formed tunnel section. As the shield machine itself is a straight rigid body, a circular curve section can only be excavated in a polyline to fit the circular curve. In order to make the axis of the shield tunnel coincide with the design axis, continuous deviation correction is required in the process of tunneling. The smaller the radius of the circular curve and the larger the diameter of the shield machine, the greater the fitting difficulty, the greater the deviation correction amount per unit distance of the tunneling, and the lower the deviation correction accuracy. Therefore, the axis control of the super-large-diameter shield tunnel on small-radius curves is more difficult due to the rising probability of quality problems (Fig. 5.64) such as segment breaking and faulting of ends. In addition, the shield tail void is likely to be uneven during the construction of a small-radius curve section, which will cause shield tail leakage.

5.6.2.2 *Targeted measures*

(1) **Design measures**

(i) *Segment form*: In order to better fit small-radius curves and meet the stress requirements of segments erected on tunnel curves, it is necessary to demonstrate the type, ring width and taper of segments during design. The taper for small-radius curve sections of super large-diameter tunnels should be reasonably determined in comprehensive consideration of the outer diameter of the shield

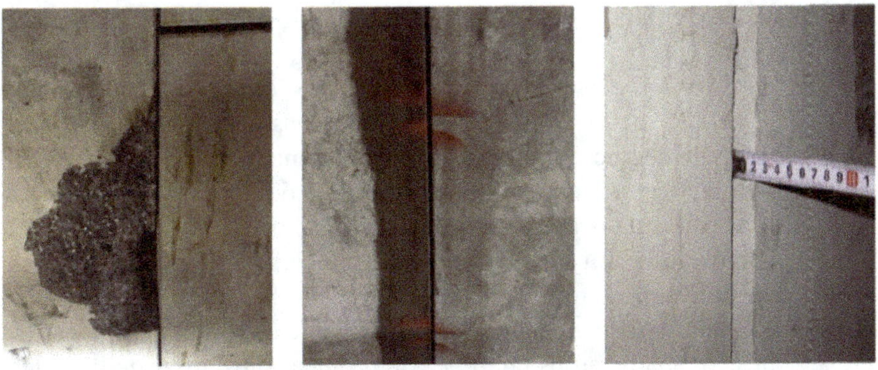

(a) Segment fra gmentation (b) Segment seepage (c) Faulting of segment ends

Fig. 5.64. Faulting of ends, fragmentation and water seepage of segments in small-radius curve section.

Fig. 5.65. Segments in small ring width for second tube of Dapu Road River Crossing Tunnel.

machine, the shield void, and the sealing capacity of the shield tail brush.

The segments erected in the *R* = 380 m small-radius curve section of the second tube for the Dapu Road River Crossing Tunnel are shown in Fig. 5.65. In this section, the segments with a 750 mm ring width and those with a 1,500 mm ring width are used together to realize the construction in the small-radius curve sections.

(ii) *Segment waterproofness*: The segment joint of large-diameter tunnels is mainly arranged with two passes of waterproof. The specific segment waterproof arrangement for the Shanghai Yangtze River Tunnel is shown in Fig. 5.66. During the construction of a small-radius curve section, in order to strengthen the waterproof ability, more waterproof layers can be adopted, such as the Sutong Power Tunnel, as shown in Fig. 5.67.

(iii) *Segment connection*: In order to enhance the overall tunnel stiffness in small-radius curve sections, shear pins are usually added between segment rings, and connectors are added longitudinally to strengthen the connection of segments.

(2) **Equipment support**

(i) *Minimum turning radius and over excavation capacity of the shield machines*: The overexcavation capacity of shield machines

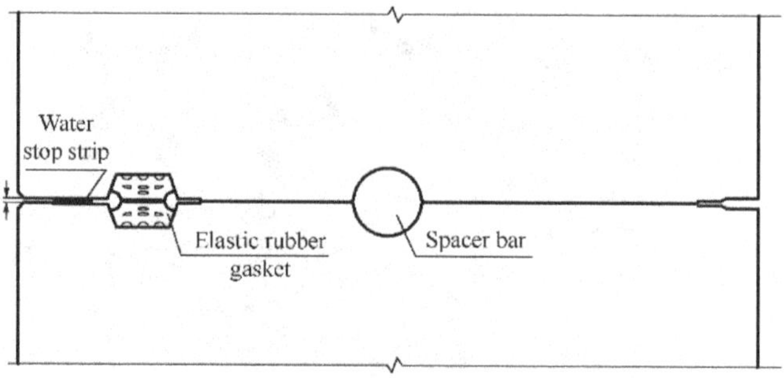

Fig. 5.66. Waterproof arrangement of segments in Shanghai Yangtze River Tunnel.

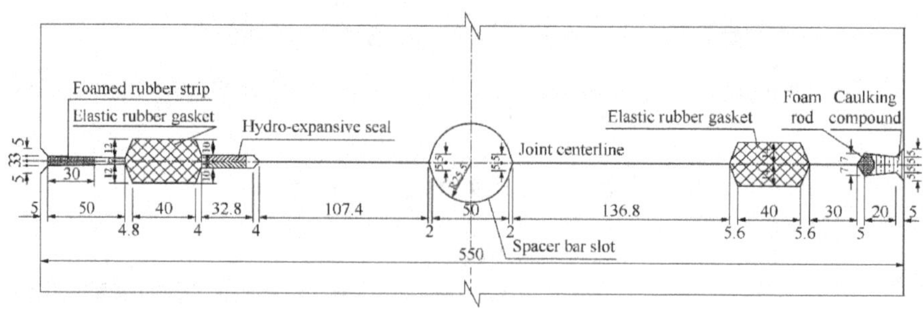

Fig. 5.67. Waterproof arrangement for segments in Sutong Power Tunnel (unit: mm).

has an important influence on small-radius curve construction. The profiling cutter on the cutterhead of the shield machine has a certain overbreak range. During curve construction, partial overbreak can be carried out according to the advance axis. The larger the overbreak, the easier the curve construction. However, overexcavation will also cause the synchronous grout to seep into the excavation face due to the looseness of the soil mass. Therefore, before using the overbreak function, it is necessary to understand its control form and capacity to control the overbreak volume to the minimum.

In addition, the articulation design of the shield machine can form a movable structure from the cutting ring to the support ring and from the support ring to the shield tail to increase the sensitivity of the shield machine, ensure control of the advance axis in the curve section, and reduce the occurrence of external arc fragmentation and water seepage of segments.

(ii) *Sealing performance of shield tail*: During the construction of a small-radius curve section, in order to prevent leakage of the shield tail, the shield tail sealing device of the shield machine can be equipped with the freezing pipe and emergency pressure bladder in addition to the steel wire brush and steel plate brush, as shown in Fig. 5.68.

As a kind of inflatable expansion seal, an emergency pressure bladder is used for sealing with an expanding bladder. The function of emergency sealing is to first stop water in case of emergency (such as a large amount of mud and sand leakage at the shield tail) and then to protect the installation area of segments

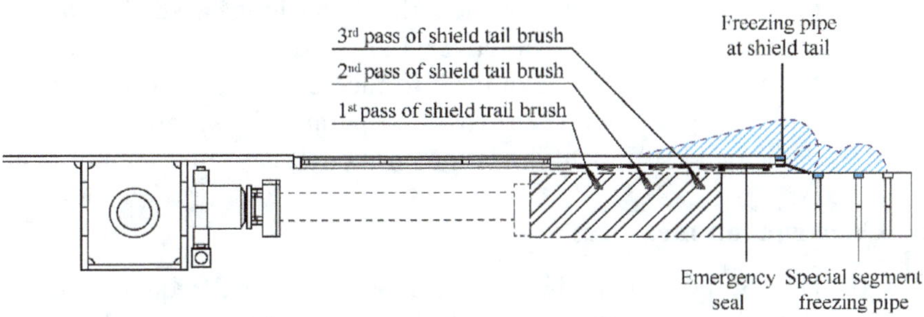

Fig. 5.68. Schematic diagram of shield tail seal and freezing pipe.

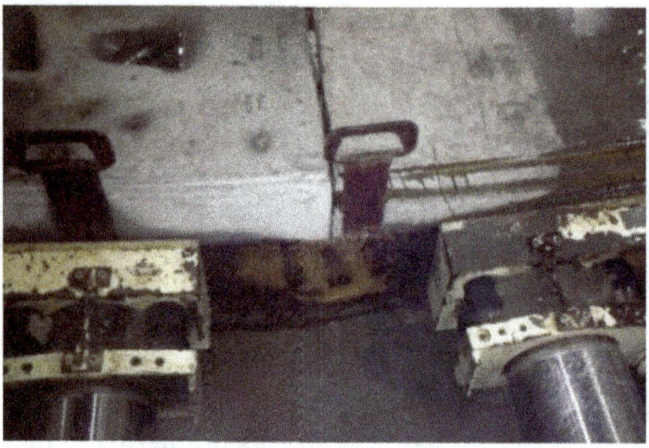

Fig. 5.69. Emergency water stop plug plate.

when replacing the first two wire brushes, so as to prevent slurry outside the shield tail from entering the shield machine. The emergency water stop plug plate is shown in Fig. 5.69.

At the end of the shield tail, a circle of ring pipes arranged along the circumference of the shield machine is provided as the freezing pipe position. Before replacing the first two wire brushes, the water and soil in this part are frozen by filling the freezing medium, and special segments with freezing pipes are used to strengthen freezing so as to ensure the safety of replacing the wire brushes.

(iii) *Frame turning performance*: In the construction of small-radius curves, the length of each section of the frame system is considered first. Each section of the segmented frame, as a rigid body, must be able to pass through the tunnel rings on small-radius curves smoothly, and the length of the longest section of the frame is calculated. Moreover, the traveling mechanism of the frame system should conform to the actual working conditions and equipped with an easy-to-operate deviation correction device and sufficient deviation correction capacity.

(3) **Construction measures**

(i) During the construction of the shield machine, frequent calibration, remeasurement and recheck should be performed against the measurement base station, and manual re-measurement should be conducted against the automatic measurement system on a regular basis. In cases of great data differences, the frequency of manual re-measurement should be increased.

(ii) The deviation correction amount should be strictly controlled in the shield machine advance to make the shield tail void around the segment uniform as far as possible and reduce the compression degree from the segment on the shield tail seal brush.

(iii) The quality control of synchronous grouting should be strengthened to timely and sufficiently carry out synchronous grouting and to timely adjust the proportional distribution of grouting among all grouting points.

(iv) The segment erection oil pressure should be reasonably set to prevent the shield machine from retreating. Bolts should be re-tightened as required, involving the segments separated from the shield tail and the frame during the advance process.

(v) High-quality shield tail grease should be used, and the shield tail grease should be injected in a timely, sufficiently and uniform manner to ensure the sealing performance. If necessary, emergency plugging grease is injected.

(vi) Emergency supplies should be prepared, such as polyurethane, plugboard and emergency grease injection.

5.6.3 Long-distance tunneling

Generally, tunnels with a shield tunneling distance greater than 3 km are defined as long-distance tunnels.

5.6.3.1 *Risk analysis*

(1) **Large wear of equipment:** In long-distance shield tunneling, the operation conditions of the equipment are critical to smooth tunneling. Long-distance shield tunneling will inevitably lead to problems such as cutter wear and shield tail brush damage. The replacement of cutters and shield tail brushes will bring great construction risks.

(2) **Low transportation efficiency:** During the shield machine long-distance construction, with the continuous increase in tunneling distance, the time and number of vehicles for transporting construction materials will increase, and the load of slurry circulating and discharge pumps will increase. Therefore, transportation efficiency and safety are particularly critical to long-distance tunneling. In the process of

shield tunneling, the main factor affecting construction efficiency is horizontal transportation.

(3) **Low measurement accuracy:** In the process of long-distance tunneling, measurement errors will continue to accumulate, resulting in the decline of construction measurement accuracy, thus affecting the overall tunnel axis deviation and holing through survey accuracy.

5.6.3.2 *Targeted measures*

(1) **Device configuration**

- (i) The cutterhead is designed to be an atmospheric pressure type. During the construction, the inspection and replacement of the cutters replaceable under atmospheric pressure are strengthened (Fig. 5.70) to improve the efficiency of cutter replacement and reduce the construction risk of operation under pressure.
- (ii) During long-distance tunneling, the shield tail brush is prone to damage due to poor control of the shield tail void or uneven grease injection at the shield tail, thus resulting in shield tail leakage. Jack length, segment erector and other equipment shall be designed to enable the replacement of at least one shield tail brush (Fig. 5.71).
- (iii) The main drive sealing system has the ability of automatic monitoring, pressure increase and pressure decrease to automatically adjust the sealing pressure according to the change in external water and soil pressure.

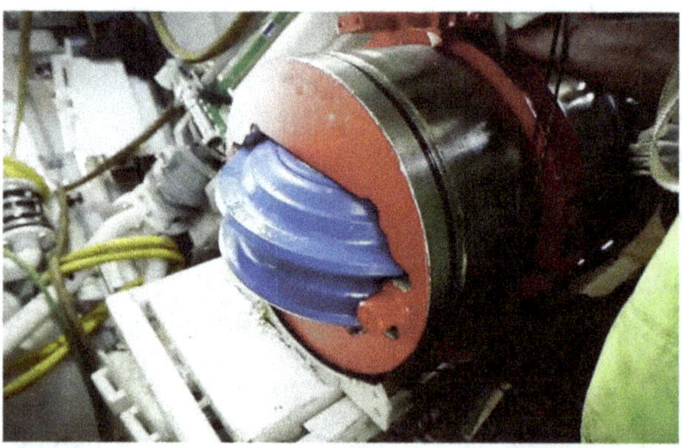

Fig. 5.70. Replacement of atmospheric pressure cutter.

(iv) All kinds of equipment are arranged reasonably and can be replaced quickly.

(2) **Transportation efficiency control**

(i) *Slurry system configuration*: The slurry delivery system consists of a slurry delivery pipe, a slurry discharge pipe, an air pressure valve, a slurry delivery pump and a slurry discharge pump. The qualified slurry passing through the slurry treatment system is sent from the regulating tank to the cutting face of the shield machine through the $P_{1.1}$ pump and via the slurry pipeline, and then discharged through the slurry tank of the shield machine. The slurry is returned to the slurry field through the $P_{2.1}$ pump on the shield machine and the $P_{2.X}$ relay pump in the tunnel, as shown in Fig. 5.72.

The slurry pipeline consists of 8–10 m pipes, and hydraulic gate valves and elbows are installed at intervals and is subject to pipeline pressure loss along tunnels sloping up and down.

Fig. 5.71. Replacement of shield tail brush.

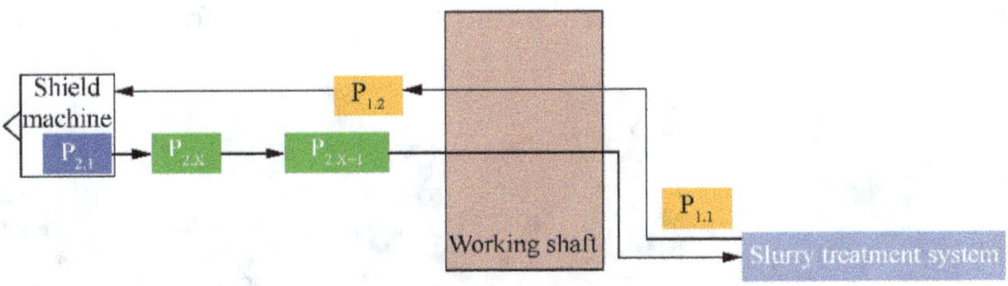

Fig. 5.72. Schematic diagram of slurry pump.

The total pressure loss of the pipeline system is the sum of the pressure loss along the linear piping sections $\Delta p\lambda$ and all local pressure losses $\Delta p\zeta$:

$$\Delta p = \sum \Delta p_\lambda + \sum \Delta p_\varsigma = \sum \lambda \frac{l}{d}\frac{\rho v^2}{2} + \sum \varsigma \frac{\rho v^2}{2}, \tag{5.21}$$

where

λ — resistance coefficient along the pipeline;
ξ — local resistance coefficient;
l — pipeline length (m);
d — pipe diameter (m);
ρ — slurry density (kg/m^3);
v — slurry velocity (m/s).

A slurry pressure sensor is placed every 500 m, and a relay pump is placed at an appropriate position according to the pressure data on the sensor.

(ii) *Horizontal transportation*

(a) *Mode of transportation*: Double-headed trucks, Steyr trucks and truck mixers (Fig. 5.73) are generally used to transport prefabricated parts, segments and grout, respectively.
(b) *Horizontal transport road structure and integrated transport design*: According to the overall construction plan, the shield tunneling and internal road structure are generally carried out simultaneously. During tunnel tunneling, the hollow rectangle members and pavement construction are carried out simultaneously. The hollow rectangle members are placed by the frame, while the pavement structure is constructed after being separated from the frame (Fig. 5.74).

Considering that the Steyr truck needs to turn around in the tunnel, a turnaround platform is made behind the shield machine frame and towed

(a) Double-headed truck (b) Steyr truck (c) Truck mixer

Fig. 5.73. Vehicles for horizontal transport in tunnels.

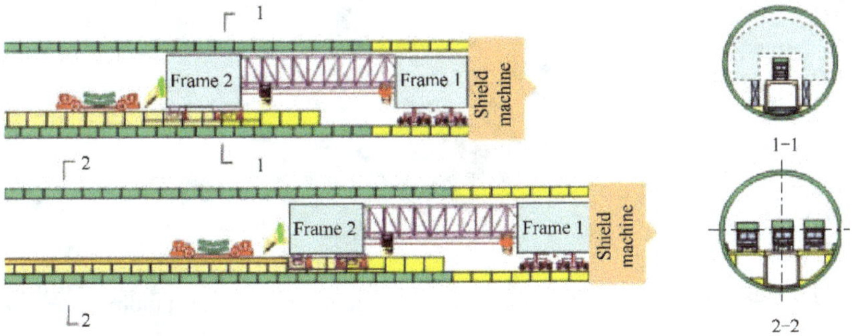

Fig. 5.74. Schematic diagram of horizontal transport.

Fig. 5.75. Turnaround platform at the rear of frame.

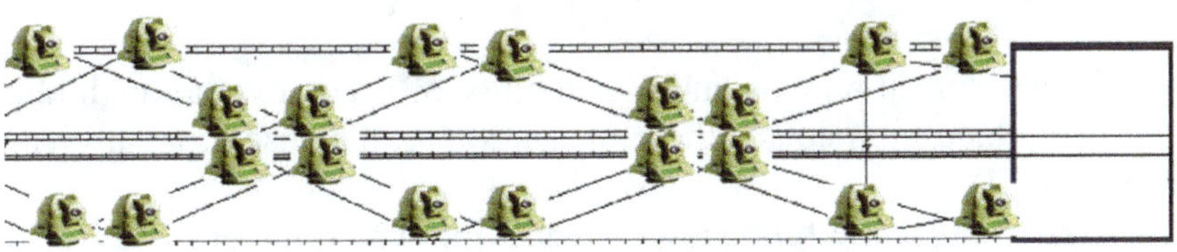

Fig. 5.76. Layout of parallel traverses.

by the frame. The turnaround platform is made of a steel structure, as shown in Fig. 5.75.

5.6.3.3 *Survey accuracy control*

(1) **Horizontal control survey in tunnel:** Two levels of branch traverses can be laid: the construction branch traverse and the control parallel traverse (Fig. 5.76). In order to ensure the integrity of traverse layout

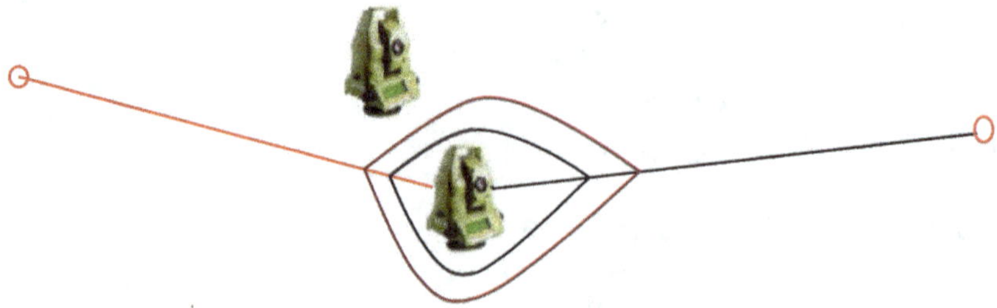

Fig. 5.77. Schematic diagram of double measurements at the same station.

Note: Double measurements at the same station means to conduct two measurements at the same station, with different instruments and by different operators.

Fig. 5.78. Vertical jacking measuring hole in tunnel.

in long-distance tunnels and the stability of accuracy, the method of "double measurements at the same station" can be used for surveying (Fig. 5.77).

The longer the tunneling distance, the worse the accuracy of its point position. It is difficult to ensure the accuracy of orientation in the tunnel. The use of a gyro theodolite can greatly improve the accuracy of tunnel connection surveys.

(2) **Vertical jacking measuring hole:** In order to ensure the accurate holing of tunnels, the construction process of a vertical jacking measuring hole can be used to transfer the coordinates of the ground control network near the receiving shaft to the tunnel so as to accurately correct the control network in the tunnel (Fig. 5.78).

5.6.4 Tunneling in river crossing section

In river crossing tunnels, the distance between the river scouring line and the tunnel roof is usually small, and the geological conditions are complex.

5.6.4.1 *Risk analysis*

(1) **Stability of cutting face:** For the shield machine advance in the middle of rivers, river water penetration is the main risk point. The instability of the excavation face will cause accidents of "tunnel roof collapse" in the river. Failure to tightly block the exploration boreholes around the section in the early stage will also form channels, causing the risk of slurry pumping in the river or river water penetration.

(2) **Leakage at shield tail:** The shield tail is the key protection object during shield tunneling in the middle of the river. The river section often has shallow overburden, high water pressure, and strong permeability of soil mass. The shield tail leakage during the shield machine advance brings risks of river bottom collapse, tunnel flooding, and even tunnel collapse.

(3) **Harmful gas prevention:** The river stratum in some areas contains harmful gases such as methane. Shield machines create a secondary confined space in which poor ventilation conditions are likely to cause serious poisoning accidents in cases of harmful gas leakage during construction.

5.6.4.2 *Targeted measures*

(i) Reasonable slurry separation system should be configured, and high-quality fresh slurry should be fed to ensure the quality of the slurry on the excavation face.

(ii) *Slurry pressure should be controlled on the shield machine excavation face.*

The slurry balance of the excavation face is a dynamic process. An changes due to some reasons may break the balance. Therefore, attention should be paid to the change in the slurry pressure both in the excavation period and the excavation suspension stage, and corresponding measures should be taken to make the slurry pressure as close to the set value as possible.

(iii) *Management and control of dry sand volume from excavation.* According to various data measured by the flowmeter and densimeter for slurry delivery and discharge, the volume of excavated dry sand contained in the delivered and discharged slurry should be calculated to judge whether the excavation face is overexcavated by comparing the volume of excavated dry sand with the theoretical value.

When the volume of dry sand is too large, it is necessary to stop the machine to analyze the data, adjust the construction parameters, make the data on dry sand volume close to the theoretical value, and reduce the possibility of soil mass collapse ahead.

(iv) *Construction process management*

(a) A reasonable advance rate should be set to ensure the advance is stable. The rate should be gradually increased after starting the advance and gradually decreased before stopping the advance.
(b) The synchronous grouting pressure should be controlled, and safety valves should be installed in the grouting pipeline to prevent the overburden from being broken due to too high a grouting pressure.
(c) If the shield machine stops due to a mechanical failure or other reasons, measures should be taken to prevent the shield machine from retreating.

(v) *Underwater monitoring measures*: Within 10 days before the shield machine advances into the middle section of the river, a comprehensive scanning (background survey) should be conducted to cover the water depth range until the river bottom along the tunnel axis and recheck the thickness of the tunnel overburden. After the shield machine advances into the middle section of the river, high-precision water depth monitoring should be started at the river bottom, and the monitoring results will be fully used to guide the construction. The operation process is shown in Fig. 5.79. In cases of large upheaval and subsidence at the bottom of the river, measures are taken in time.

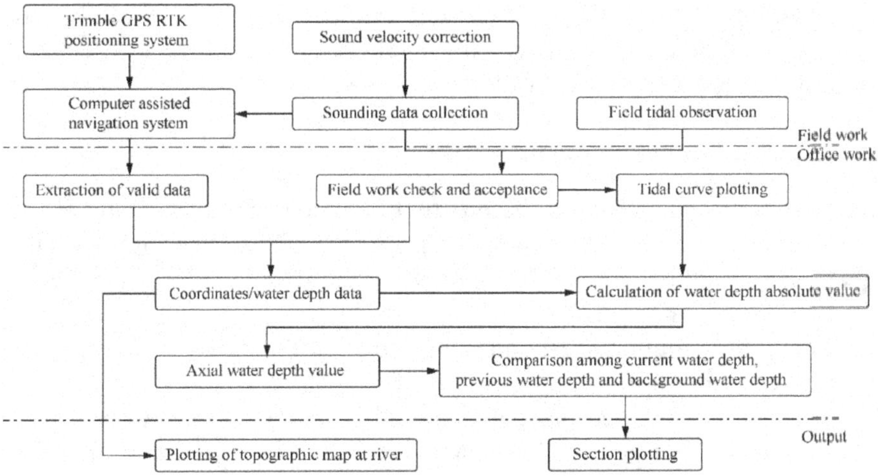

Fig. 5.79. Operation process.

5.6.4.3 *Leakage at shield tail*

(1) **Water pressure stabilization on cutting face:** In the advance process, the cutting water pressure is kept stable to prevent the fluctuation of the cutting water pressure due to equipment failure and human error. After each operation to raise water pressure, specially-designated personnel shall be arranged to observe the grout leakage at the shield tail.
(2) **Quality improvement of synchronous grouting:** Before the advance for each ring, it is necessary to conduct a small-sample test on the synchronous grout. During the synchronous grouting process, it is necessary to reasonably control the grouting pressure so that the grouting amount, grouting rate and tunneling parameters are coordinated to present the best performance.
(3) **Preparation of standby pump and emergency supplies:** In order to prevent the shield machine equipment from being flooded due to huge slurry accumulation after the shield tail leakage, a set of standby pumps shall be added to facilitate the discharge of accumulated slurry, and emergency rescue supplies shall be prepared at the construction site.
(4) **Shield tail grease injection:** The shield tail grease shall be injected regularly, quantitatively and at a fixed position. When slight slurry leakage is found at the shield tail, the slurry leakage part shall be timely supplemented with grease.

(5) **Erection of segments:** The erected segment shall be centered to prevent excessive voids between the shield machine and the segment and reduce the sealing effect of the shield tail.

5.6.4.4 *Harmful gas control*

(1) In order to ensure the safe construction of the shield machine in the harmful gas section, the harmful gas in the area to be crossed by the shield machine should be depressurized and released in advance before construction.
(2) Ventilation equipment should be selected with a reasonable type and proper configuration.
(3) Specially-designated personnel should be assigned to monitor harmful gases in the tunnel. Explosion-proof lighting and transport vehicles should be used. Fire management should be strengthened during construction. No open fire is allowed when entering the tunnel, and smoking is prohibited in sensitive areas.
(4) Special safety disclosure to deal with harmful gases should be made to the construction personnel in advance; emergency gas masks must be provided; and highly sensitive automatic alarm detectors for harmful gases must be provided in each important construction area in the tunnel. Once harmful gases are detected to exceed the standard, the personnel present should be organized to evacuate immediately. After a period of ventilation, the monitoring personnel shall wear gas masks and carry out monitoring gradually from the outside of the tunnel inward. If the methane concentration in the tunnel reaches or exceeds 1.5%, the monitoring personnel should stop moving on. Other construction personnel should not enter the tunnel to resume construction until the methane concentration in the tunnel is detected to be less than 0.5%.

5.6.5 Tunneling in composite stratum

Tunneling is considered in the composite stratum when the excavated section contains soft soil and rock with greatly different geotechnical characteristics.

5.6.5.1 *Risk analysis*

The construction using a slurry balanced shield machine in the composite stratum often suffers from abnormal cutter wear and mud caking on the cutterhead, which are mutually affected.

The main function of the cutter is to break the rock and soil on the working face into muck. It is necessary to break both hard rock and cut soft soil. The hard rock is generally broken and cut with a hob, and the soft soil is cut with a cutting tool. For composite formations with soft soil and rock, the hob is used for cutting, which is prone to eccentric wear, as shown in Fig. 5.80.

Some special composite formations, such as those containing strongly weathered silty mudstone, are low in natural compressive strength but high in the content of argillaceous particles. For a shield machine with a cutterhead of panel type, the opening ratio is not high, and with the wear of the cutter, muck cannot quickly enter the opening through the diversion groove of the cutterhead panel, resulting in mud caking around the cutterhead panel and the cutter. As shown in Fig. 5.81, mud caking greatly reduces the cutting ability of the cutter and affects the tunneling efficiency of the shield machine.

5.6.5.2 *Targeted measures*

(1) **Supplementary geological exploration**

(i) Identify the formation lithology, geological structure and adverse geological processes within the survey scope, and generate geological drilling logs.

(ii) Ascertain the distribution range, size and law of boulders.

(iii) Perform natural uniaxial compressive strength test and RQD value statistics of rock samples within the tunneling scope.

(iv) The geological boreholes within the tunneling scope must be blocked.

(a) New hob (b) Hob with serious eccentric wear

Fig. 5.80. Eccentric wear of hobs.

Fig. 5.81. Mud caking on cutterhead panel.

(2) **Pre-treatment measures before construction:** The characteristic of granite strata is that uneven, weathered bodies often occur in completely weathered and strongly weathered strata, showing a spherical weathering effect. When shield machine construction is carried out in the composite stratum, an appropriate rock stratum (boulder) exploration scheme shall be selected first according to the specific engineering conditions, and the distribution of rock surface lines and the size and location of boulders shall be accurately judged by geophysical exploration technology. Generally, a combination of drilling and geophysical exploration is adopted.

(i) *Borehole exploration*: Before the shield machine construction, drill more holes. Based on the previous geological exploration data, judge the possible area of boulders, and arrange boreholes in this area within the tunneling scope or on both sides.
(ii) *Geophysical prospecting*: Cross-hole CT, transverse wave reflection method, seismic wave scattering method and other geophysical prospecting methods may be applied to determine the distribution, spatial location and shape of rock strata in the tunnel section.

After accurate positioning of the rock stratum in the composite stratum, appropriate pre-treatment measures shall be taken to clear obstacles before the shield tunneling. The commonly used composite formation pretreatment technologies are as follows:

(i) *Artificial rock breaking*: High-pressure air or hydraulic hammers are used to crush boulders, or static blasting is applied to break rocks manually, and the broken rocks are transported away from the shield machine. The precondition for using this method is that the working face must be stably maintained.
(ii) *Rock breaking by a shield machine*: Grout is injected horizontally from the shield machine or from the ground to reinforce the soil layer around the boulders so that the boulder can be immobilized and then directly broken by the shield machine cutter.
(iii) *Ground drilling and underground blasting*: Through ground drilling, the boulder is blasted to break the rock mass, increase the rock mass fracture and reduce the rock strength, so as to reduce cutter wear on the cutterhead of the shield machine. The main difficulty of this method lies in the accurate control of blasting so as to blast boulders into relatively uniform, small pieces.
(iv) *Stone removal with steel casing*: Through a steel casing driven from the ground, first suck out soil around the boulder, and then take out the boulder directly with the lifting equipment.
(v) *Rock breaking by punching*: After the location of the boulder is identified, it is broken by a punch drill before the shield machine tunneling.

(3) **Equipment management**

(i) *Replacement of cutter*: The configuration of the cutterhead and cutters is very important for the construction in composite strata. Different cutters are suitable for different strata. In order to ensure cutting efficiency, cutters often need to be checked and replaced to adapt to changes in strata. Moreover, in the composite formation construction, the cutter wear is very serious. At present, there are two ways to replace cutters: under pressure and under atmospheric pressure. The cutter replacement under pressure means that in the high-pressure environment, the operator enters the earth chamber or slurry tank to replace the cutter. There are two ways to replace the cutter under normal pressure: one is to reinforce the soil around the excavation face, and then the operator replaces the cutter under normal pressure; the other is to check and replace the cutter replaceable under atmospheric pressure directly in the atmospheric working chamber with the corresponding tooling under atmospheric pressure.

A total of 93 groups of cutters replaceable under atmospheric pressure have been provided for the tunneling of the Sanyang Road River Crossing Tunnel in Wuhan, including 41 groups of cutters interchangeable between hobs and shell cutters and 52 groups of side scrapers replaceable under atmospheric pressure.

(ii) *Wear detection*: As shown in Fig. 5.82, the cutterhead and cutter wear detection device are set mainly to give an alarm when the cutter is worn to the extent that the cutterhead is about to be worn. The principle of wear detection is to set a probe head or detection slot with a certain height on the cutter rest panel and cutterhead panel replaceable at atmospheric pressure, which is filled with hydraulic oil at a certain pressure, and connected to the shield machine via a pressure detection device. When the probe head or detection slot is worn and damaged, hydraulic oil leaks and pressure loss occur, triggering an alarm.

(iii) *Advance detection system*: During the construction in composite stratum, the shield machine is equipped with advanced detection technology (Fig. 5.83) to predict the obstacles ahead of the shield machine excavation face. An SSP advanced detection device is installed on the shield machine to continuously detect the geological conditions ahead of the working face. After automatic data collection and geological evaluation, the prediction results in a certain range ahead of the shield machine are displayed in real time, which is conducive to making quick judgments on the site.

(a) Physical device

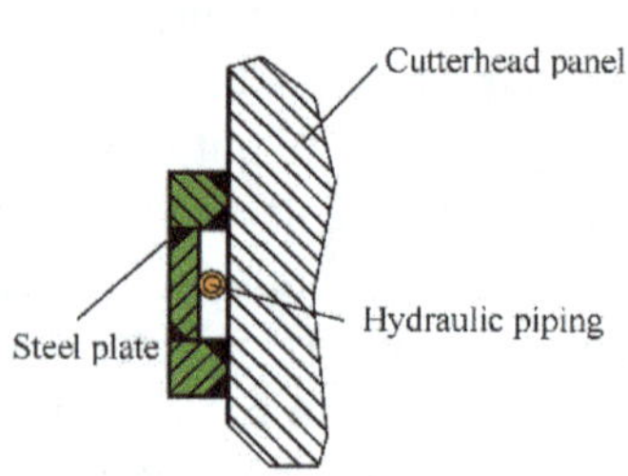

(b) Structural sketch

Fig. 5.82. Cutterhead and cutter wear detection device.

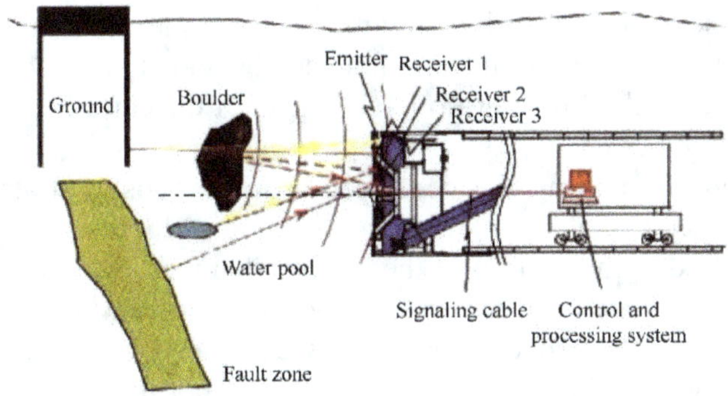

Fig. 5.83. Schematic diagram of SSP advanced detection device.

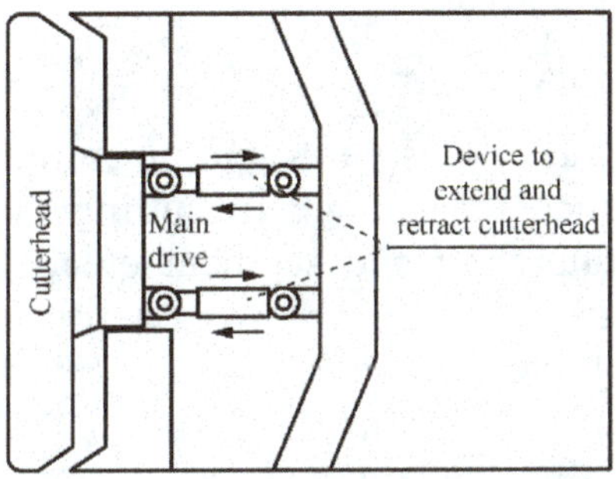

Fig. 5.84. Schematic diagram of telescopic cutterhead.

(iv) *Telescopic main drive*: During the construction in composite strata, there are risks of cutterhead jamming and slurry discharge pipe plugging. As shown in Fig. 5.84, the telescopic drive can extend and retract the cutterhead within a certain range to expand the space of the excavation face, improve the efficiency of slurry circulation and muck carrying, and reduce the risk of mud caking by the cutterhead.

(v) *Flushing on the cutterhead panel*: During construction in a stratum with a high viscosity, failure to discharge the muck cut means a high probability of mud caking on the cutterhead panel. The mud cake on the panel will wrap the cutter to reduce its cutting ability, resulting in further effects on the muck flow and triggering a vicious cycle.

The cutterhead panel is provided with flushing holes and a P0.1 slurry pump, raising the slurry flow rate and pressure to flush the cutterhead panel and prevent mud caking on the panel.

If the center is equipped with a cutter replaceable under atmospheric pressure, a flushing hole can be added on the cutter rest to increase the flushing flow rate ahead of the cutterhead, as shown in Fig. 5.85.

(vi) *Crusher*: During the construction in composite strata, a slurry balanced shield machine must be equipped with a stone crusher, located at the bottom of the bubble chamber and behind the connecting gate between the slurry tank and the bubble chamber, mainly to crush the large rocks excavated before entering the slurry discharge pipe and ensure the normal operation of the slurry circulating system. Its structure is shown in Fig. 5.86.

(4) **Shield tunneling control**

(i) *Thrust parameters*:

(a) Cutting face pressure: Keep the cutting face pressure stable, reduce the change in cutting face pressure during shield tunneling and segment erection, and prevent the upper soil layer from flowing and even collapsing under repeated disturbance.
(b) Advance rate: It is matched with the synchronous grouting amount, shield tail grease injection amount, slurry circulation deviation flow rate, etc.
(c) Penetration: The penetration in composite formation (rock strength ≤ 40 MPa) is generally controlled within 15 mm/r. When the rock strength is greater than or equal to 40 MPa, the penetration is gradually reduced to protect the cutting cutters.

(ii) *Shield machine attitude*: The shield machine advancing in composite strata is prone to attitude instability. Great importance should be attached to the attitude trend control, and the profiling cutter may be turned on when necessary. The angle of the shield machine should be controlled within ±1° and may be adjusted by adjusting the angle of the tunneling jack under special circumstances.

(iii) *Slurry circulation*: The quality inspection of slurry should be strengthened on the excavation face and in the slurry delivery and discharge pipelines. The slurry pipeline to the front cabin should be opened normally, while the pipeline flushing the bottom and suction port should be opened normally in the rear

Fig. 5.85. Cutter rest flushing.

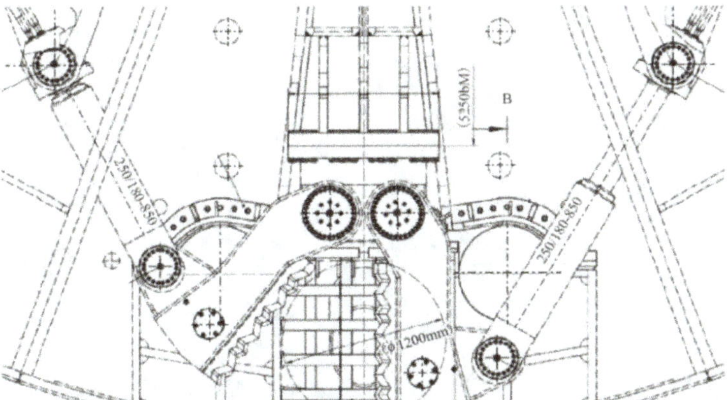

Fig. 5.86. Structural diagram of crusher.

cabin. The shear pump and grout barrel are installed on the frame to mix fresh grout, enabling direct injection into the excavation chamber when necessary.

Before advancing to the next ring, slurry shall be circulated in the whole system until the specific weight of the slurry fed is close to that of the discharged to prevent the sediment in the excavation chamber from silting down to the bottom, causing secondary wear of the cutter.

(iv) *Central scouring*: For the atmospheric pressure cutterhead, washing should be strengthened in the central area of the cutterhead and the central cutter. If the flushing pipe of the central cutter rest is blocked, the central rest may be drawn out for cleaning, and the flushing pipe can be dredged. The central flushing pipeline should be opened in turn to prevent blockage. During the construction of the shield machine, detection should be strengthened for the central flushing pipeline to judge whether the pipeline is blocked according to the flow rate and pressure.

Chapter 6

Internal Structure and Connecting Passage

Large-diameter shield tunnels are usually built for multiple purposes, such as highways, railways, utility pipelines, escape and evacuation passages, requiring more complex and diversified forms and functions of the internal separated structure. In the construction process of large-diameter shield tunnels, for the sake of a short construction period and easy construction, the internal structure is often constructed synchronously with the shield tunneling, which puts forward higher requirements for the construction project management. In addition, in order to meet the safe evacuation requirements of large-diameter shield tunnels, connecting passages are often built between tunnels, applying special construction technologies. This chapter mainly introduces the construction of the internal structure and connecting passages of large-diameter shield tunnels, including the structure forms, construction processes and key technologies.

6.1 Internal structure construction

6.1.1 Forms of internal structure

The internal structures of circular tunnels accommodate highways, rail transit, railways, electrical equipment pipelines, smoke exhaust and ventilation, accident handling in the operation stage and escape passages, giving them integrated functions. The cross-section layout of tunnels with a diameter of more than 14 m is relatively flexible. Up to now, there are four types of tunnels in China: single-deck highway tunnels, road-rail tunnels, double-deck highway tunnels and railway tunnels. See Table 6.1 for the-cross section layouts of circular tunnels and corresponding examples.

Table 6.1. Cross section layouts of large-diameter shield tunnels.

Type	**Sketch of cross-section**	**Tunnel examples**
Single-deck highway tunnel	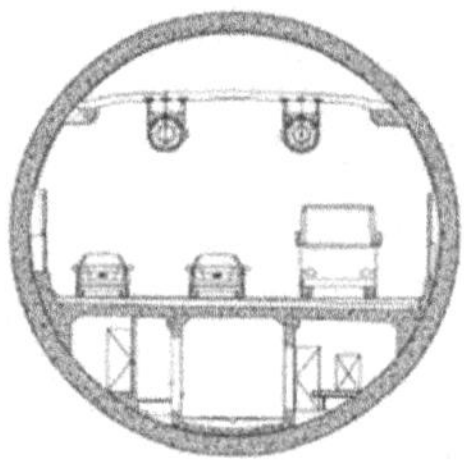	Changjiang West Road Tunnel, Shanghai; Hengqin No. 3 Tunnel, Zhuhai; Qianjiang Tunnel, Hangzhou; Weiqi Road Yangtze River Tunnel, Nanjing; Wuhu South River Crossing Tunnel; Jiluo Road Yellow River Tunnel, Jinan; Shantou Bay Tunnel; Tuen Mun — Chek Lap Kok Subsea Tunnel, Hong Kong;
Single-deck highway tunnel	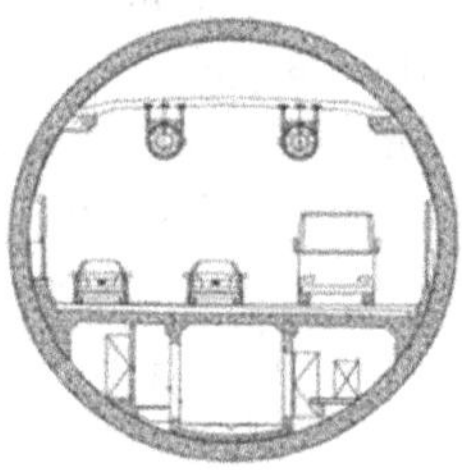	Caltanissetta Highway Tunnel, Italy; M-30 Loop Tunnel in Madrid, Spain; Fourth tunnel of the Elbe River in Hamburg, Germany; Lefortovo Tunnel in Moscow, Russia; Waterview Highway Tunnel in Auckland, New Zealand; Tokyo Bay Submarine Highway Tunnel, Japan
Road-rail tunnel	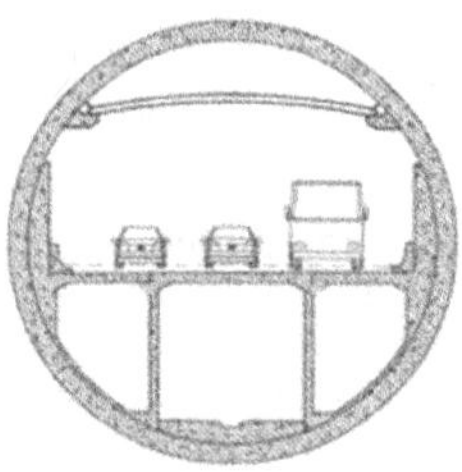	Shanghai Yangtze River Tunnel; Sanyang Road River Crossing Tunnel, Wuhan
Double-deck highway tunnel	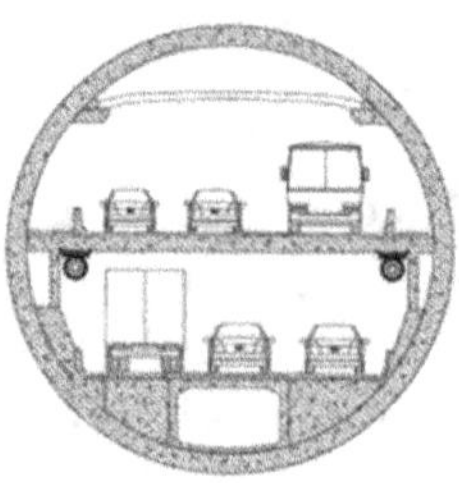	Shangzhong Road Tunnel, Shanghai; Weisan Road Yangtze River Tunnel, Nanjing; Slender West Lake Tunnel, Yangzhou; Heping Avenue South Extension Tunnel, Wuhan; Chunfeng Road Tunnel, Shenzhen Tunnels in Silverpine Forest (Silberwald), Russia Alaskan Way Tunnel, USA

Table 6.1. (*Continued*)

Type	Sketch of cross-section	Tunnel examples
Railway tunnel	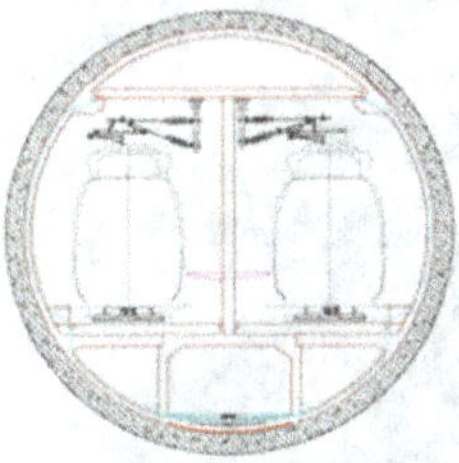	Phase I of Line S2 of the Wenzhou Rail Transit; "Green Heart" Double-track Railway Tunnel, Netherlands

Fig. 6.1. Vertical transport of members.

6.1.2 Transport for internal structure

Transport for the internal structure of circular tunnels mainly includes the hollow rectangle members (or π-shaped members) to be mounted on the tunnel floor, the pavement slabs, the binding reinforcement for the corbels on both sides, the anti-collision curb, concrete and the flue sheet to be mounted on the tunnel roof.

6.1.2.1 *Lowering methods of materials for internal structure*

There are two main modes of transporting materials for the internal structure. The first is vertical transport by a crane down to the shaft bottom and then horizontal transport to the tunnel by a transport vehicle, as shown in Fig. 6.1; the second is horizontal transport directly from the prefabrication yard to the tunnel, as shown in Fig. 6.2. The former is generally used

Fig. 6.2. Horizontal transport of members.

Fig. 6.3. Material transport vehicles for internal structure (unit: mm).

when the lower ramp is not built at the launching stage and direct ground transport cannot realize the tunnel.

6.1.2.2 *Selection of transport vehicles*

There are several common choices of material transport vehicles for tunnel internal structures:

(1) rail electric locomotive (Fig. 6.3(a)),
(2) double-headed truck (Fig. 6.3(b)),
(3) flat-bed trailer (Fig. 6.3(c)).

As shown in Fig. 6.3, a rail locomotive has a strong single transport capacity and can transport segments, hollow rectangle members (or π-shaped components) for internal structure, slurry barrels, various pipelines, shield tail grease, bolts and other miscellaneous materials necessary for the advance of each ring of the shield. However, due to the speed, the layout of rails and turnouts and the restrictions on the maximum gradient of transport, the transport efficiency of a rail locomotive is generally low.

The single transport volume of a double-headed truck is small, but since there is no need to lay rails and turnouts and traffic routes can be flexibly arranged in circular tunnels, transport efficiency can be maximized by reasonably arranging the frequency of vehicle transport and transport capacity allocation.

A flat-bed trailer can realize two-way transport in tunnels without turning around, but it is towed by a truck, so it has poor flexibility and is only used as an auxiliary measure for the transport of large-sized materials.

6.1.3 Construction process of internal structure

During the circular tunnel excavation, the internal structure construction lags behind the tunneling, imposing restrictions on the overall construction period of the tunnel to a certain extent. Among the available construction processes, the common processing method is to adopt an assembly-line-style process combining prefabrication and cast-in-situ for the internal structure while the shield machine advances to ensure synchronous progress with the shield tunneling. The prefabricated members to be mounted on the floor are hoisted synchronously with the shield advance to build a transport lane for floor materials while stabilizing the tunnel segments for rapid construction. After the road structure construction is completed, the flue layer structure at the roof is built by prefabrication or cast-in-situ mode. After the internal structure is built and the tunnel settlement is stabilized, the elevation of the pavement structure is measured fully, according to which the slope adjustment scheme, the anti-collision wall for construction pavement, and the road pavement layer is defined.

6.1.3.1 *Construction process combining prefabrication and cast-in-situ*

The Shanghai Yangtze River Tunnel is the largest road-rail tunnel completed and open to traffic in the world. The tunnel spans the Yangtze River and connects Pudong and Changxing Island in Shanghai, with a total length of 8.9 km. The circular tunnel section was excavated using a slurry balanced shield machine, the largest in diameter (15.43 m) and highest in all indicators in the world at that time, including the distance excavated in one shot reaching 7.5 km and the maximum burial depth at the river bottom of 55 m.

As shown in Fig. 6.4, the Shanghai Yangtze River Tunnel is designed as a double-tube, single-layer, two-way, six-lane arrangement. Its internal structure is divided into three layers from bottom to top: the rail transit layer (including cables, pipelines and escape passages), the road layer and the flue layer, respectively serving as the passage for rail transit and road vehicles and for handling of emergencies such as fire accidents in the operation stage. The tunnel internal structure was built in an assembly-line style, making full use of the characteristics of the Shanghai Yangtze River Tunnel with a large diameter and long distance; that is, the construction of the internal road structure was organized by combining prefabrication and cast-in-situ during shield tunneling. With the synchronous hoisting of prefabricated hollow rectangle members as the core, the structural platform above the hollow rectangle members is used to transport materials while playing an anti-floating role, so as to achieve the

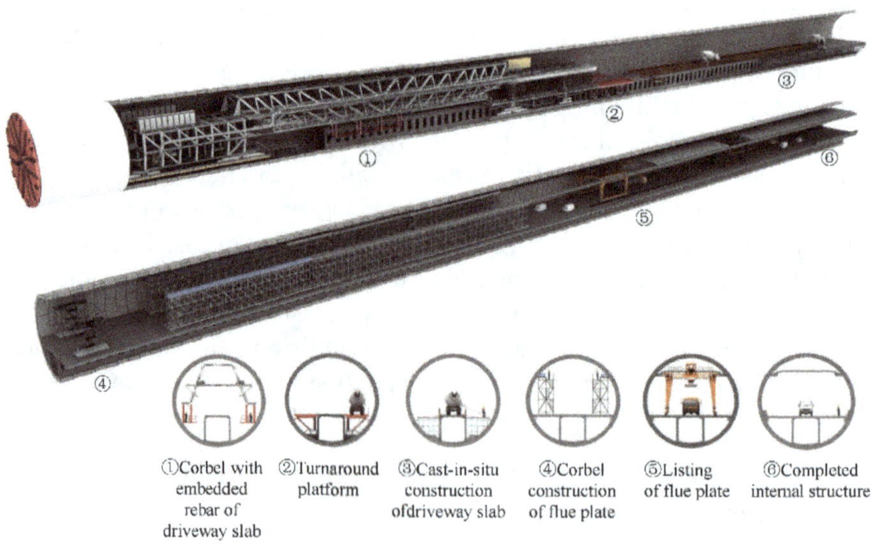

Fig. 6.4. Internal structure construction of Shanghai Yangtze River Tunnel.

goal of rapid construction. During this period, manual roughening and rebar embedment are carried out horizontally on both sides of the platform at the top of the hollow rectangle member, and the corbel of the pavement slab on both sides and the top pavement slab is poured. After the construction of the road structure is completed, the flue layer structure is constructed by means of prefabricated flue plate hoisting and cast-in-situ flue opening.

During the construction of the internal structure, the cast-in-situ operation of the ballast weights, corbels and driveway slabs on both sides of the hollow rectangle members was carried out at 30 m as a construction unit, to reasonably arrange the assembly-line-style process and the corresponding working face. The top flue structure was built after the completion of the pavement structure and before the construction of the pavement layer. The flue corbel was connected with the segment by embedded reinforcing bars, and the customized mobile formwork was used for construction in the tunnel; after the corbel of the top flue is completed, the prefabricated 1.2 m wide flue plates were erected in place with a specially customized crane. The top smoke exhaust outlet and jet fan positions were built with movable gantry and fixed formwork in an integral steel beam and cast-in-situ concrete composite structure.

6.1.3.2 *Construction process of prefabrication and assembly*

The new Zhuguang Road Tunnel is located in Minhang District and Qingpu District of Shanghai, starting from the ground point at Beiqing Road in the north and ending at the ground point at Convention and Exhibition Ring Road in the south of Songze Elevated Road in the south, with a length of about 2.8 km. The shield method and open cut method were used for construction. An earth pressure balanced shield machine with a diameter of 14.45 m was used for shield tunneling, for 1,390 m total and in 695 rings in total. The tunnel lining structure has an outer diameter of 14 m, an inner diameter of 12.8 m, a ring width of 2 m, and a thickness of 0.6 m. The maximum longitudinal slope along the main line of the shield tunnel is −48‰, and the minimum radius of curvature along the plane axis is 700 m.

The internal structure of the Zhuguang Road Tunnel was built by the prefabrication and assembly process, except for the cast-in-situ structures in the filling on both sides of π-shaped prefabricated members, the post-cast beams on both sides of the lower base (including the lower anti-collision curb) and the upper prefabricated driveway slab, and the prefabricated members include the driveway slabs, columns and cover plates.

The assembly-line-style construction was formed in the circular tunnel along the shield advance direction. The construction procedures are as follows:

Procedure 1. Installation of π-shaped prefabricated members: The π-shaped prefabricated members are installed simultaneously with the shield tunneling to form a construction lane for the transport of segments and other related materials.

Procedure 2. Pouring of the lower anti-collision curb and base: After the deformation and settlement of the shield tunnel are stabilized, the lower anti-collision curb and base are fabricated, and the column dowels are embedded, with an allowable deviation of ± 2 mm for the dowel center position.

Procedure 3. Installation of a prefabricated column: After the base strength meets the requirements, the prefabricated column is installed. The prefabricated column is connected to the dowel provided in the base through the grouting sleeve.

Procedure 4. Paving of driveway slab: After the strength of the joint between the prefabricated column and the base meets the requirements, the prefabricated upper driveway slabs are paved. At this stage, the prefabricated upper driveway slabs are in a four-point simply-supported stress state and only bear the dead weight load.

Procedure 5. Post-cast beam construction.

Procedure 6. Paving of the prefabricated cover plate and anti-collision curb, and pouring of anti-collision curb at the post-cast joint.

The internal structure construction of the Shanghai Zhuguang Road Tunnel is shown in Fig. 6.5.

6.2 Construction of connecting passage

6.2.1 Functions of connecting passage

Long and large double-tube tunnels are generally equipped with cross-connecting passages. In case of emergencies, such as a fire in the tunnels, traffic crew and passengers can be safely evacuated to another tube

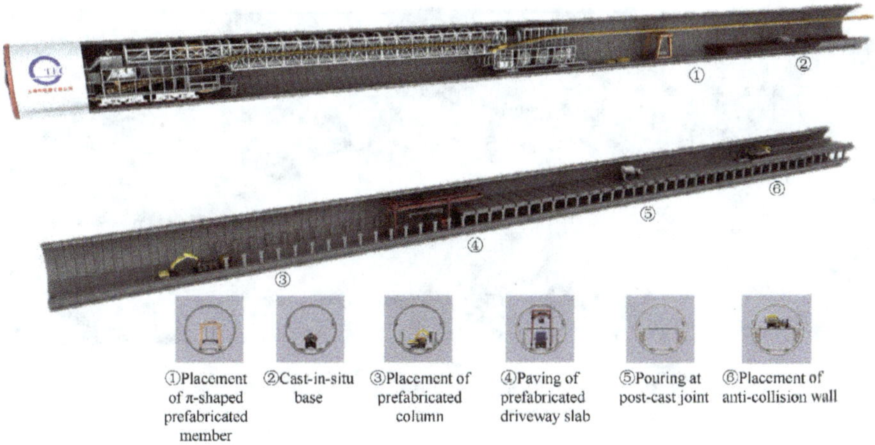

Fig. 6.5. Internal structure construction of Shanghai Zhuguang Road Tunnel.

through the cross-connecting passages, and rescue personnel can quickly reach the accident site through such passages.

In tunnel design, various forms of escape and refuge can be adopted, such as a cross passage, an underground pipe gallery and an evacuation lane, among which the pedestrian cross passage and pedestrian longitudinal evacuation passage are commonly adopted in tunnels at present. The pedestrian cross passage is set perpendicular to the length direction of two tubes to connect two adjacent tubes. When an emergency occurs in one of the two tubes, evacuation can be conducted to the adjacent tube through the pedestrian cross passage. The longitudinal pedestrian evacuation passage is set under the tunnel pavement and directly leads to the outside of the tunnel. When a disaster accident occurs, occupants in the tunnel can enter the passage to escape.

There are two ways to evacuate vehicles in the tunnel: one is to set up a cross passage between two tubes, and the other is to set up a longitudinal special evacuation passage between two tubes. The former is widely used in engineering practices due to its small workload and low cost; the latter is more reliable and safe, but its application is limited due to its high cost. Taking the Tuen Mun — Chek Lap Kok Subsea Tunnel in Hong Kong as an example, the two tubes are provided with connecting passages every 100 m, which are arranged for pedestrians and vehicles alternately. The vehicle cross passage and special vehicle evacuation passage between two tubes can not only be used for vehicle evacuation in the tunnel but also for patrol, maintenance, rescue and vehicle direction change.

Fig. 6.6. Structural diagram of connecting passage.

To sum up, the main function of the connecting passages is to meet the needs of tunnel disaster prevention, escape of occupants, vehicle evacuation and rescue during operation. The structure of the connecting passage is shown in Fig. 6.6.

6.2.2 Construction technology of freezing method

Artificial ground freezing (AGF) is a special construction process which applies artificial refrigeration technology to freeze water in the stratum, turn the natural rock and soil into frozen ground with increased strength and stability and isolate groundwater from the underground works, facilitating underground engineering construction under the protection of the frozen wall. Without restrictions on the support range and depth and with little impact on the surrounding environment, it is widely used in the construction of shield tunnels, especially in the construction of connecting passages between formed tunnels, to effectively prevent water inrush and deformation of soil in the boring process and therefore has attracted the attention of the engineering industry.

The indirect freezing method is generally adopted for the freezing construction of connecting passages. Ammonia, freons or other substances are used as the refrigeration medium. The refrigerant is compressed, throttled and expanded repeatedly by the refrigeration compressor to refrigerate the brine water until it reaches negative temperature, and this negative-temperature brine water is used as the medium to transfer cold flow to the ground for local freezing.

6.2.2.1 *Construction process of freezing method*

The construction process of the freezing method is as follows: carry out horizontal drilling first after the preparations are in place and install the freezing system in parallel; freeze the soil after the drilling is completed; excavate the soil after the freezing index meets the design requirements, erect the temporary supports and then place concrete and pave the waterproof layer, and bind reinforcing rebar, erect the formwork support and pour concrete; remove the formwork support after the concrete strength meets the design requirements; finally conduct forced unfreezing and grouting against thawing settlement, as shown in Fig. 6.7.

(1) **Construction preparations:** Before freezing construction, the steel pipe sheets are fully welded at the connection of the connecting passage, and the rectangular steel support is set, on which the emergency protective door is installed. The protective door should be agile in opening and closing without affecting the subsequent construction, on which the air compressor, air duct, drainage pipe, grouting pipe, control valve and other equipment are mounted to meet the design requirements. When the protective door is closed and the air compressor is started, the air tightness of the protective door and the retention of its internal air pressure should meet the design requirements. In case of any danger of water seepage and sand burst, the protective door is closed immediately, with effective prevention of water and soil outflow, to control construction danger in time and reduce construction loss.

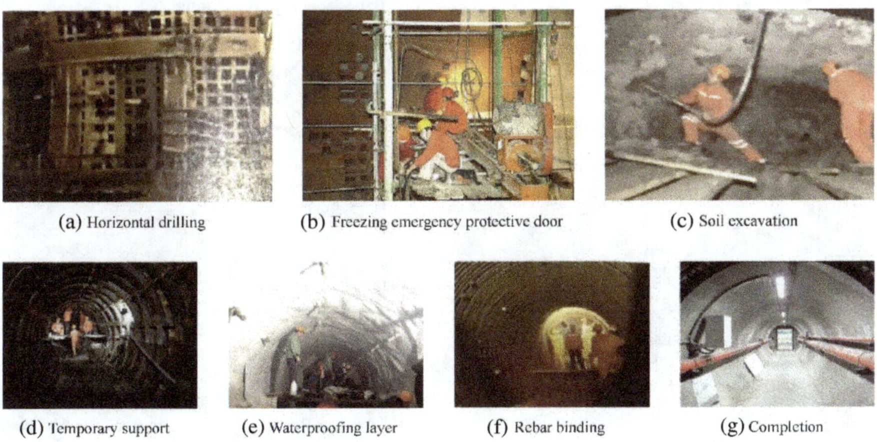

(a) Horizontal drilling (b) Freezing emergency protective door (c) Soil excavation

(d) Temporary support (e) Waterproofing layer (f) Rebar binding (g) Completion

Fig. 6.7. Construction flow chart of connecting passage.

(2) **Drilling:** After surveying and positioning, a drilling rig is installed for drilling operations, during which the deviation measurement and correction are carried out in a timely manner. After the drilling depth meets the requirements, the deviation measurement and leakage detection are carried out. The secondary hole forming construction process is adopted for the drilling construction, and a collar sealing device is installed to prevent sand burst when the freezing hole is drilled penetrating the tunnel segment or in the subsequent operation by a pipe drill, as shown in Fig. 6.8.

After the casing construction by jacking, the horizontal through-hole freezing construction is completed by drilling through the segment on the opposite side with a drill in the casing, as shown in Fig. 6.9.

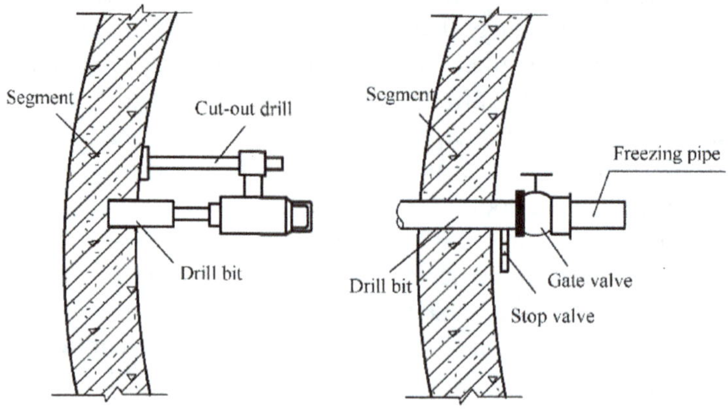

(a) Hole cutting at one shot (b) Secondary hole cutting

Fig. 6.8. Drilling operation.

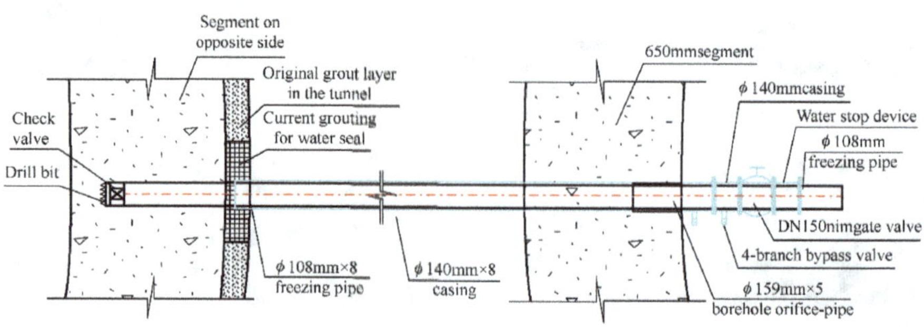

Fig. 6.9. Schematic diagram of through-hole construction in casing.

(3) **Installation and commissioning of freezing system:** The freezing station is set in the tunnel, close to the connecting passage. The equipment in the station mainly includes a refrigeration unit, brine water tank, brine water pump, submersible pump, cooling tower and power distribution control cabinet.

The pipes are connected by flanges, and the brine pipes in the tunnel are laid on the tunnel segments through pipe racks so as not to affect the normal traffic in the tunnel. Valves, thermometers, pressure gauges and other test components are set on the brine pipeline and cooling water circulation pipeline. Thermal insulation measures are taken for the brine pipeline after leak detection and cleaning.

Thermal insulation of the segments on both sides of the connecting passage: Since concrete and steel pipe sheets have strong heat dissipation capacity relative to soil, in order to strengthen the cementation capacity between the freezing curtain and the segments, thermal insulation measures should be taken on the surface of the segments on both sides of the connecting passage, for a thermal insulation range of 2 m beyond the design frozen wall boundary.

Commissioning and test run are carried out after equipment installation. During the test run, the operating parameters are stabilized by pressure regulation so that the unit can operate under the technical parameters specified in the relevant technological procedures and equipment instruments.

(4) **Soil freezing:** Soil freezing is divided into two stages: active freezing and maintenance freezing.

During the active freezing period, the temperature of brine water ranges from −28°C to −30°C, the flow rate of a single freezing hole is not less than 3 m^3/h, and the temperature difference between the feed and return circuits is not greater than 2°C. During construction, the detection is strengthened in the freezing process, and temperature measuring holes and pressure relief holes are arranged in the frozen soil curtain to release the frost heaving pressure, accurately measure the thickness of the frozen soil curtain and judge whether the frozen soil curtains are closed. During excavation, the temperature of brine water is reduced to below −28°C. The average temperature at the interface between the frozen wall and the tunnel segment shall not be higher than −5°C, and the design average temperature of the frozen wall at other parts shall not be higher than −10°C and may be reduced to a range from −13°C to −15°C for special formations.

During the maintenance freezing period covering the excavation of the connecting passage and the construction of the main structure, the temperature of brine water ranges from −20°C to −25°C, and the flow rate of a single freezing hole is not less than 3 m^3/h.

(5) **Soil excavation and support:** The whole section is excavated in one shot with an excavation step of 0.5–0.8 m. The section at the bell mouth on both sides is large. In order to reduce the influence of excavation on tunnel deformation, the excavation step is controlled at 0.5 m. During excavation, the excavation step is strictly controlled to prevent over excavation, and the principle of "exploration before excavation, exploration when in doubt, and support while excavation" is followed. In order to control deformation, the frozen wall is supported promptly after the excavation of frozen soil. The support layer is used not only as an important technical measure to maintain ground stability and ensure construction safety but also as a part of the structural layer, which is the most critical step of the support technology. During the construction process, two supports are provided: the first support is made of steel arch frame and shotcrete with mesh; the second support is made of cast-in-site reinforced concrete.

6.2.2.2 *Forced unfreezing and thawing settlement control*

Forced unfreezing of artificially frozen soil is realized through heat conduction and phase-change heat transfer based on the basic principles of heat transfer theory. Through the freezing pipe originally used to freeze soil, the cold brine in the freezing pipe is replaced with hot brine to replace the original cold brine circulation system with hot brine circulation, so that the temperature of the frozen soil gradually rises until final thawing to achieve forced unfreezing of the soil.

The layout of the forced unfreezing circulation system is shown in Fig. 6.10. The circulating system is mainly composed of a circulating pump, a heater and an unfreezing pipe. Heat is transmitted to the stratum to be unfrozen through the circulation of hot brine water (or other heat medium) and then returned to the heater for reheating.

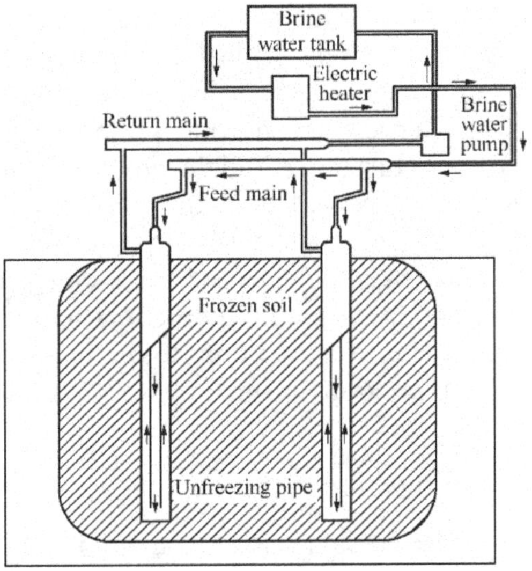

Fig. 6.10. Layout of forced unfreezing circulation system.

The operation process is as follows: Hot brine water is delivered by the circulating pump to the heater for heating to a suitable temperature, injected into the unfreezing pipe to transfer heat to the formation, and then recycled back to the heater for heating. The formation is unfrozen through the above circulation. The heater can be set on the ground surface of the formation to be unfrozen, and the heating mode and unfreezing spacing are determined according to the specific conditions of the construction site. The freezing pipe buried for the previous freezing construction can be directly used as the unfreezing pipe, and the arrangement of the unfreezing pipe and the temperature of the hot brine water can also be adjusted according to the specific requirements.

Generally, the thawing of frozen soil generates excess pore water pressure, which greatly reduces the strength of the soil mass. The soil mass suffering from freezing and thawing is lower than the undisturbed soil, resulting in the thawing settlement amount generally being greater than the frost heaving amount. Thawing settlement of soil will affect the surrounding environment and produce hazards of different degrees. The treatment of thaw settlement generally adopts the grout filling method to compensate for the thawing consolidation settlement of the stratum with grout and minimizes the impact of thawing settlement on the environment.

Grouting against thawing settlement is carried out through the embedded grouting hole, and the method of clean water washing at normal temperature is usually adopted as an extension of the grouting pipe beyond the frozen wall. Grouting against thawing settlement is carried out in coordination with forced unfreezing, and the grouting sequence is consistent with the unfreezing sequence, that is, the floor slab of the connecting passage first, then the two sides, and finally the roof slab. The dual-liquid grout of cement and water is mainly used for grouting against thawing settlement, supplemented by single-liquid grout. The mix proportion and grouting pressure is determined according to the engineering hydrogeological conditions and monitoring results.

6.2.2.3 *Freezing construction monitoring*

The whole construction process needs to be monitored in order to timely feed back the construction information of the connecting passage, eliminate the hidden dangers, monitor the weak links of freezing, and ensure construction safety and stability after completion. The monitoring system consists of a hardware system, a software system and a database. The hardware system is shown in Fig. 6.11. The software system is mainly composed of a recording module, a control module and a display module. The Access database is adopted, which is composed of data tables and data views (queries). The monitoring contents include freezing parameters, stability of adjacent segments and connecting passages.

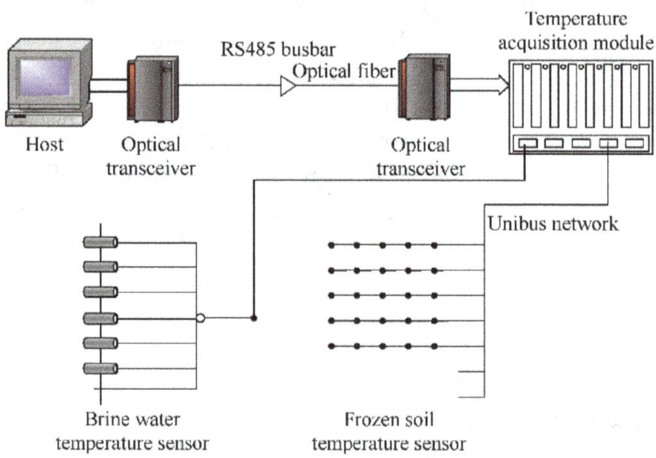

Fig. 6.11. Monitoring hardware structure.

6.2.3 New technology of mechanized construction

With technological development, mechanized construction has begun to emerge in the construction of cross passages. Compared with the freezing method + mining method, mechanized construction is characterized by good economic efficiency, a short construction period and high safety, which are embodied in Ningbo Metro Line 3 and the Tuen Mun — Chek Lap Kok Subsea Tunnel in Hong Kong.

Taking the Tuen Mun — Chek Lap Kok Subsea Tunnel in Hong Kong as an example, the entire dual-tube subsea tunnel section is about 5 km long, with 46 connecting passages of about 13 m in length each. The stratum where the cross passages are located is composed of relatively soft alluvium and a completely weathered zone, subject to water pressures of up to 5.5 bar. The geological profile is shown in Fig. 6.12.

The standard size of a sidewalk is 2.1 m in height and 2.0 m in width. Considering the allowable error of construction, the passage of a circular section is 3.0 m in inner diameter. The schematic diagram of the longitudinal section of the passageway is shown in Fig. 6.13.

As shown in Fig. 6.14, the multifunctional concrete outer frame is designed to ensure the watertight connection between the connecting

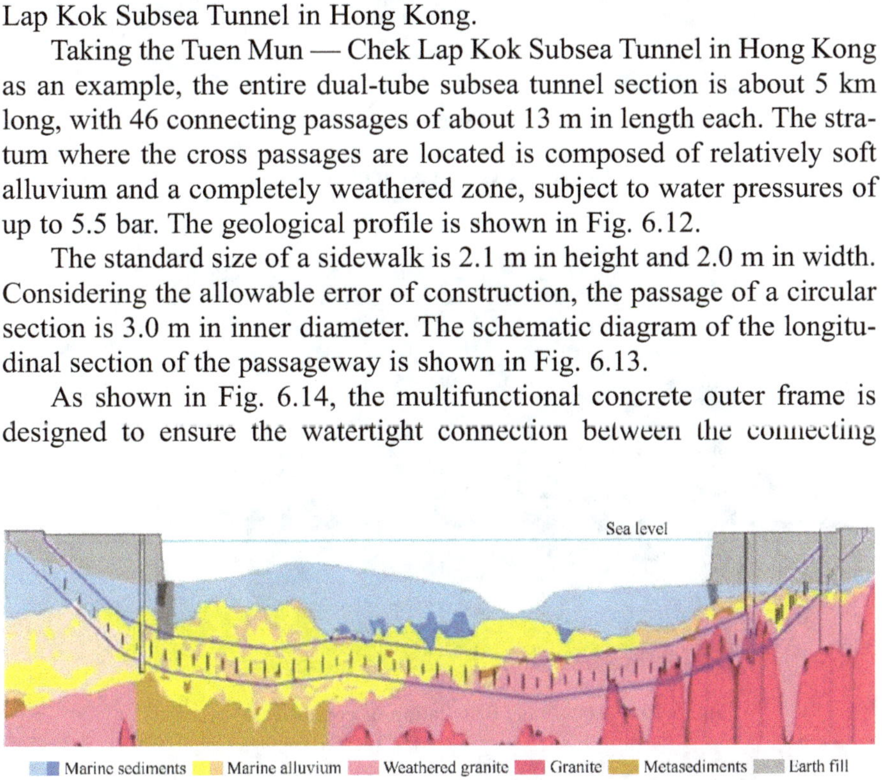

Fig. 6.12. Geological profile of Tuen Mun — Chek Lap Kok Subsea Tunnel in Hong Kong.

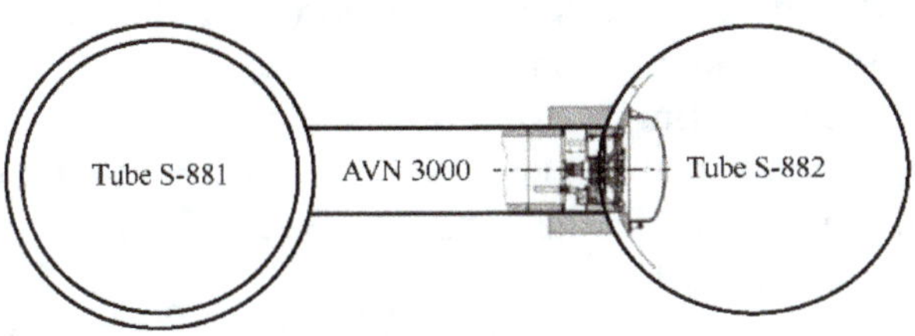

Fig. 6.13. Schematic diagram of longitudinal section of Tuen Mun — Chek Lap Kok Subsea Tunnel in Hong Kong.

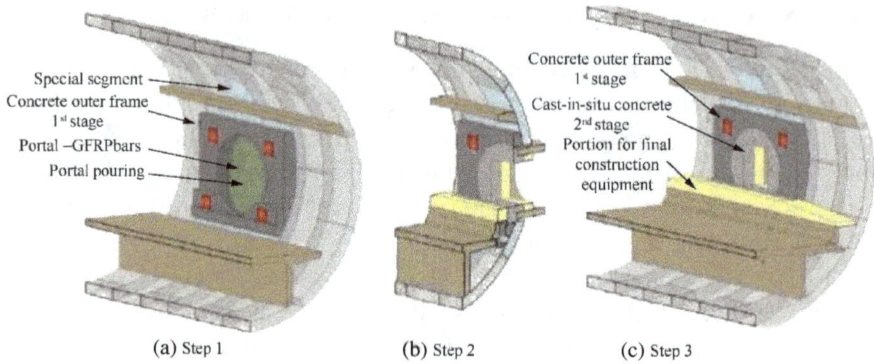

Fig. 6.14. Outer frame construction of connecting passage.

Fig. 6.15. Anchorage installation of connecting passage.

passage and the main tunnel, bear the load of the pipe jacking machine and serve as a permanent structure. The composite segments with steel bars and GFRP bars are erected at the portal of the main tunnel and painted in color to avoid erection errors.

Before the pipe jacking construction, the outer frame was built, and the pouring was conducted in the upper and lower parts. Before pouring the lower part, the metal flange used for pipe jacking was installed in place, as shown in Fig. 6.15.

The connecting passage was constructed by a pipe jacking machine with a diameter of 3 m. The tunnel portal seal was connected to the

Fig. 6.16. Launching of pipe jacking machine for connecting passage.

Fig. 6.17. Receiving of pipe jacking machine for connecting passage.

Fig. 6.18. Photo of connecting passage completed.

launching flange, and the four anchors were fixed on the outer frame by the post-tensioning method.

As shown in Fig. 6.16, the launching seal and the receiving structure were tested to check tightness before launching. The pipe jacking machine was advanced under jacking by the inserted segments and relay jacking stations. After putting the new segments into the jacking trailer, the segment brake was applied to maintain pressure.

After the pipe jacking machine arrived at the receiving device, the jacks at the relay jacking stations at both ends of the connecting passage were started to uncover the expandable seals and inject grout into these seals as temporary seals.

The outer ring void was grouted, and tests were conducted to check the pressure of the receiving device and the launching seal and verify whether pressure could be reduced and the equipment could be disassembled. As shown in Fig. 6.17, after pressure verification, the pipe jacking machine was pushed into the receiving device through the last relay jacking station and was dragged away through the open cover plate of the receiving device, and then the remaining parts were dismantled before transferring the entire set of equipment to the next connecting passage.

The connecting structure was constructed after removing the pipe jacking machine: first pour the invert slab to provide access for workers, and then install the reinforcement with embedded connectors.

For the circular tunnel section of the Tuen Mun — Chek Lap Kok Subsea Tunnel in Hong Kong, the 42 connecting passages were built by the mechanized construction method under high water pressure and soft soil geological conditions, with less than 13.5 months, that is, three passages per month, and a good construction effect. A photo of the connecting passage completed is shown in Fig. 6.18.

Chapter 7

Information Monitoring

For large-diameter shield tunneling, parameter control is involved in many aspects of the excavation process, and it is obviously impossible to meet the construction requirements only through conventional and simple machine acquisition and processing. Therefore, the development of information-based construction technology is of very important research significance for shield tunneling and also represents an important trend for the development of this technology. As the basic configuration to realize modern information-based shield machine constructions, a shield machine monitoring system can not only ensure the safety of tunnel construction and improve the quality of tunnel products but also provide an information basis for improving the level of fine tunnel construction and promoting the progress of underground engineering technology. This chapter focuses on the latest technology of big data information monitoring of large-diameter shield tunneling, mainly including data processing technology, site information management and remote intelligent management and control.

7.1 Data processing technology in shield construction

Accurate shield construction data are the basis of information analysis for shield construction. In order to generate stable, all-round, and accurate shield construction data, it is necessary to effectively solve the technical difficulties in data source processing in consideration of the characteristics of construction data, such as multi-source heterogeneity, multiple dimensions and transmission data link construction environment.

7.1.1 Shield construction data sources

7.1.1.1 *Classification of shield construction data sources*

The correct classification method for shield construction data is the basis of information standardization for shield tunneling. According to their attributes, shield data are classified into analog data or switching data. According to its role, the shield system is divided into more than 10 shield configuration systems, such as a cutterhead system, thrust system and grouting system. According to their source, shield construction data are divided into shield equipment data, shield attitude data, surface settlement monitoring data and manual recording data.

(1) **Shield equipment data:** The following shield equipment data are acquired through a shield programmable logic controller (PLC) communication interface:

 (i) current and voltage of the power supply and distribution system and the working conditions of all motors and solenoid valves;
 (ii) cutterhead speed and total torque of the cutterhead system;
 (iii) hydraulic pressure and jack stroke of each section of the thrust system;
 (iv) speeds of pumps, flow rates, density and pressure of slurry delivery and discharge, air pressure and liquid level of the cutting cabin of the slurry delivery balanced system;
 (v) grouting flow rate and pressure at each injection port of the synchronous grouting system;
 (vi) physical quantities and working conditions of other auxiliary systems.

Due to the requirements of equipment maintenance, shield machine suppliers have made the PLC communication interface and ladder logic control program available to users, which not only facilitates equipment troubleshooting by users but also provides them with a data acquisition communication interface for the acquisition of real-time shield machine information.

The analog and switching quantities in Table 7.1 represent the shield machine equipment operation information collected through the PLC communication interface. The analog data reflect the information within a certain numerical range and in physical units during

Table 7.1. Main technical parameters of large-diameter shield machine data communication interface.

Shield machine type	PLC brand	Communication protocol	Analog quantity (No.)	Switching quantity (No.)
15.53 m slurry-balanced shield machine	Siemens, Germany	PROFIBUS	709	198
14.27 m earth-balanced shield machine	Mitsubishi, Japan	CCLINK	399	220

shield machine construction, and the switching data reflect the opening and closing status of various devices or components of the shield machine.

(2) **Shield attitude data:** The shield attitude information is presented in the design tunnel axis (DTA) coordinate system, that is, the path position information of the DTA in local urban coordinates. The shield attitude information reflects the current position and attitude of the shield machine, including five main variables: cutting plane deviation, cutting elevation deviation, shield tail plane deviation, shield tail elevation deviation and shield tunneling mileage.

The "VMT" (Company abbreviation) shield construction alignment system in Britain and the "ENZAN" (The Guidance System for Japanese Enterprise Development) shield construction alignment system in Japan represent the internationally advanced level of shield attitude data acquisition technology at present. The common feature of the two technologies is that the total station is used to automatically detect the current mileage, elevation and other position parameters of the shield construction, and a two-dimensional inclinometer is used to measure the tilt angle and rotation angle. The difference between the two technologies lies in the measurement principle of shield attitude azimuth.

The British "VMT" shield construction alignment system automatically measures the shield plane angle using the laser grating method. When the grating plane inside the target rotates to be parallel to the incidence angle of the laser total station, the laser signal received by the photosensitive sensor triggers a recording of the rotation angle of the grating, and this angle record can be converted into the data of the shield plane angle. The operation interface of the "VMT" shield construction alignment system is shown in Fig. 7.1.

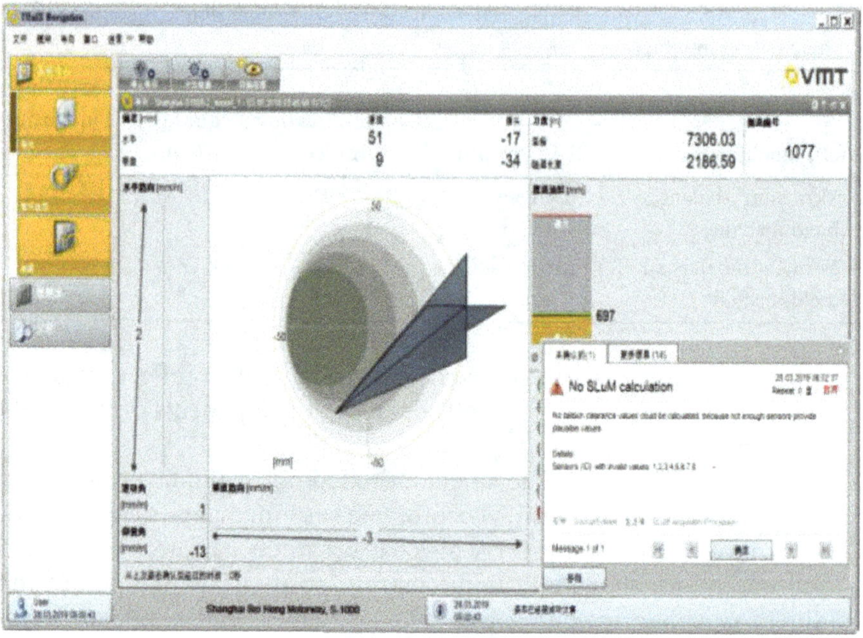

Fig. 7.1. Operation interface of "VMT" shield construction alignment system.

The Japanese "ENZAN" shield construction alignment system automatically and sequentially measures the distances and angles of the three target prisms arranged at different positions on the shield plane with a total station and generates the plane angle data of the shield by solving the geometric relationship. The operation interface of the "ENZAN" shield construction alignment system is shown in Fig. 7.2.

According to the current ring number of the shield and the measured mileage, plane angle, elevation angle, inclination angle, rotation angle and other data, the computer generates the position and attitude data of the shield through calculation and displays the real-time shield attitude information to guide the shield driver in implementing deviation corrections.

(3) **Surface settlement monitoring data:** The surface state monitoring of shield tunneling is mainly conducted by professional survey agencies. Data are collected and sorted out manually from the survey

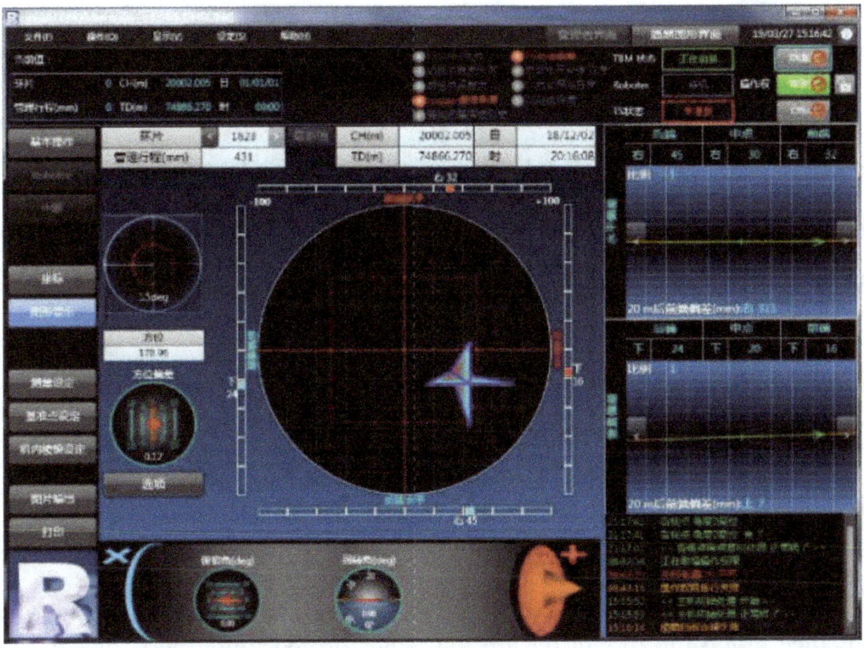

Fig. 7.2. Operation interface of "ENZAN" shield construction alignment system.

sensors and sent to relevant units in the form of electronic and hard-copy documents, covering the dominant survey information of construction environment quality, such as surface settlement and soil displacement.

(4) **Manual recording of construction data:** During the construction process, the information reviewed or measured manually shall be manually filled out in a statement form for each ring according to the operation system, which mainly records the segment location information.

7.1.1.2 *Multi-source heterogeneity of shield construction information*

The surface settlement monitoring information and the manually recorded construction information are derived from different data files with their own characteristics of information organization structure and recording triggering methods, which are different from the real-time data acquisition and recording methods of shield tunneling. The key to shield construction information management is to systematically establish the shield construction standard database for data search and analysis. Especially for information management in large-diameter shield construction and large-scale cluster construction, it is necessary to establish a reference model of data organization structure that conforms to the multi-source heterogeneity characteristics of shield construction information.

7.1.1.3 *Multiple dimensions of shield construction information*

Generally, construction projects involve fixed sites. The two-dimensional keyword (time and event) recording method can basically meet the users' demands for information query and analysis, so that the construction managers can receive timely notifications about when and what events occur on the site.

In the mobile operation of shield machine construction by continuous advance along the tunnel design axis, the formed tunnel axis or surface settlement related control parameters of the main objects of the shield construction quality control can only be queried according to the corresponding location information (such as the advance mileage or ring number), which enables a spatial analysis corresponding to the geological conditions of the location or the tunnel design alignment. Therefore, it is necessary to provide the shield construction information in at least three dimensions (time, space and events).

The real-time data come from the PLC communication interface of the shield machine control system, but the current PLC communication interfaces of most shield machines do not provide shield position (ring number or mileage) data. Poor environmental conditions at the shield construction site, failure of communication optical fiber lines, failure of the control power supply, weak functioning of the application software and other problems will cause the shield information management system of the construction enterprises to lose the reference datum of the shield position after troubleshooting and restarting. Without manually entering the current position data of the shield, the subsequently recorded shield construction data without location (ring number or mileage) is of almost no practical significance.

7.1.2 Data collection standardization

Since the shield machines supplied by different suppliers are equipped with control systems of varied configuration, the shield machine information management system of the construction manager faces the problem of adapting to the various shield machine information sources. It is necessary to formulate corresponding data acquisition coding rules which can meet the general application requirements of various non-standard shield machine information interaction and information sharing. The coding rules are as follows:

(1) The total number of coding characters is four, selected from the 26 English characters of "A–Z" and the 10 natural numbers of "0–9".
(2) The first character reflects the "system" to which the data belong in the shield machine, such as the cutterhead system, thrust system, erection system, slurry conveying system or electrical system.
(3) The second character reflects the type of sensor used for shield data detection, such as stroke, angle, oil pressure, current, voltage and density. Each type of detection sensor is designated with a specific English character code according to the rules.
(4) The third character is a spare bit, which is user-defined to increase the flexibility of rule application and is set to "0" when it is not used.
(5) The fourth character is the "serial number" of the same kind of data. For example, the characters "1–9" and "A–Z" can be used to represent the natural number sequence of the same kind of data. The data code "BA0D" of the construction data acquisition system of Shanghai Beiheng Passageway represents the jack stroke of the shield machine bottom area D of the shield thrust system.

Generally, the existing shield construction information management system can be quickly expanded for application in different shield types as long as the variables are appropriately increased or decreased or the positions are changed on the display screen, which improves the application conversion efficiency of the shield construction information management system. In addition to the data from the shield machine interfaces, data from other sources can also be coded according to this rule.

7.1.3 Data transmission security

Data transmission security mainly refers to the security between the PLC communication interface and the data switch in the ground control room at the construction site, including data transmission media, data transmission hardware and software isolation security, and preventing the impact of abnormal network information on the normal operation of PLC programs.

7.1.3.1 *Reliability measures for data transmission media*

Optical fiber is used as the transmission medium of the data switch in the ground shield monitoring room and the shield control room, and a photoelectric converter is used to realize signal conversion between optical fiber and an Ethernet router. The attenuation coefficient of optical fiber is related not only to the relay distance of optical fiber communication but also to optical fiber bending and optical fiber splicing technology.

The attenuation coefficient of optical fiber is defined as the attenuation value of optical signal power per kilometer of optical fiber (unit: dB/km), which is expressed in the following formula:

$$a = 10\log\frac{P_i}{P_o}, \tag{7.1}$$

where P_i is the input optical power value (W) and P_o is the output optical power value (W).

The optical cable reel with the Optoelectronic/Electro-optic conversion function should be used for large-diameter shield tunnels to prevent winding faults in optical cable laying over a long distance and avoid optical signal attenuation defects.

7.1.3.2 *Security isolation of PLC communication protocols in different types*

PLC manufacturers apply different communication protocols. For example, the bottom layer of a Siemens PLC in Germany is applied with the PROFINET communication protocol, and the bottom layer of a Mitsubishi PLC in Japan is applied with the CCLINK (Control & Communication Link) communication protocol. The data conversion

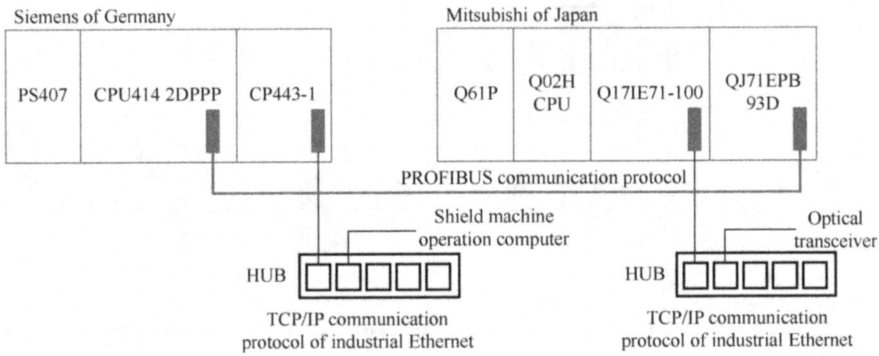

Fig. 7.3. Schematic diagram of PLC communication module hardware isolation.

between the two types of protocols is realized through a dedicated PLC communication hardware module. In general, special PLC communication hardware can only transmit heterogeneous PLC interface data without changing the logic program, so different types of PLC communication hardware modules can effectively achieve security isolation.

As shown in Fig. 7.3, the shield machine is configured with a Siemens PLC from Germany. The user's shield construction information management system is added with a set of Mitsubishi Q series PLCs from Japan before the communication interface with the Siemens PLC. The Mitsubishi PLC is configured with a PROFIBUS DP communication module compatible with the Siemens PLC. The shield construction data are obtained from the Siemens PLC system of the shield machine through the PROFIBUS communication protocol to realize the security isolation of the communication interface between the user and the shield manufacturer.

7.1.3.3 *Security isolation of unidirectional data transmission computer*

As shown in Fig. 7.4, a special unidirectional communication industrial computer is used, which has the functions of data acquisition, recording and backup, not only improving the safety of data isolation but also the reliability of data acquisition.

An RS232 communication line enables PC1 to send shield data acquisition information to PC2 only according to the conventional variable

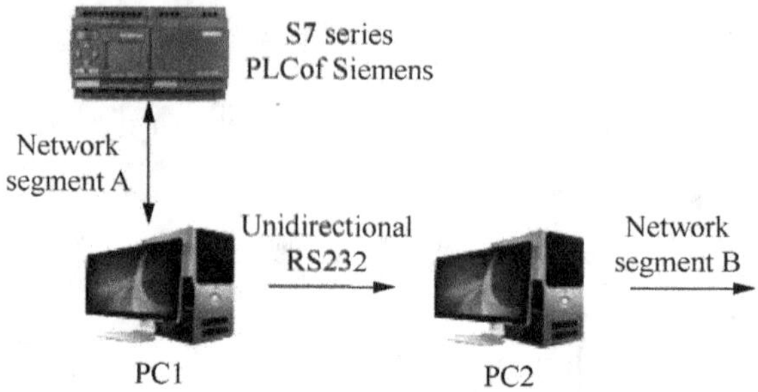

Fig. 7.4. Schematic diagram of security isolation configuration of unidirectional transmission data computer.

sequence and communication protocol, so as to achieve security isolation of the shield PLC communication interface.

7.1.4 Database design management

7.1.4.1 *Relational database structure*

The relational mathematical model describes data in the form of two-dimensional tables. In a relational database, information is stored in two-dimensional tables, and a relational database can include multiple two-dimensional tables.

In the shield construction information system, relational databases are divided into three categories: data definition table, key field table and historical data record table.

The data definition table mainly describes the engineering information of the shield machine, the information coding, engineering unit, maximum value, minimum value, data type and the shield machine subsystem.

The key field table contains time and the pointer values of historical data record tables, mainly to establish a retrieval relationship by connecting historical data record tables.

The historical data record table mainly includes time, ring number, advance mileage, net stroke and the variable data of shield machine systems. All historical data are saved in the corresponding record table. If the number of variables exceeds a certain limit, the historical data record table will be divided into multiple tables. The historical data record tables are

Local time	Ring	Mileage	Net stroke jxc	Thrust cylinder group A, position qid_1	Thrust cylinder group B, position qid_2	Thrust cylinder group C, position qid_3	Thrust cylinder group D, position qid_4	Thrust cylinder group E, position qid_5
2017-02-09 21:53:33	16	2348.288	1	761.270000	761.547000	753.269000	707.637000	701.976000
2017-02-09 21:53:43	16	2348.289	2	761.270000	762.610000	754.410000	708.640000	702.980000
2017-02-09 21:53:53	16	2348.290	3	762.380000	763.650000	755.470000	709.850000	704.000000
2017-02-09 21:54:02	16	2348.291	4	764.650000	764.840000	756.530000	710.910000	705.060000
2017-02-09 21:54:07	16	2348.292	5	765.680000	765.940000	757.550000	710.910000	706.060000
2017-02-09 21:54:17	16	2348.294	7	767.750000	768.080000	758.700000	713.110000	708.070000
2017-02-09 21:54:25	16	2348.295	8	768.840000	770.220000	760.950000	715.290000	709.270000
2017-02-09 21:54:27	16	2348.296	9	769.880000	770.220000	760.950000	715.290000	710.440000
2017-02-09 21:54:39	16	2348.298	11	770.940000	771.230000	762.050000	716.310000	711.510000
2017-02-09 21:54:53	16	2348.300	13	773.240000	773.520000	764.120000	718.400000	713.660000
2017-02-09 21:55:03	16	2348.301	14	775.400000	775.850000	766.440000	720.520000	714.820000

Fig. 7.5. Schematic diagram of two-dimensional table data recording.

connected in series by the index pointer table, and a retrieval relationship is established among them, as shown in Fig. 7.5.

The data recording characteristics of two-dimensional tables contained in relational databases are as follows. During shield construction, including advance, erection or shutdown, the shield construction data must be recorded in real time and at a frequency of 1 time/s, and a row of data must be generated that includes all variables in the two-dimensional table for each recording. The total number of variables for 15 m shield machines is about 1,000.

7.1.4.2 *Linked list database structure*

(1) **Structure design of linked list database**

(i) *Note area*: The note area summarizes the engineering parameter information, mainly including the total number of variables, project name, file start time, file end time and the number of historical data lines recorded in the file.

(ii) *Index area*: The index area includes the Chinese description of variables, information code of variables, data type, unit, display format, minimum and maximum values, change record threshold, initial value, final value and pointer to the last record.

(iii) *Data area*: To protect the privacy of data, this area is encrypted. After decryption, this area is displayed in the format of line number, time, net stroke, code of variable 1, value of variable 1, line number of last record variable 1, code of variable 2, value of variable 2, line number of last record variable 2 and so on.

(iv) *Query area*: *There are three query methods*: query by time, query by ring number and query by advance mileage. The query flow chart is shown in Fig. 7.6.

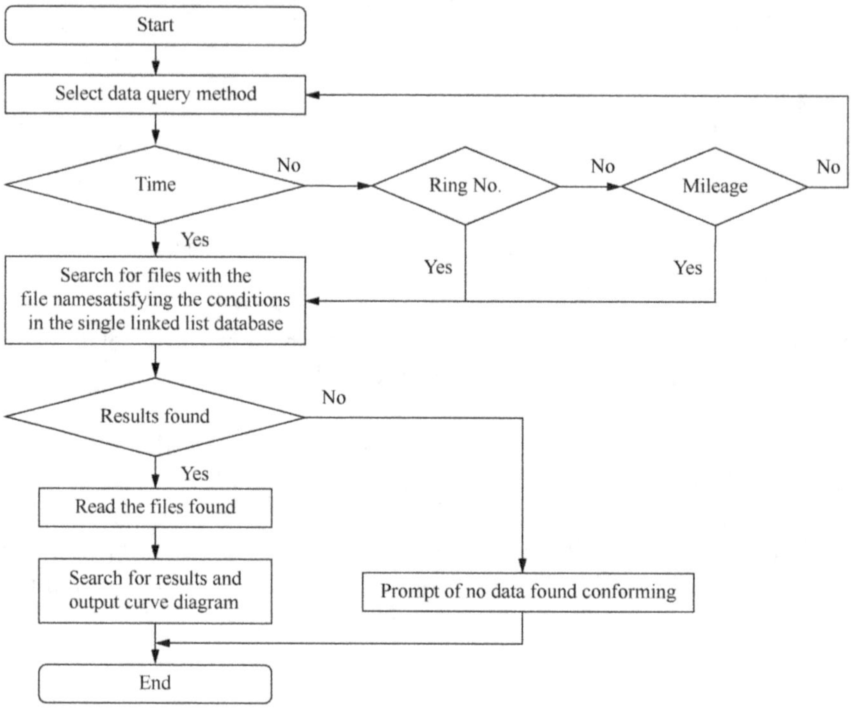

Fig. 7.6. Process of three different query methods.

In addition, due to the data encryption function against unauthorized modification, the linked list structured data have good security and confidentiality.

After opening the linked list structured data file found according to the search conditions, the data variable definition part is first parsed, and then the historical data part is decrypted. If the pointer of the retrieved variable in the data variable definition part is displayed as "00000", it means that the variable has not changed in the recording time period of the file.

If the pointer of the variable in the data variable definition part is, for instance, "*d*", instead of 0, it means that the variable was last recorded in line *d*. The logic of the historical data parsing process of the linked list structure is shown in Fig. 7.7. First, the line *d* of the historical data part is found, the variable and the following value are found, and then the line number *c* of the last record is displayed following the value of this variable. In this way, the last recorded value of the variable can be found in line *c*, and so on until the end.

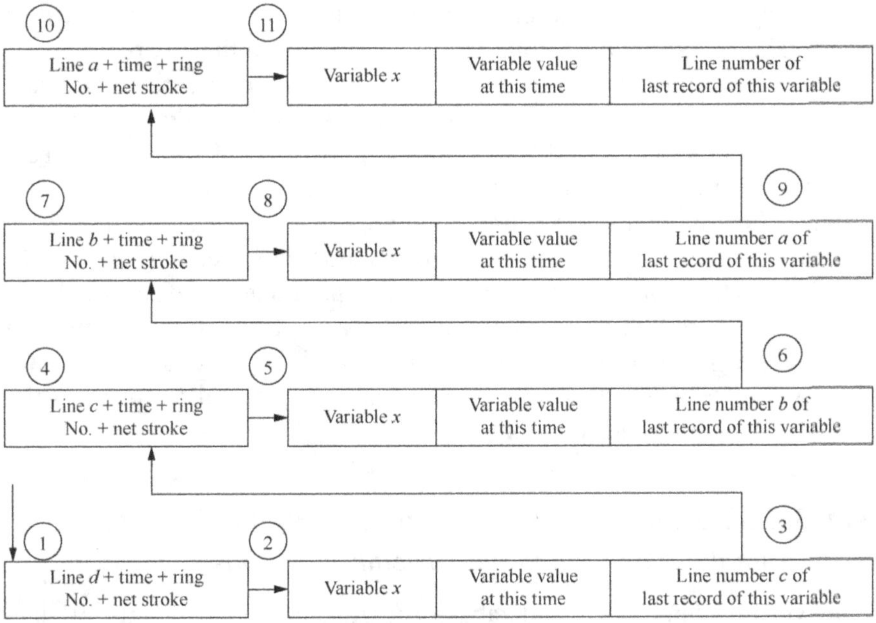

Fig. 7.7. Logic of historical data parsing process of linked list structure.

(2) **Characteristics of linked list database:** The data storage technology is developed in the linked list structure to record data through the event triggering method.

One file represents the historical data of the shield construction for a certain period of time. The file name consists of shield construction status, ring number, mileage and file generation time. For example, the file name S1088_2186_201101191941 represents the historical data of the 1088th ring under the thrust state, an advance mileage of 2,186 m, recorded at 19:41 on January 19, 2011. The whole file consists of three parts: the engineering definition part, the data variable definition part and the historical data recording part.

The engineering definition part mainly includes the total number of variables of construction information, the time of variable parameter modification, the description of engineering works, the sampling period, the start mileage, the end mileage, the start net stroke, the end net stroke, the start and end times, and the total number of rows of historical data.

The data variable definition part mainly includes the Chinese description of variables, information code of variables, data type, unit, display format, minimum and maximum values, change record threshold, initial value, final value and pointer to the last record.

The historical data recording part records the historical data during the shield construction process for retrieval, including the instantaneous value of a variable at a time, the position pointer at the current time, the mileage at the current time and the net stroke.

The linked list database only records the variables that have changed from the previous time for an amount exceeding the threshold value. All changes that involve state parameters are recorded.

Compared with the data volume of a relational two-dimensional table, the capacity of a linked list structured database is significantly lower. Due to the small data size, it is very helpful to improve the speed of query and retrieval.

7.1.4.3 *Bidirectional conversion between linked list structured database and relational structured database*

The two-dimensional table database records massive amounts of data when the shield machine is shut down for a long period of time, which is compressed into the linked table database. The linked list structured data are optimized in the data recording mechanism when the shield machine is shut down for a long period of time. The mechanism of recording any change exceeding the threshold value is adopted, and recording is suspended when there is no change or when the changes are small, so as to reduce the size of data transmitted remotely.

The linked list structured database contains engineering definition information to identify the engineering data of different shield constructions and realize the data integration of multi-source heterogeneous shield construction information. The linked list data structure can be converted into a two-dimensional table data record structure through a specific decompression method, so as to achieve integration with the existing shield construction information relational database.

7.2 Site information management in shield construction

The site information management of shield construction is necessary in the configuration of modern shield machines, and the primary task of information management is to build an appropriate database. The shield machines produce mass data during construction, which are different along with the types of shield machines. Many factors need to be considered in database design. Each kind of database structure has its own advantages and disadvantages, and factors in all aspects should be considered in the design.

7.2.1 Shield construction site information

In order to maintain the integrity of shield data information, detailed data information shall be provided as much as possible. In addition to the data recorded during shield construction, relevant information about the shield machine shall also be included, such as the necessary data information for the user to operate and use the system and the basic parameter information of the shield machine.

7.2.1.1 *Basic parameter information*

Basic parameter information consists of data definition, shield information and system code.

(1) **Data definition:** Data definition mainly includes the basic physical meaning, maximum value, minimum value, unit, minimum display precision, table name and field name in the historical database, PLC address and other contents used to record variables in the system.
(2) **Shield information:** As the shield machine information is fixed, it is stored in a file format similar to ".ini" in Windows, mainly including the mechanical and physical parameters used to record the shield machine, such as the diameter and length of the shield machine, the number of thrust jacks and the position of articulation jacks.
(3) **System code:** The system code includes the shield machine subsystems and their codes, such as system name, switch quantity and analog quantity in the system.

7.2.1.2 *Construction data*

According to their characteristics, the shield construction data are classified into analog quantity and switching quantity for differential treatment. Taking Shanghai Qingcaosha Tunnel as an example, it has 793 pieces of

Table 7.2. Categories of data number (unit: piece).

System name	Switching quantity	Analog quantity	Total
Cutterhead system	20	120	140
Thrust system	10	42	52
Erection system	12	6	18
Frame traveling system	23	2	25
Shield tail system	115	124	239
Grouting system	13	27	40
Lubrication system	2	21	23
Slurry conveying system	84	81	165
Slurry treatment system	4	12	16
Water system	2	2	4
Auxiliary system	0	1	1
Attitude system	0	17	17
Electrical system	11	3	14
Balance system	20	15	35
Others	0	4	4
Total	316	477	793

construction data (Table 7.2), mainly including data groups and relevant indexing information during shield construction, which are readable and writable during system operation.

7.2.2 Considerations for database design

When building a shield construction site database, there are disadvantages to storing all construction data in one worksheet of an Excel file, as follows:

(1) **High data redundancy and low query efficiency:** A data collector works to take samples per second, no matter whether the data changes or not. In fact, only a few of the hundreds of construction data change every second, and sometimes there is even no change at all. This recording method not only increases the size of data stored in the database but also results in high data redundancy, reducing

query efficiency since a backward search operation is required for querying data.

(2) **Unstable and non-universal table structure:** The construction data items generated by different types of shield machines are different. To collect and store the construction data of different types of shield machines, the database structure needs to be modified, and some data items need to be added or deleted. This will seriously affect data security and bring many unfavorable factors to the development of information management software.

(3) **Lack of a large-capacity data storage structure for sampling per second:** In the mode of sampling per second, the shield data collector contains mass data, so large-capacity data storage must be considered to support high sampling rates. Engineering information, data definition, system code and other basic information are critical to data analysis and must be stored in the database system together with construction data for query.

In view of the above considerations and in combination with the specific shield construction data, the factors to be considered in designing the shield construction database include:

(i) *Quick query*: The shield construction database is operated mostly for data query, so the query speed shall be mainly considered in database design. There are many factors that affect the quick query performance of the database, among which the database structure is an important aspect. A complete set of shield construction data consists of hundreds of data items, covering integer, factual, Boolean and other data types. In order to truly reflect the shield machine status during construction as much as possible, data shall be recorded at a frequency high enough. In the extreme case, the database memory can reach 40G, so it is necessary to divide the data modules reasonably and classify the data scientifically to avoid query performance being affected due to the data growing too fast in the database system.

(ii) *Structural stability*: The main purpose of establishing a database is to realize data integration and data sharing, as well as plan and coordinate information resources throughout various relevant application fields through data integration, so as to maximize data sharing. Changes in the database structure will cause a series of reactions, which will make it necessary to modify all application systems based on the database. Therefore, when designing the shield construction database structure, the types of data collectors, management software requirements, etc., should be considered to make the database structure relatively stable.

(iii) *Data redundancy*: Duplication of data in the database is called data redundancy. In database design, for the sake of operational business and technology, certain data redundancy can be retained to improve query speed, instead of eliminating all data redundancy. However, data redundancy caused by an unreasonable design of the database table structure should be avoided as much as possible.

(4) **Easy maintenance:** The database is used to store the shield tunneling parameters and status during the construction process, so the original data should be protected against addition, modification or deletion to ensure the integrity, authenticity and objectivity of the shield tunneling data records for further analysis and study by professional technicians afterward.

7.2.3 Access database

7.2.3.1 *Data table structure*

(1) **Basic data table:** The basic data table consists of three parts: the system code table, the data definition table and the shield information file. See Tables 7.3–7.5 for specific formats. The basic data table is generated in one shot during the system definition process, is only readable and cannot be altered during system operation.

Table 7.3. Table of system codes.

Field name	Data type	Data length	Keyword	Note
Id_system	Char	1	Y	System code
Name_system	Char	20	N	System name
Num_D	Int	4	N	Switching quantity in the system
Num_A	Int	4	N	Analog quantity in the system

Table 7.4. Table of data definitions.

Field name	Data type	Data length	Keyword	Note
Sno	Int	4	N	Serial number
Name_field	Char	5	Y	Field name (i.e. information code)
Description	Char	30	N	Chinese name
Type_var	Char	8	N	Type of variable
Min_var	Float	4	N	Minimum
Max_var	Float	4	N	Maximum
Units	Char	8	N	Unit
Precision	Float	4	N	Display accuracy (10,1, 0.1, 0.01,...)
Id_system	Char	1	N	System code
Table	Char	6	N	Data table
Source	Char	20	N	Data source (such as PLC address)
Query	Bool	1	N	Query ID
Id_used	Bool	1	N	Use ID

Table 7.5. Shield machine information file.

Outer diameter = 14.88 m
Total length = 13.08 m
Whether articulation shield = *N* (not)
Number of thrust jacks = 22
Position of No. 1 jack = 0
Numbers of jacks in zone 1 = 21, 22, 1, 2, 3
...

(2) **Data record sheet:** The data record table is actually a table set, which is composed of several single sheets. Each sheet has a pointer field and several real-time data record fields. The pointer field is used to associate with the pointer record table. When the on-site real-time data change, a record will be added to the relevant single sheet, and the current time serial number will be used as the pointer field data. See Table 7.6 for the specific data recording method. Each sheet has a time field, and the index pointer in the index pointer table is associated with this field.

Table 7.6. Index pointer table.

Field name	Data type	Data length	Keyword	Note
RinaNo	Int	4	N	Ring number
Time	Date/time	8	Y	Time
Distance	Float	4	N	Advance distance
PA1	Date/time	8	N	Index point of Table I
PA2	Date/time	8	N	Index point of Table II
PA3	Date/time	8	N	Index point of Table III
……				

Table 7.7. Structure of historical data record table A1.

Field name	Data type	Data length	Keyword	Note
Time	Date/Time	8	Y	Time
AAA01	Bool	1	N	Left turning of cutterhead
AAA02	Bool	1	N	Right turning of cutterhead
ADA01	Bool	1	N	No.1 cutterhead motor running
ADA02	Bool	1	N	No.2 cutterhead motor running
….				

According to the characteristics of different construction data generated by different shield machines, several historical data record tables are set, that is, tables with the same structure as the historical data record table A1 in Table 7.7. The number of data tables may vary depending on the type of shield machine. For example, there are no slurry conveying and treatment systems in the EPB shield machine, so the corresponding database is unnecessary.

(3) **Index pointer table:** The real-time data record is divided into several tables, and an index pointer table is established according to the characteristics of different tables so that all tables can be associated. PA1, PA2, PA3,… in the index pointer table are pointers to each historical data record table, and the specific number is consistent with the number of historical data record tables. The index pointer table is mainly used to record the index pointer in the data record table, including the main fields of serial number, time, ring number, advance distance,

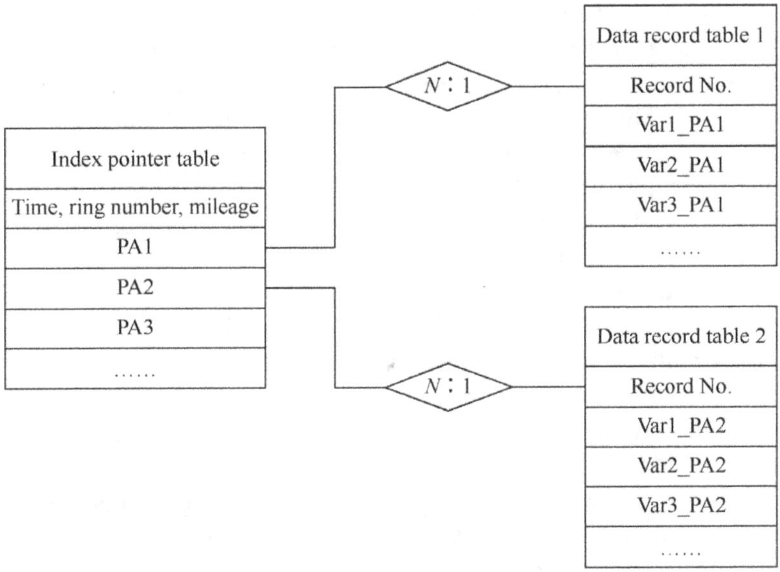

Fig. 7.8. Relationship between index pointer table and data record table.

mileage and several pointer fields. Through the corresponding association between the pointer field and each data record table variable, the relevant historical data can be easily queried, and the corresponding relationship is shown in Fig. 7.8. For example, index pointer PA1 corresponds to variable Var1_ PA1, Var2_ PA2, Var3_ PA3, etc., in data record Table 1, and for querying the variable Var2_ PA2, Var3_ PA3 in a specific range, the first step is to find the corresponding pointer PA1 on the index pointer table, and the second step is to find the Var2_ PA2 and Var3_ PA3 values in the data record table corresponding to PA1 that meets the query conditions.

7.2.3.2 *Recording method*

According to the overall system requirements, the basic operating parameters of the database system are defined as follows:

(1) **Time resolution accuracy of the system:** Record data once every 1s.
(2) **System data accuracy:** For analog quantities, the change threshold is 0.5%, that is, any change in the collected data greater than 0.5% is

considered a data change. For switching values, any data change from 0 to 1 or from 1 to 0 is considered a data change.

(3) **System data recording strategy:** Any data change that occurs is recorded. The analog quantity reflecting any thrust status parameter is only recorded during the thrust process; the analog quantity reflecting any segment erection status is only recorded during the erection process; and others are recorded during the whole process; all switching quantities are recorded.

(4) **System data recording process:** The system contains variables (Var_PA1, Var_PA2,...) of the same type as the index pointer in the index pointer table. When a datum changes in the system, a record will be automatically added to the corresponding data record table, and the corresponding variable Var_PA in the table changes to the time value recorded in the data table.

When the system is running, the index pointer table is added at one record per second, in which the time, ring number and tunneling distance are recorded according to the actual value of the system, and the PA1, PA2, PA3,... pointers are recorded according to Var_PA1, Var_PA2, Var_PA3,....

(5) After the data collector is started, the database system records data according to the process as shown in Fig. 7.9.

7.2.3.3 *Advantages of database*

(1) **Optimize database structure:** Most data in the system do not change. The method based on saving storage space is to add a data record when a data change occurs, but this recording method will bring difficulty to the query operation. In database design, the basic principle of generating data records is adopted to record the relevant generated data when a state change occurs. Moreover, a record pointer table is introduced, which is associated with the corresponding data table through the pointer field. The uncertainty in data queries and backward search operations can be completely avoided.

(2) **Avoid data inconsistency:** In order to improve the system speed and reduce the number of complex operations, this database retains a small amount of data redundancy. The data record operation

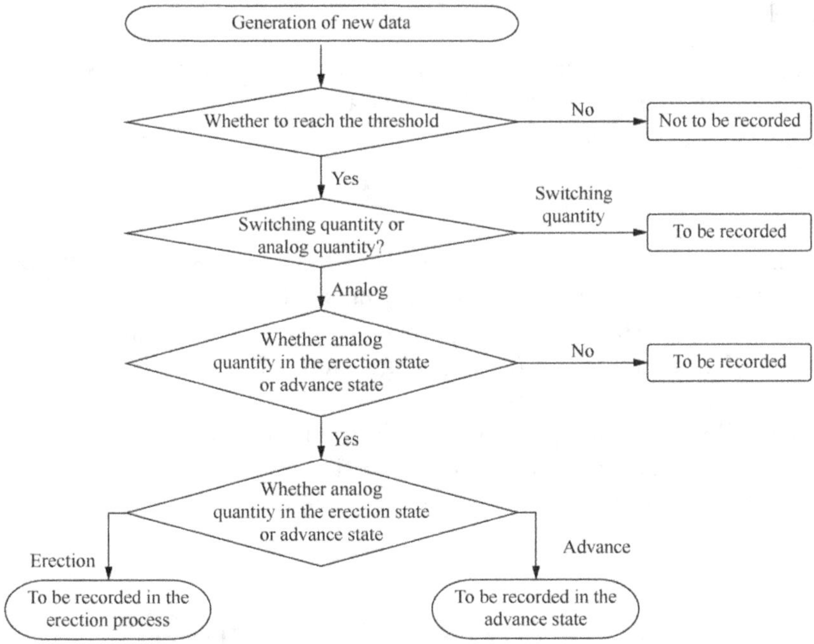

Fig. 7.9. Data recording process.

only performs two basic operations: addition and retrieval, without modification or deletion, to fundamentally avoid the problem of data inconsistency caused by data redundancy.

7.2.4 Text database based on linked list structure

The users' queries are generally concentrated in the range from around the first to tenth rings. Considering the mass data of shield construction and the characteristics of multiple engineering applications, saving data in the form of files is more convenient for the management of construction data. Based on this idea, a text database is designed. When the data recording is triggered, only the change value is recorded to realize the real data change record.

7.2.4.1 *File name definition*

In order to improve query efficiency, the original data file is divided into multiple text files according to the characteristics of the recorded data. The data file is named as follows:

Ring number_ Mileage_ Time.txt

The ring number is composed of one letter (S, T, or A) and four characters (0–9) (integer number). S represents the erection status, T represents the thrust status, and A represents the shield attitude. Mileage consists of four characters (0–9) (integer numbers), and time consists of 12 characters (0–9) (integers).

The file name designed using this method consists of the overall information of the data contained in the file. Reading the file name can realize quick retrieval of files for data queries, which greatly improves query efficiency.

7.2.4.2 *Reading method of linked list structured data*

Text databases are also triggered for recording by data changes, but only for recording the changed data. The number of columns in each row in the record section is not necessarily the same. A data record is saved as a group variable. There are three columns of data in each group, meaning variable code, variable value and variable pointer. Each variable recorded is represented by a code. The value changed is a variable value. The next position of the variable is saved in the variable pointer.

When reading the text database, the required information code is found according to the data index part at the beginning of the file, and the initial value, final value and pointer of the final value of the variable are read. The final value is the value of the variable when the recording is stopped, and the final value pointer is the data row of the previous value of the variable; the variable code of the variable is found in the data row of the previous value. The variable code is followed by the variable value and the variable pointer. The variable pointer is the data row of the previous value of the variable. These steps are repeated up to the initial value, as shown in Fig. 7.10.

If the end condition has not been reached at the end of a search in one file, the search is continued to the next data file until all the files have

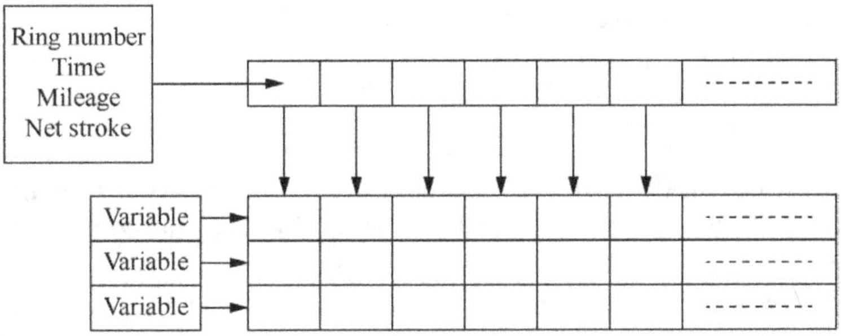

Fig. 7.10. Text database reading mode.

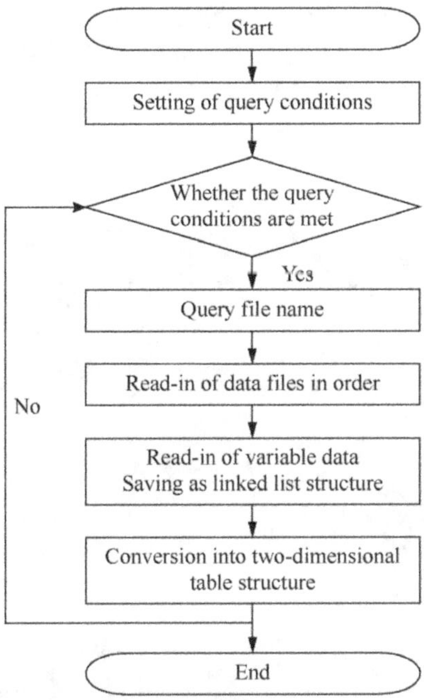

Fig. 7.11. Data query process of text database data.

been searched. The basic idea of retrieval and conversion of linked list structured data files is to retrieve a linked list structured data file according to conditions and convert it into a temporary file of two-dimensional data structure for processing and application of various data. The whole query process is shown in Fig. 7.11.

7.2.5 Site information management software of shield construction

7.2.5.1 *Basic process*

The real-time data for the shield construction information management software are obtained from the shield PLC through the communication link and interface. Its basic idea is to use the database as a data support and convert the data into corresponding curves for display so that users can quickly make reasonable, scientific judgments on the shield construction, display the corresponding data in graphs of variables according to the construction habits, print the ring report of the variable curve or generate the required data files, and display various data states in real time, so as to achieve real-time management of the shield machine. The main interfaces of the system are shown in Fig. 7.12.

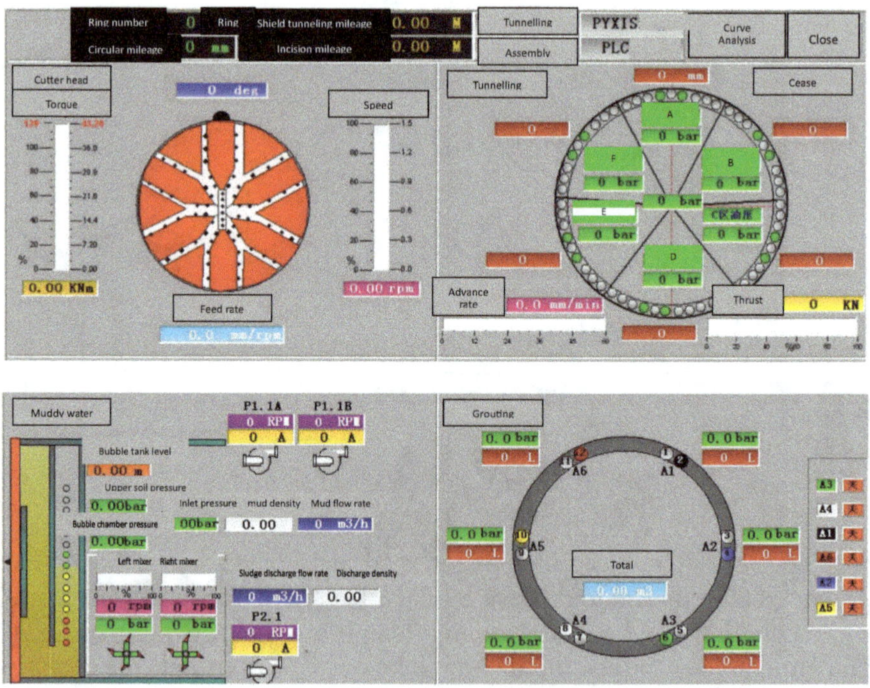

Fig. 7.12. System main interfaces.

7.2.5.2 *Main software modules*

(1) **Real-time monitoring:** The construction parameters are transmitted back to the system when the shield machine is advancing, and the staff monitors the real-time operation of the shield machine according to these parameters. The real-time monitoring module displays these parameters separately for different equipment systems. The specific display items depend on the data acquisition PLC form to visually reflect the change in this characteristic variable and help the staff make a more accurate judgment on the operation of the shield machine, so as to achieve the purpose of effectively monitoring its operation. The real-time monitoring interface is shown in Fig. 7.13.

(2) **Data analysis:** A flexible sampling method is adopted for curve plotting. When there are sufficient data, the data involved in interaction with the data layer are kept within an appropriate volume, which is conducive to accurate and rapid curve plotting and display and can

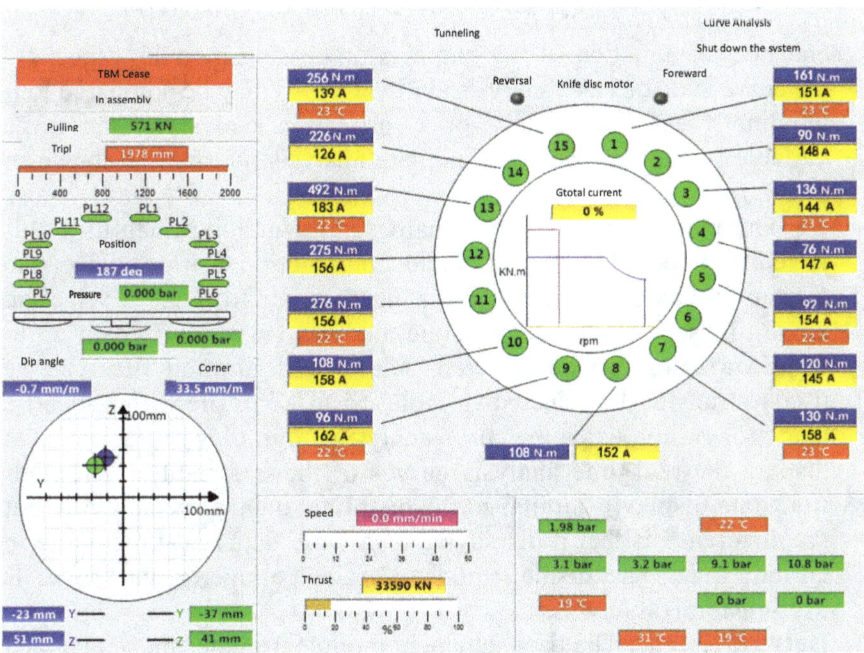

Fig. 7.13. Real-time monitoring interface.

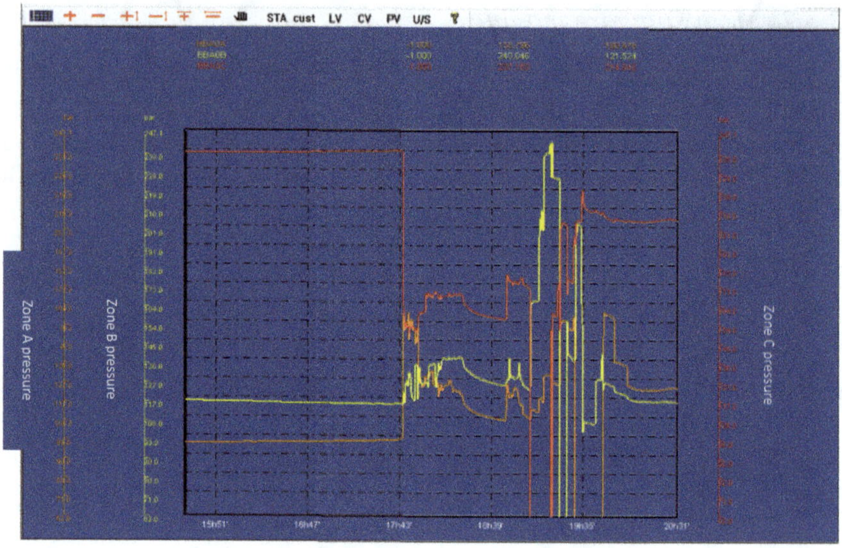

Fig. 7.14. Data analysis interface.

enhance the function of the graphics processing module, especially the curve statistics and analysis ability. Curves may be compared longitudinally and horizontally, or a correlation analysis of the curve variables can be conducted. The data analysis interface is shown in Fig. 7.14.

The module provides the analog and switching quantities collected and recorded during shield construction to the user in the form of a database system, allowing the user to make a selection among these variables to generate analysis curves, of which up to eight variables can be selected. These configuration files can be used to transfer the selected variables to an XML file with a specific name for calling when the curve analysis program is started, so as to display the real-time analysis curves of these variables. The real-time data display is similar to the historical data curve analysis, but the difference is that it displays the variable data real-time curve in current time. The display mode of each parameter and curve is user-customizable.

(3) **Data statement:** The data statement module provides the analog and switching quantities collected and recorded during the shield construction to the user in the form of a database system and generates

statements according to some parameter information of the selected ring for preview and printout by the staff. The ring statement is used to plot the change in variable value in each ring, with the shield machine advancing according to the variable value in the database, and simulate the jack status and various integrating results. In order to clearly know the attitude information, the ring statement also includes the graphic and numerical displays of the cutting plane deviation, shield tail plane deviation, cutting elevation deviation, shield tail elevation deviation and shield advance. The generated ring statement can be viewed on the screen or output with a printer. Users can select up to 10 variables for display. The print format of the ring statement is shown in Fig. 7.15.

The data module for the ring statement can also be set with a data range and density according to the file parameters, and the data can be exported in the specified attribute column to the specified Excel file. In order to realize centralized management of shield construction data, remote data export is also available.

The data generated in the shield construction process is obtained through PLC and saved in the local database. It is convenient for users to generate data curves and statements and view and analyze the shield construction status with the site information management software.

7.3 Central management and control of shield machine cluster

Shield tunneling is usually applied in the construction of large or even super-large tunnels, featuring a relatively complex construction process and numerous shield construction information.

Without effective information communication and data interaction, rapid and effective management cannot be achieved. The pyramid management structure requires top-down passing of decision-making instructions and bottom-top passing of work reports level by level. In the process of transmission, information may shrink, be distorted or even be lost, and this also delays information transmission.

"Internet + big data" has been developed on the basis of integrating distributed computing, virtualization technology, network computing and web services to provide users with access to resources on the internet anytime, anywhere and on demand via the internet without time and space

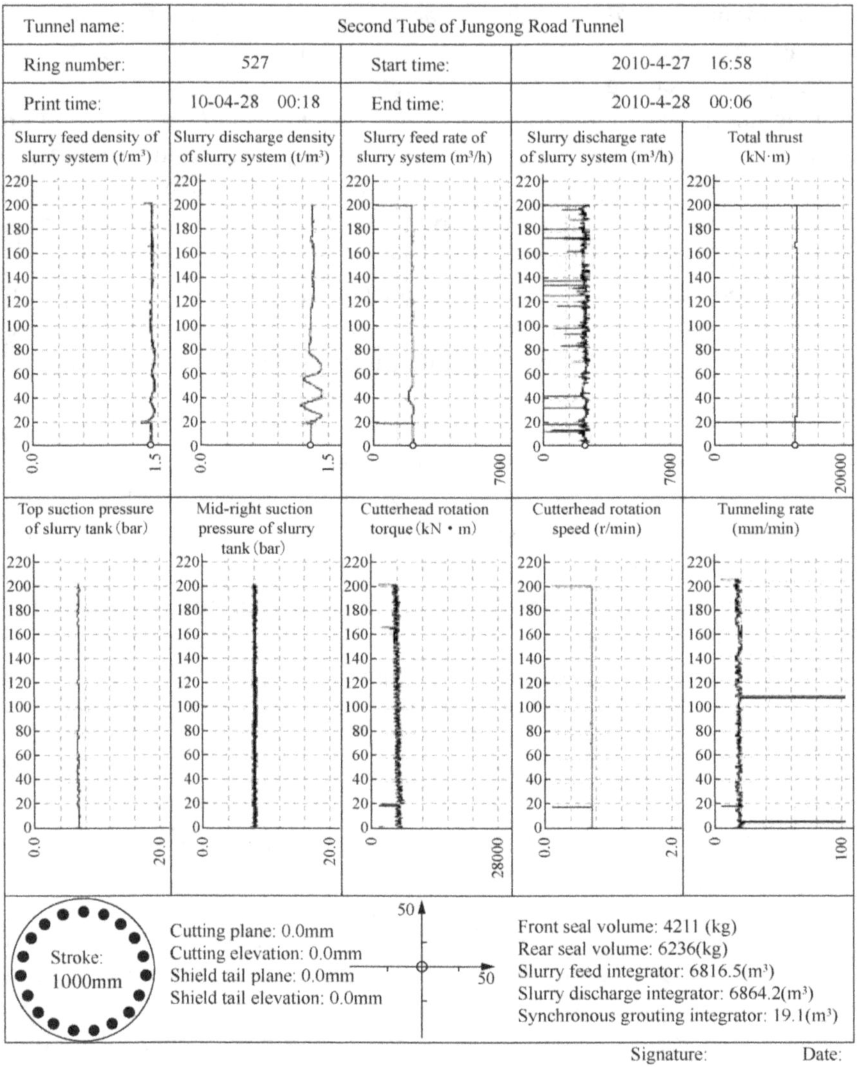

Fig. 7.15. Print format of ring statement.

constraints so as to deal with massive data and calculations. The application of "Internet + big data" to shield construction can break the barriers that solid structures cannot overcome and enable users to apply data resources in a better way, representing the development trend of information-based management for shield construction.

Fig. 7.16. Centralized management and control.

7.3.1 Management and control architecture

Drawing on the practical experience of the "Internet+" model in other industries of the national economy and combining the current development of the shield tunnel industry, the central management and control of the shield machine cluster have transformed the conventional single and scattered supervision mode into systematic and intelligent centralized supervision through the "Internet + big data" technology, providing a decision-making basis for shield tunneling.

Based on the concept of base management, the centralized display of shield tunneling enables managers to more systematically and visually know the current progress of the engineering construction, achieve centralized monitoring and strengthen the dynamic process management of the engineering construction, as shown in Fig. 7.16. The central management and control of shield machine cluster construction can realize remote management and dynamic analysis of a shield machine group, real-time risk warning and timely intelligent decision-making to improve enterprise management level and efficiency and reduce production costs in the central management and control system of shield machine cluster construction.

7.3.1.1 *Data type*

Tunnel engineering data are divided into six categories: Basic engineering data, engineering management data, tunnel construction data, tunnel quality data, risk data, and equipment management data.

(1) **Basic engineering data:** Project overview data, the data of surrounding buildings, design axis data, ground characteristics data,

monitoring point data, tunnel section data and shield machine equipment data.

(2) **Engineering management data:** Management document data, collaboration process data, mail data, log data and management trace data.
(3) **Tunnel construction data:** Shield machine attitude data, thrust data, cutterhead data, mucking data, synchronous grouting data, grouting supplement data and shield machine slurry treatment data.
(4) **Tunnel quality data:** Tunnel axis data, segment attitude data, segment deformation data and segment disease data.
(5) **Risk data:** Tunnel section risk source data, axis surface settlement data, building settlement data, patrol inspection data and video monitoring data.
(6) **Equipment management data:** Equipment damage data, equipment repair data and equipment maintenance data.

There are processing difficulties in four major aspects of tunnel engineering data, as listed in Table 7.8.

Table 7.8. Processing difficulties and requirements of tunnel engineering data.

S. No.	Data processing difficulty	Problem arising	Analysis of processing requirements
1	Date of varied sources	Saving format	Data governance is required after data acquisition
2	Massive data	Contradiction between slow processing in ordinary saving architecture and high processing rate required in tunnel engineering management and control	Quick data processing is required in the system architecture
3	High data sampling frequency	Increased difficulties raising probability of data node failures	High redundancy and capacity expansion of memory system
4	Complex and diversified data structures	Saving and analysis difficulties for data in different structures	Processing platform to be designed for data in different structures

7.3.1.2 *Overall architecture*

The data generated during the whole tunneling process are acquired by the data acquisition equipment at the tunnel construction site and manual records uploaded by the construction personnel. The topology of the central management and control system is shown in Fig. 7.17. The shield

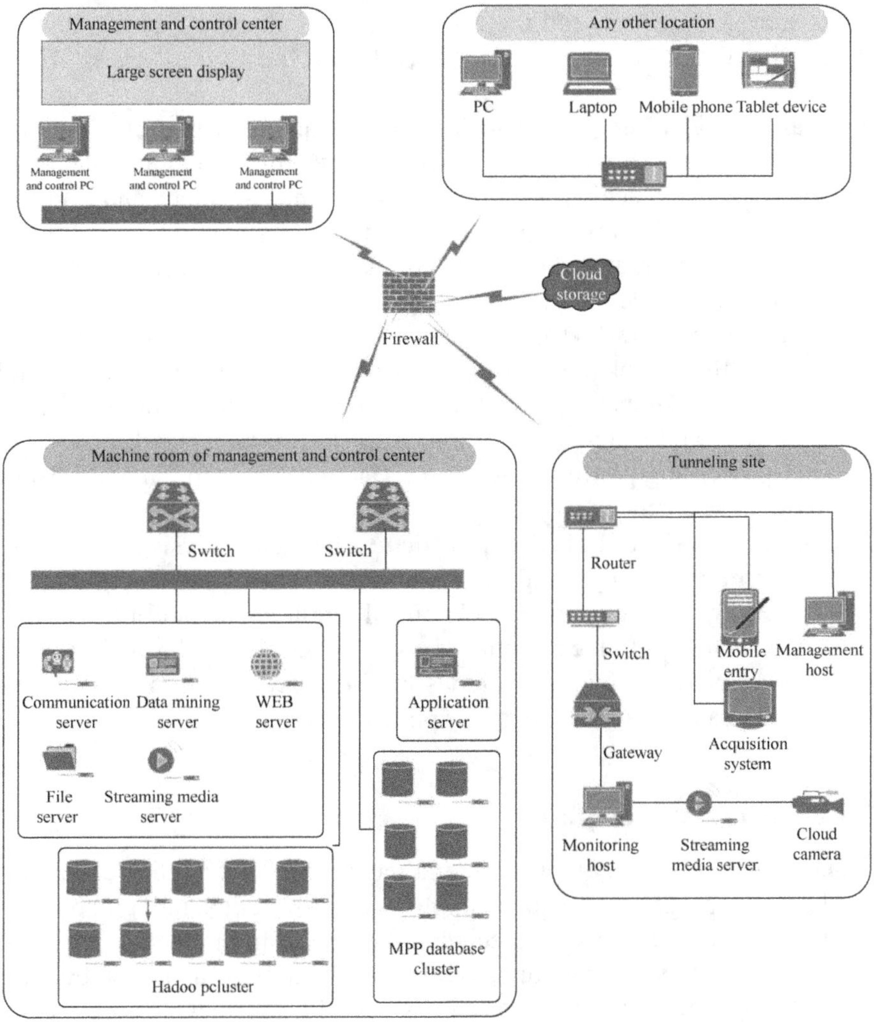

Fig. 7.17. Topology of central management and control system.

machine acquisition system, monitoring equipment, camera monitoring equipment and other automatic data acquisition equipment at the construction site record the tunnel construction-related data in real time and upload them automatically.

According to the data flow direction, the system is divided into two main parts: background data processing and foreground data display. The background data processing part includes a data acquisition platform and a data storage platform, and the foreground data display part is a tunnel engineering data analysis and display platform.

(1) **Data acquisition platform:** The data acquisition platform mainly serves as an access point for all the collected tunnel engineering data to reach the machine room server at the management and control center via various types of interfaces on the data source layer. It preliminarily integrates the tunnel engineering data in preparation for subsequent processing and storage.
(2) **Data storage platform:** The data storage platform is mainly used to process the tunnel engineering data and then effectively store it in the appropriate database system for a long period of time. It preliminarily processes and analyzes the stored data. The core of the data storage and processing platform includes a Hadoop platform, an MPP database and a memory database.
(3) **Data analysis and display platform:** The data analysis and display platform is mainly used to analyze the tunnel engineering data according to management and control requirements and present the analysis results to the system users in a visual form. The analysis and display platform mainly includes the logic layer, presentation layer, big data analysis and a unified application portal.

7.3.2 Visual data display

Visualization is a theory, approach and technology that applies computer graphics and image processing technology to convert data into graphics or images, display them on the screen and then perform interactive processing. It can visually convey key aspects and features so as to achieve in-depth insights into complex data sets. Its basic idea is to express each data item in the database as a single image and constitute a data image with a large number of data sets. Moreover, each attribute value of data is expressed in the form of multi-dimensional data to achieve data observation and in-depth analysis in different dimensions.

Through the visual display of various problems, monitoring curve analysis, periodic statement generation and other data visualization means, the management work is made scientific, standardized and institutionalized, which is conducive to a more efficient, comprehensive and systematic improvement of the level of safety risk monitoring and management during engineering construction.

7.3.2.1 *Project overview*

In order to enable centralized control and management and facilitate personnel intuitively and clearly know the progress of engineering construction in all aspects at the management and control center, the project overview module displays the progress of all aspects, construction indicator alarms and alarm processing progress in the form of graphics.

The personnel at the management and control center can also quickly view the engineering construction in specific details and visually display the statistical results of the normal section, the focus section, the warning section, the project completion rate, the number of shield machines along the whole line and the number of rings in daily advance.

7.3.2.2 *Real-time monitoring*

The real-time monitoring module is used to display the parameters of the shield tunneling on a large screen at the management and control center in real time so that the management and control personnel can visually learn the current construction status and real-time operation parameters of the shield machines, as shown in Fig. 7.18. These real-time parameters can help the management and control personnel analyze the construction status of the shield machines so as to make better management and control decisions.

7.3.2.3 *Attention*

The attention module mainly displays Levels I and II risk sources and summarizes and describes the project risks so that management and control personnel can quickly learn the risks of all items under attention.

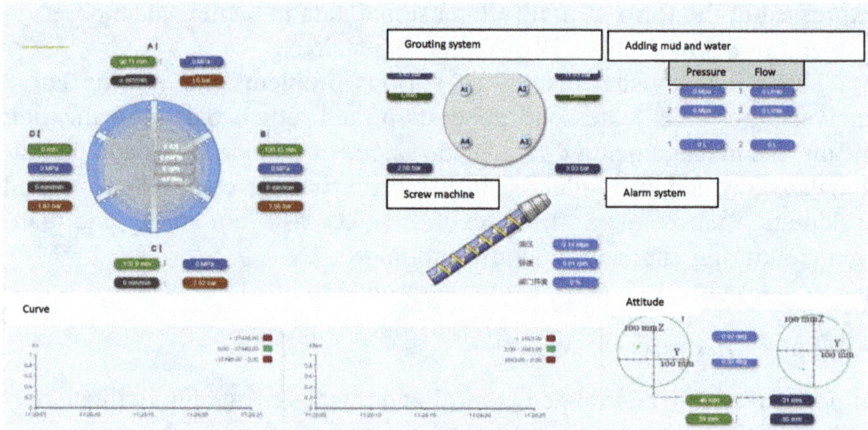

Fig. 7.18. Real-time monitoring interface.

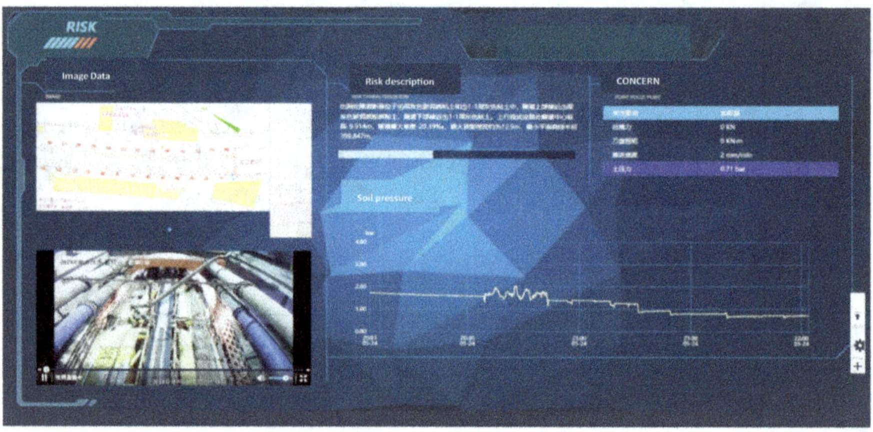

Fig. 7.19. Risk interface.

For the attention items, the specific risks are presented, including the basic risk information, the construction process, the key construction parameter values and data curves, as well as the plan and profile of the risk point and the surrounding environment map and video, as shown in Fig. 7.19.

7.3.2.4 *BIM (Building Information Modeling) model*

The BIM model allows the management and control center personnel to visually learn the internal structure of the tunnel, the alignment of the

Fig. 7.20. BIM model interface.

tunnel axis and the surrounding environment of the tunnel. This module can display the corresponding construction data of different parts (tunnels, buildings, monitoring points, etc.) in the model, the building settlement data during shield tunneling and the corresponding ring number, thrust parameters, axis parameters, settlement monitoring and other information. For a single monitoring point, the module can display a single change polyline and a cumulative change polyline. The BIM model interface is shown in Fig. 7.20.

7.3.2.5 *Tunneling quality*

The tunneling quality module is used for tunnel axis control, segment disease and segment ovality. The axis control displays the tunnel axis points in rings, in which a blue dashed box indicates that the axis is within ±50 mm, a red dashed box indicates that the axis is within ±100 mm, and the axis points display information such as ring number and axis deviation. Nearly 10 rings or the latest ring data are available for view. The segment disease type shows the number of segment diseases (including fragmentation, leakage, crack and faulting of ends) in the tunneling section and is updated monthly. The segment ovality shows the ovality of nearly 20 rings of tunnel segments in the tunneling section; when the axis deviation and ovality exceed the preset warning values, the alarm details will be displayed. The tunneling quality interface is shown in Fig. 7.21.

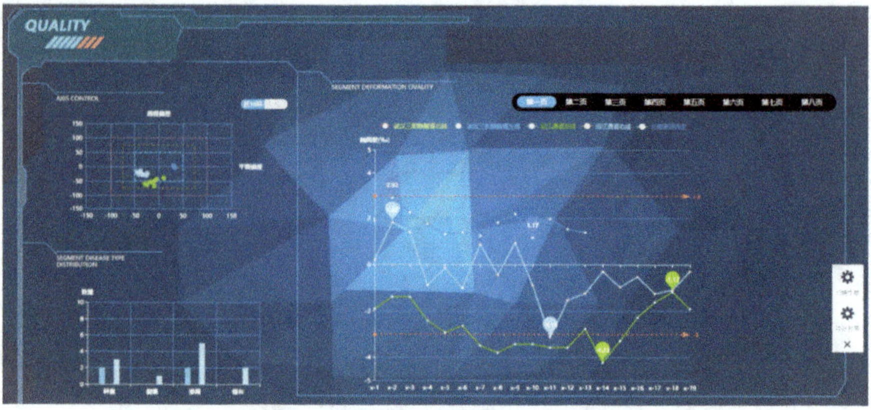

Fig. 7.21. Tunneling quality interface.

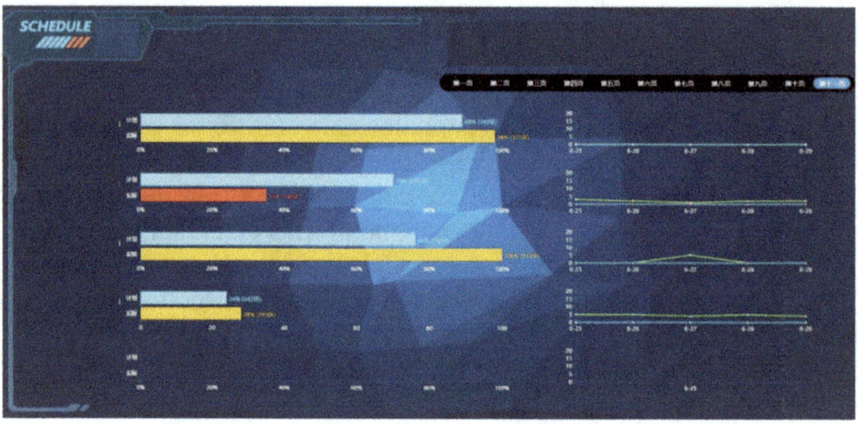

Fig. 7.22. Progress analysis interface.

7.3.3 Specialized data analysis

7.3.3.1 *Progress analysis*

The schedule and actual progress histograms are used to display the daily tunneling schedule, the cumulative number of rings actually advanced and their progress percentage; when the actual cumulative number of rings falls behind the schedule, the actual histogram becomes darker and redder gradually. The progress polyline shows the planned number of rings, actual number of rings and remarks of the corresponding tunneling section in the last five days. The progress analysis interface is shown in Fig. 7.22.

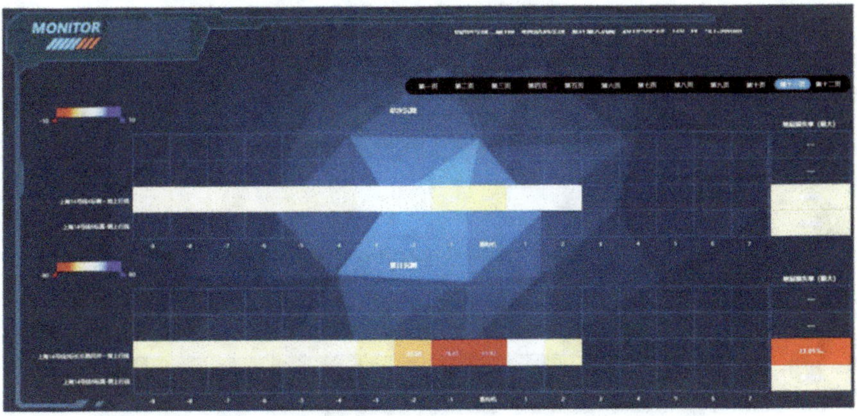

Fig. 7.23. Settlement analysis interface.

7.3.3.2 *Settlement analysis*

The settlement monitoring data show the single settlement value and cumulative settlement value of the shield machine at different locations for different tunnel projects. Different settlement values are distinguished by different colors, as shown in Fig. 7.23. The horizontal axis represents the distance from the shield machine location, and the vertical axis represents different tunnel projects.

7.3.3.3 *Comprehensive analysis*

For the tunnels under supervision, the management and control expert team uses the system data visualization platform to conduct comprehensive analysis of construction data based on shield construction experience and provide various professional management and control consulting reports for engineering construction, including project daily reports, project weekly reports, project monthly reports and special reports. The management and control consultation report interface is shown in Fig. 7.24.

7.3.4 Automatic data alarms

7.3.4.1 *Real-time data alarm*

There are two kinds of data collected from the construction site through the PLC: analog quantity and switching quantity. The switching quantity

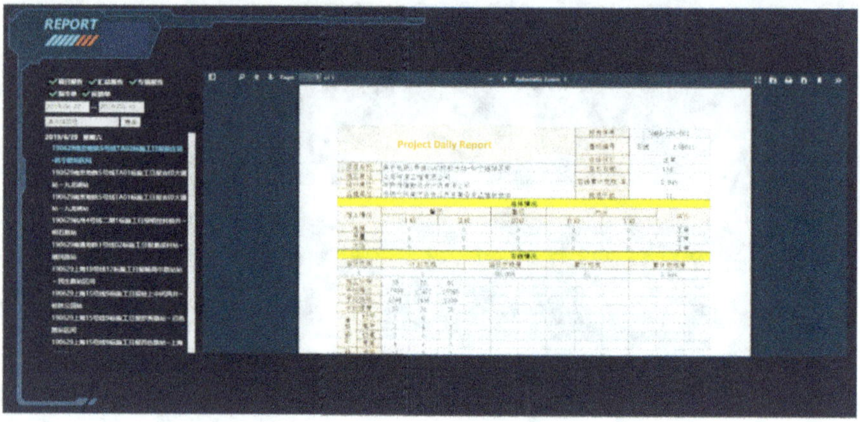

Fig. 7.24. Report interface of management and control consultation.

itself contains alarm data. For an analog alarm, the threshold values — the maximum and minimum values — are preset. The real-time data page of the part exceeding the threshold value is displayed in red, and the part within the range is displayed in blue. The shield machine attitude is displayed in three levels: the normal range is within ±50 mm, and the attitude data are shown in blue; the caution range is between −100 and −50 mm and between 50 and 100 mm, and the attitude data are shown in yellow; the warning range is beyond ±100 mm, and the attitude data are shown in red boxes.

7.3.4.2 *Analysis and prediction alarm*

The shield tunneling process inevitably causes ground loss. The surface settlement is closely related to the stability of the cutting face and the quality of shield grouting. The stability control of the cutting face involves construction parameters such as earth pressure, tunneling rate, the total thrust of the shield, the torque of the cutterhead, the rotation speed of the cutterhead and shield attitude. The quality control of the shield grouting involves construction parameters such as synchronous grouting pressure and synchronous grouting amount. Through big data analysis of these parameters and settlement information, a surface settlement prediction and alarm model is established. Once the model predicts that the construction parameters would cause large ground deformation,

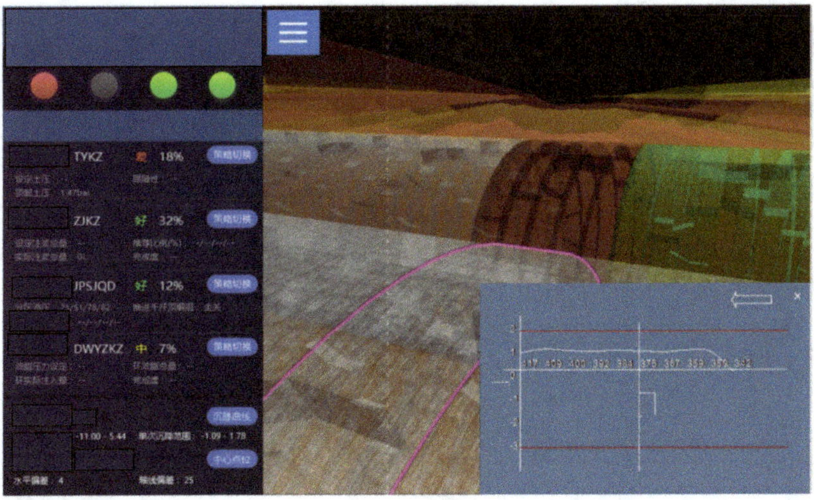

Fig. 7.25. Analysis and prediction alarm interface.

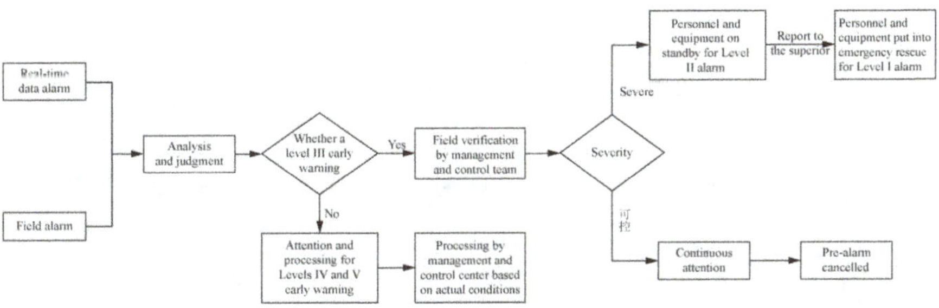

Fig. 7.26. Leveled alarm logic sequence.

an alarm will be raised. The analysis and prediction alarm interface is shown in Fig. 7.25.

7.3.4.3 *Alarm processing in levels*

The severity is graded according to the three major indicators of progress, quality and risk, with a total of five levels. Different levels of early warning data are pushed to the relevant management personnel in real time by the system platform, and the automatic and manual data are automatically processed and displayed in levels from I to V so as to establish an effective early warning management system. The logic sequence of prediction and alarm in levels is shown in Fig. 7.26.

In the case of Levels IV and V early warning, the person on duty shall initiate and push collaborative tasks to the project construction personnel, prompting them to take corresponding measures. When the abnormality reaches Level III early warning, the management and control team shall rush to the site for further confirmation. If the situation is not under control, a Level II alarm shall be issued, and relevant rescue personnel and equipment shall be on standby at the base. If the situation continues to deteriorate, a Level I alarm shall be issued, and all personnel and equipment shall be put into emergency rescue mode.

Chapter 8

Risk Management

During the construction of large-diameter shield machines, the cutterhead is in direct contact with a large range of soil mass ahead of the excavation face. Boulders, fracture zones and other unfavorable geological conditions in the soil mass seriously affect the service life of the cutters and even cause cutterhead failure. Therefore, it is necessary to conduct early exploration and risk investigation on the soil mass ahead of the excavation face and replace the cutters in time to adapt to the changes in stratum conditions. Moreover, due to the large overall size of the shield machines and the operating conditions in the river crossing tunnels with sufficient groundwater, shield tail leakage becomes one of the biggest risk points in the construction process of large-diameter shield machines, and its seal control is particularly important. This chapter mainly introduces the management of main risks and countermeasures for large-diameter shield construction, including advanced geological prediction, cutter management and shield tail seal management.

8.1 Advance geological prediction

8.1.1 Unfavorable geological hazards

Shield tunneling is a complex task in the underground environment, where it is difficult to clearly know the geological conditions and simple mistakes may lead to major accidents, resulting in incalculable losses. In particular, the geological prediction in the composite stratum has always been an issue puzzling the engineering community.

A large number of engineering practices reveal that the most common unfavorable geological conditions in shield tunneling mainly include fault fracture zones, karst, boulders and special lithologic intrusions, which are potential hazards to tunneling safety.

8.1.1.1 *Fault fracture zone*

Fault fracture zones, especially active faults, overthrust faults, tensile faults, torsional faults, uncemented structures and secondary structures, have the common characteristics of loose structure, developed fractures, significantly reduced strength and rate, and proneness to water enrichment. Therefore, there are obvious geological differences between the fault fracture zones and the rock masses on both sides. There are obviously strong reflection wave groups and low P-wave velocity in these sections. Tunneling in the stratum rich in water suffers from frequent occurrences of instability and collapse.

8.1.1.2 *Karst*

Karst refers to the carbonate-containing rocks (dolomite, limestone, etc.). The constituent calcium carbonate generates slightly soluble calcium bicarbonate under the action of fissure water and carbon dioxide. Karst caves grow as the result of the long-term dissolution of groundwater in limestone areas. The karst geological landscape is formed under the following conditions: soluble rock strata; water with dissolution capacity and sufficient flow in rock; recharge water flowing or penetrating into the underground rock mass and flowing water in the rock mass. Such three factors play an integral role.

Karst is distributed in varied shapes, and it is difficult to accurately identify its locations, occurrence, water inflow and other properties, so the design and construction of karst tunnels are often difficult. For example, for a tunnel passing through high-pressure and water-rich ground, lining design and shield tunneling are difficult problems.

It can be seen from the engineering cases in China and abroad that karst problems in tunnel construction mainly fall into two categories. First, it is difficult to control the large-diameter shield attitude and maintain pressure on the excavation face; second, the tunnel structure, under excessive deformation, is prone to fragmentation, water leakage and other defects.

How to avoid karst disasters is a key concern for construction personnel. Karst caves and underground rivers have different shapes and irregular spatial distributions, which are the main disaster-causing factors in tunnel construction. These geological bodies are characterized by low density and low P-wave velocity. Generally, when seismic waves encounter karst caves in the process of propagation, they will generate strong reflections. Therefore, the prediction of karst strata using the seismic wave method is the main research subject in China and abroad.

8.1.1.3 *Boulder and other special lithological strata*

Boulder, also known as spherical weathered rock, tends to cause cutter damage and even cutterhead deformation, resulting in the breakdown of the passing shield due to the large difference in strength with the surrounding soil. The uncertainty of boulders has a great impact on the shield construction and imposes a risk of stranding the shield. Therefore, in the early geological supplementary exploration, it is required to carry out an in-depth and detailed investigation to accurately ascertain the distribution, quantity, size and other specific conditions of boulders, generally by the following three methods:

(1) Drill holes to detect the distribution of boulders along the tunnel.
(2) Detect boulders in the tunnel range with ground penetrating radar.
(3) Monitor the geological conditions ahead of the tunneling face at all times with an ultrasonic detection system installed in the shield earth chamber to reveal the boulders ahead in advance.

8.1.2 Stratum exploration technology

The advance exploration and prediction technology that emerged in the 20th century was first applied in the field of coal mine production. With the development of road engineering, water conservancy engineering, military fortification engineering and urban underground engineering construction, the geological advance exploration and prediction technology has been applied in underground tunnel construction wider and wider, and geological experts and researchers in related fields have carried out a lot of experimental research and practical exploration. Multiple methods have been successively applied in the research of underground rock mass engineering in various countries, including the down-hole DC method, infrared thermometry method, ground penetrating radar method, seismic wave reflection method and electromagnetic wave method (including radio wave perspective method, ground penetrating radar method and transient electromagnetic wave method), mainly aimed at various geological disasters encountered in underground production and construction safety, solving many practical difficulties in production and construction safety, and providing scientific and effective guidance for detecting and predicting underground geological disasters.

8.1.2.1 *Main methods of exploration and prediction*

(1) **Down-hole DC method:** In the 1950s, former Soviet scholars began to study the application of the DC method to the underground exploration of coal mines and successfully solved the geological problems related to the production safety of mines, such as the exploration of small coal seam structures and scour zones, the investigation of hydrogeological conditions in mines, the evaluation of roof stability, and the monitoring of roadway deformation.

    DC exploration is a kind of earth exploration method. Based on the study of the differences in the electrical properties among various rocks and ores in the crust, the spatial and temporal distribution laws of electric or electromagnetic field are used to learn the geological structures or look for minerals. This technology achieves the purpose of exploration by establishing a full-space stable artificial electric field in the rock stratum around the underground roadway, identifying the variation law of the electric field by measurements, calculating the apparent resistivity of the rock stratum and plotting the apparent resistivity curve or profile.

    The down-hole DC exploration method is mainly used to solve the problems of water-flowing and water-bearing structures in mines. As the rock resistivity mainly depends on the water abundance and pore space characteristics in the voids, the exploration purpose can be achieved by taking advantage of this electrical property difference, combining the full space electric field theory and the corresponding data interpretation and processing system.

(2) **Infrared thermometry:** The geothermal field of the rock mass where a tunnel is located is mainly affected by the burial depth of the tunnel. Generally speaking, the rock mass temperature of the tunnel with a large burial depth is greatly affected by the geothermal field. The greater the burial depth, the higher the surrounding rock temperature.

The infrared thermometry technology is used to measure the temperature field of the geological body in a certain range ahead with the detection instrument. If there is a water-bearing body in the geological body ahead, the abnormal fields of convection, conduction and radiation will be superimposed on the normal radiation field, and there will be sudden changes in the measured data. Therefore, through the analysis of the collected temperature field data, it is possible to predict the unfavorable geological bodies in the tunnel ahead, especially the underground water bearing bodies, such as the karst caves in limestone areas, which can be well detected by this method.

(3) **Ground penetrating radar method:** Ground penetrating radar (GPR) is a nondestructive method to detect geological distribution using high-frequency broadband electromagnetic pulse technology, usually composed of a radar host, transmitting antenna and receiving antenna. When electromagnetic waves transmitted from the transmitting antenna to the ground ahead of the tunneling face encounter different impedance interfaces (differences in electrical properties on both sides of the interface), reflected waves and transmitted waves will be generated. The echo is received by the antenna, and the received electromagnetic wave signal is processed and analyzed. The location, structure, geometry, etc., of the abnormal geological body are judged according to the waveform information, reflection intensity, two-way travel time and other parameters.

Derived from aerospace-sounding radar technology in Europe and America, the GPR method has a long history of application in geological exploration based on the radar principle. However, its practical application scope had been limited until the 1970s. GPR was initially mostly used for engineering quality detection or site investigation, but it has also been used for geological advance exploration and prediction in recent years. The advantage of this method lies in its very high identification ability for fault zones, especially water-bearing zones and fracture zones, and its advantages in application in water-rich strata and karst cave development areas. Due to its disadvantage that the detection distance is relatively short, the exploration and prediction of long distance tunnels need to be carried out in sections for multiple operations, and data processing is inconvenient. In addition, radar records are vulnerable to interference, so their accuracy is not stable.

(4) **Transient electromagnetic wave method:** The transient electromagnetic wave method is used to study the distribution of rock conductivity by receiving the strength of the secondary field and analyzing its variation with time. If the conductivity of rock mass is strong, the secondary field is strong; if the conductivity of rock mass is poor, the secondary field is weak. The early secondary field reflects the conductivity of the shallow rock mass, while the late secondary field reflects the conductivity of the deep rock mass. Without using a grounding electrode, the transient electromagnetic wave method is applicable to the exploration of areas with high resistivity on the surface. Up to a survey depth of 1 km, it is sensitive to the distribution of high-conductive bodies, so it can be used to predict the water-bearing property of tunnels. At present, transient electromagnetic has been incorporated in the research on its application in geological advance prediction on tunneling faces. Certain effects have been proven in the tests with small coils and high current, but the supporting hardware and software are still to be developed.

(5) **Seismic wave reflection method:** The seismic wave reflection method is to artificially excite seismic waves with a small amount of explosives, which propagate in three-dimensional rock and soil media, to identify the geological structure ahead based on the echo characteristics. The seismic wave propagating in rock and soil mass generates a reflection wave when encountering unfavorable geological bodies such as karst, faults or fracture zones. The reflected wave is received by a highly sensitive detector, and the recorded signal is amplified, collected and processed to realize the geological prediction ahead.

8.1.2.2 *Prediction principle of seismic wave reflection method*

The Tunnel Seismic Prediction (TSP) system and Seismic Scattering Profile (SSP) system based on wave reflection exploration are the main technical means used in advance geological exploration of shield tunnels in China. The prediction principle of the seismic wave reflection method is introduced in the following, focusing on the currently mature geological advance prediction technology — the seismic wave transmission method.

(1) **Concept of seismic wave**

(i) *Types of seismic wave*: Seismic waves are classified into body and surface waves. Body waves propagate in the medium, while

Table 8.1. Classification of seismic waves.

Seismic wave	Propagation mode	Classification
Body wave	Propagation in media (different propagation features)	P-wave S-wave
Surface wave	Propagation along the free surface and interface of media (different natures)	Rayleigh surface wave Love surface wave

surface waves propagate along the free surface or interface of the medium. Body waves are further classified into P-waves and S-waves. Surface waves include Rayleigh surface waves and Love surface waves. The classification of seismic waves is shown in Table 8.1.

(a) P-wave: The seismic wave generated by elastic deformation (i.e. tensile and compression deformation) of the medium is called the P-wave. P-wave propagation in the medium generates compression zones and sparse zones alternately. Therefore, a P-wave is also called a compression or tensile wave in a propagation direction consistent with the vibration direction of the medium.

(b) S-wave: It refers to the seismic wave generated by shear deformation of an elastic medium, with the particle vibration direction perpendicular to the wave propagation direction. The S-wave is composed of vibrations in both vertical and horizontal directions. The SV wave is the shear wave component of vibration in the vertical plane. The SH wave is the shear wave component of vibration in the horizontal plane.

(c) Rayleigh surface wave: It refers to the surface wave propagated along the medium and mountain surfaces, and its particle vibrates counterclockwise along the trajectory in the vertical plane along the propagation direction.

(d) Love surface wave: This type of wave propagates along the interface of two kinds of elastic media and has little impact on seismic exploration.

(ii) *Propagation of seismic wave*: Waves represent the propagation process of vibration, and vibration is the source of waves. In the process of studying elastic seismic waves, particles in the medium interact through the action of elastic forces. For example, if a particle A leaves its original position under disturbance, the surrounding particles will produce an elastic force on A, making A return to the equilibrium position and vibrate near the equilibrium position. Moreover, the surrounding particles of point A are subjected to the force of A, making particles vibrate near the equilibrium position. Therefore, vibrating particles in the medium cause adjacent particles to vibrate, thus resulting in waves propagating outward in all directions.

The propagation process of a P-wave is classified into incident wave, reflected wave and transmitted wave (transmission in seismology is equivalent to refraction in physics) according to the characteristics of propagation paths. The seismic wave propagation diagram is shown in Fig. 8.1.

In seismic exploration, the amplitude of the incident wave is related to the wave impedance of the medium on both sides of the interface. As shown in Fig. 8.1, ρ_1 and V_1 represents the density and velocity of the medium on the reflecting interface respectively; ρ_2 and $V_{2,}$ represent the density and velocity of the medium under the reflecting interface, respectively. The product of density and velocity is called wave impedance. The condition for reflection is that the wave impedance on the reflecting interface is not equal, and the greater the difference in wave impedance, the stronger the reflection. In the process of seismic wave propagation, when encountering karst caves, underground rivers and sinkholes, strong reflection will be generated; especially for water-filled and mud-filled karst caves, the reflection coefficient can generally be as high as 0.5; in the case of a cavity, the reflection coefficient is higher. Therefore, for large-scale karst caves, the seismogram shows that the event is relatively continuous, while for small-scale karst caves, the seismogram shows obvious arc diffraction. See Table 8.2 for wave impedance, density and velocity in common rock and soil.

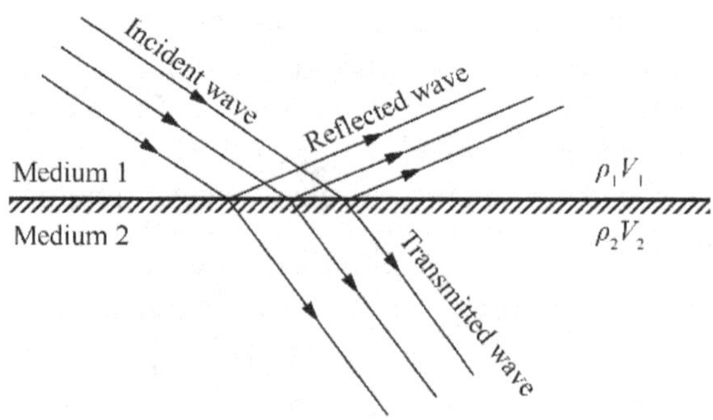

Fig. 8.1. Schematic diagram of seismic wave propagation.

Table 8.2. Wave impedance, density and velocity in common rock and soil.

Rock and soil	**Density (g/cm^3)**	**Velocity (m/s)**	**Wave impedance (g/m·cm^2 × 10^4)**
Soil	1.1–2.0	200–800	2.2–16
Sand	1.4–2.0	300–1,300	4.2–26
Clay	1.5–2.2	1,800–2,400	27–52.8
Sandstone	2.1–2.8	2,000–4,000	42–112
Limestone	2.3–3.0	3,200–5,500	73.6–165
Salt rock	2.0–2.2	4,500–5,500	90–121
Crystalline rock	2.4–3.4	4,500–6,000	108–204

(2) **Influencing factors of elastic wave velocity**

(i) *Physical properties of rocks*: The propagation speed of an elastic wave in rock mass depends on the rock density and elastic parameters, while the rock elastic parameters depend on the mineral composition and the properties, temperature and pressure of the fluid filled in the rock pores. Porosity, fluid in pores and environmental factors affect the elastic parameters and density of rock mass, thus affecting the propagation speed of elastic waves. This makes the propagation velocity of the elastic wave in rock overlap in range. Therefore, it is generally believed that it is more accurate to use the ratio of P- and S- waves to reflect the impact of rocks with different lithologies on elastic wave velocity.

(ii) *Porosity of rock*: When the influence of formation pressure on elastic wave velocity is negligible, the elastic wave velocity depends on the porosity and mineral composition of rocks. In the 1950s, Wyllie proposed that the elastic wave velocity can be correlated with the elastic parameters of rocks in a relatively wide range of porosity. It was found that elastic wave velocity (V) and porosity (ϕ) are correlated by the propagation velocity (V_F) of the elastic wave in the rock matrix and its fluid velocity (V_M) in the rock pores, as follows:

$$\frac{1}{V} = \frac{\phi}{V_F} + \frac{1-\phi}{V_M}. \tag{8.1}$$

(iii) *Saturation and temperature of fluid in pores*: In the mid-70s of the 20th century, Gregory found that the elastic wave velocity in the rock fluctuates to a certain extent with the different fluid saturation in the pores under a certain pressure, and the P-wave changes with amplitudes up to 25%. Moreover, under the pressure kept unchanged, when the temperature condition changes from 24°C to 750°C, the change in the P-wave velocity is about 63% in dry rocks and about 33% in saturated rocks.

Domenico also found through research that when the rock saturation is low, with the increasing saturation, the density of the rock increases and the V_p decreases; however, in the case of a high water saturation, the porosity of the rock increases significantly, making the speed change exceed that is caused by the increase in density, thus making V_p increase significantly.

(iv) *Rock density*: There is a positive correlation between elastic wave velocity and rock density. In general, the greater the density of the medium, the faster the elastic wave travels in it.

(3) **Absorption and attenuation of elastic waves:** In the process of seismic wave propagation, the rock mass absorbs seismic waves to varying degrees, making the elastic wave attenuate continuously. In general, the main causes of seismic wave attenuation are:

(i) *Geometric diffusion*: The seismic wave propagates from the source to the surroundings, with the wave front getting bigger and bigger, and the amplitude of the wave propagating headmost getting smaller and smaller. The spherical wave strength and energy flow density both decay inversely with the square of the distance, which is caused by the geometric diffusion of the seismic wave. Seismic waves generate energy when passing through a medium, which represents the characteristics of waves.

(ii) *Medium absorption*: In fact, the stratum is often not completely elastic and homogeneous, which makes the energy of seismic waves gradually decrease when passing through the medium. The friction among vibrating particles transforms part of the energy of the wave into heat energy, which is called medium absorption. It can be seen that the better the elastic property of rock, the smaller the energy loss.

When the seismic wave propagates in the rock mass, the amplitude caused by absorption decreases exponentially with distance, which is specifically expressed as follows:

$$A = A_0 \cdot e^{-\alpha r}, \tag{8.2}$$

where

A_0 — initial amplitude;

r — propagation distance;

α — absorption coefficient, the factor of exponential attenuation of amplitude with distance due to absorption. The absorption coefficient is related to frequency, which is generally linear and sometimes squared. The magnitude of absorption and diffusion depends on the distance and frequency of seismic wave propagation.

(iii) *Transmission loss*: Theoretically, when the ray energy is reflected, transmitted or diffracted from the interface, no energy loss will occur. However, for data processing, if only reflected wave energy is involved, transmission and diffraction will cause attenuation of reflected energy.

If the seismic wave is vertically incident and passes through a geological interface twice, the two-way transmission coefficient T_r is expressed as follows:

$$T_r = \frac{4\rho_1\rho_2V_1V_2}{\left(\rho_1V_1 + \rho_2V_2\right)^2}. \tag{8.3}$$

The relational expression between the transmission coefficient T_r and the reflection coefficient of the interface is

$$T_r = 1 - R^2. \tag{8.4}$$

For two layers of media, the reflected wave at the interface R_2 observed on the ground is shown in Fig. 8.2. The wave passes through the interface R_1 twice, downward and upward, respectively. Energy loss occurs at each transmission, so the observed reflected energy decreases. The transmission loss factor of interface R_1 is $(1-R_1^2)$.

For multiple layers of media, the transmission loss of the ith interface is the energy loss of seismic waves after passing

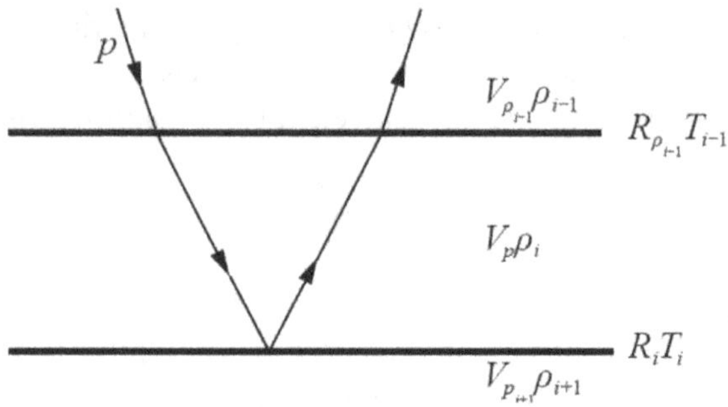

Fig. 8.2. Schematic diagram of transmission loss.

through the upper ($i-1$) interface. Therefore, the attenuation of reflected energy is the product of the two-way transmission coefficient of wave passing through all interfaces and the reflection coefficient of interface i. As shown in Fig. 8.2, the wave incident amplitude is A_0, and the reflection coefficient of each layer interface is R_i. Then, the reflection amplitude of the first interface is $A_1=A_0R_1$, the reflection amplitude of the second interface is $A_2=A_0\ (1-R_1^2)\ R_2$, and the reflection amplitude of the nth interface is $A_n = A_0\ (1-R_1^2)\ (1-R_2^2) \cdots (1-R_{n-1}^2)\ R_n$.

It can be seen that the transmission coefficient of the n–1th layer medium and the reflection coefficient of the nth layer medium determine the reflected wave energy of the nth interface.

In general, seismic wave velocity and seismic wave attenuation are closely related to rock properties. The medium acts as a filter, filtering out the higher-frequency part and retaining the lower-frequency component. On this principle, the application of seismic waves can be expanded in the exploration field.

8.1.3 Tunnel engineering application

8.1.3.1 *Seismic wave reflection method*

The seismic wave reflection method is applied as an advance exploration technology based on the classical elastic wave theory. Its exploration process flow is shown in Fig. 8.3.

It is necessary to clearly know the overall geological environment before carrying out tunnel geological prediction, specifically including the

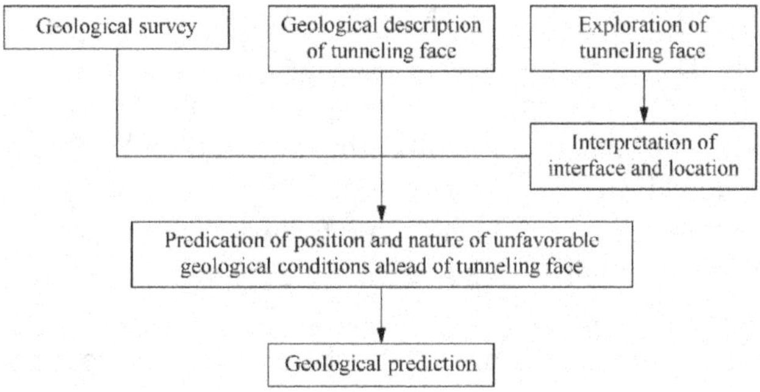

Fig. 8.3. Geological exploration process of seismic wave reflection method.

collection, sorting and analysis of geological data; geological survey; geological description of tunneling face; selection of tunnel geological exploration methods (selection of geophysical exploration methods); exploration implementation, etc. The collection, collation and analysis of geological data provide proofs for the prediction of various geological conditions and possible geological disasters that may occur during the tunnel construction, providing a scientific basis for tunnel engineering geological exploration. Through the work at this stage, the key work sections of geological advance prediction during tunnel construction can be identified.

In combination with the geological survey, the key sections of geological advance prediction are determined for tunnel construction, including the distribution positions of faults and their fracture zones, the location where karst may develop and the distribution location of boulders.

Geological survey in tunnel section and geological description of tunneling face mainly include: stratigraphic and lithological description; determination of occurrence and location of stratum interface; determination of the location, occurrence and scale of geological structures and description of their properties; joint fissure statistics; and description of special geological conditions.

The principles to be followed in the selection of geophysical exploration methods include: less occupation of the construction time of the tunneling face; strong applicability; simple operation; high adaptation to tunnel construction requirements; and strong pertinence to the geological problems faced by tunnel construction.

The exploration starts with the method selected to focus on the following problems:

(i) Shield tunneling and exploration should not be carried out simultaneously.
(ii) During the interpretation, a clear forward image concept must be established for the object to be predicted, and a clear geological concept must be established for the image. Attention should be paid to the multiple interpretations of the geophysical exploration data, and the different possibilities reflected by the data should be eliminated one by one through multiple comparisons and analyses.
(iii) In the geological interpretation of physical exploration data, the interpretation results should be verified and compared with the existing geological data.

The application of two geological advance exploration technologies based on the principle of seismic wave reflection in tunnel engineering, i.e. the TSP and SSP systems, is introduced as follows:

(1) **TSP system:** In the TSP method, as shown in Fig. 8.4, a 1.5–2 m deep borehole is generally drilled far behind the tunneling face, filled with water and placed with a three-component detector; in the direction of the tunneling face, boreholes with a drilling depth of about 1.5 m are evenly arranged, filled with water and placed with high explosives, and then blasting is carried out in turn. The elastic wave generated by blasting is received by a geophone. The elastic waves in direct propagation toward the geophone are received by the geophone and, on the other side, propagate forward, crossing the tunneling face. Waves encountering different wave impedances in the rock mass are reflected. The reflected waves return to each geophone successively

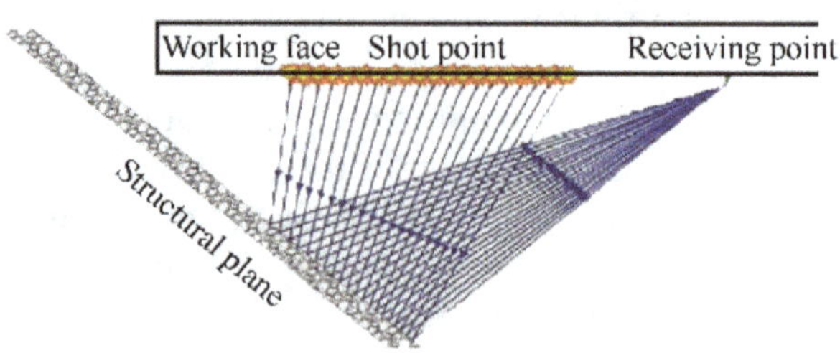

Fig. 8.4. Schematic diagram of advance geological exploration in TSP system.

and are received. After data processing, plotting and image interpretation, geological prediction conclusions can be drawn.

The TSP seismic wave method-based advance exploration data processing system consists of nine processing modules: establishment of the observation system → frequency filtering → shot time base correction → energy balance → wave field separation → inverse Q filtering → P and S wave separation → depth migration imaging → comprehensive geological interpretation.

In the Zhegushan Tunnel of Wenchuan-Maerkang Expressway, the TSP-win software was applied to process and analyze the data collected using a TSP200 PLUS instrument in the phyllite section, and the resulting data were generated, including the strength of the surrounding rock ahead of the tunneling face, the intersection angle between the seismic interface and the tunnel axis and the distance from the tunneling face, preliminary judged physical and mechanical parameters of rock (elastic modulus, density, Poisson's ratio, etc.), as well as the time profiles of the P, SH and SV waves ahead of the tunneling face, depth migration profiles, rock reflection horizon, energy size of each reflection layer, etc. It was finally predicted that there are possible unfavorable geological bodies (fault fracture zone, weak structural plane, lithologic change zone, water rich zone, karst cave, etc.) ahead of the tunnel heading, and the location, strike, scale, shape, etc., of the above unfavorable geological bodies were calculated and inferred. In the process of excavation, the actual conditions on site are basically consistent with the results of advance prediction; that is, the method can accurately predict the weak interlayer and small fracture zone ahead with high resolution and reliability.

The TSP203-PLUS geological advance predictor was used in the Lijiaying Tunnel to ascertain and clearly identify the changes in geological conditions ahead in advance, and technical measures were taken accordingly to ensure the safety of tunnel construction. This method not only accurately detected the unfavorable geological interface ahead of the working face but also provided parameters such as wave velocity, Poisson's ratio, rock integrity coefficient and Young's modulus. In the geological interpretation of the exploration data, the rock stratum is divided mainly depending on P-wave data. In the actual excavation process, the grade of the surrounding rock ahead is predicted according to the parameters generated and the contents related to engineering geology, which provided a basis for the tunnel excavation plan.

(2) **SSP system:** The SSP seismic scattered wave exploration system can be used as a geophysical exploration method to conduct a preliminary geological survey or be installed on the shield machines to continuously detect the stratum ahead with the shield tunneling.

As a geophysical exploration method for a preliminary geological survey, a spark source is used to excite seismic waves in the SSP system. When the seismic wave propagates underground and encounters a body with an abnormal wave velocity, such as the stratum interface or a boulder, reflected and scattered waves are generated. Then, the reflected and scattered waves are received by the geophone cable arranged on the ground surface, and the geological structure, location and shape of boulder are judged after processing.

The principle of seismic scattering is shown in Fig. 8.5. A small array of receiving systems is used for seismic scattering exploration, with a length of 1/3–1/2 of the exploration depth, to improve the resolution of lateral changes in geological structure. The technology is characterized by high resolution and wide applicability which can be used for homogeneous strata, complex geological structures and isolated bodies with drastic changes.

In the construction of the Zhuhai Maliuzhou Traffic Tunnel, the distribution of bedrock in the water area was detected by the SSP seismic wave scattering method, and the coincidence rate between the detection results and the actual borehole detection is up to 86%. The field implementation is shown in Fig. 8.6.

SSP installed on the shield machine as an advance exploration system emits elastic seismic waves under control by the PC system, and the transmitted wave signals are then received by the receivers, so

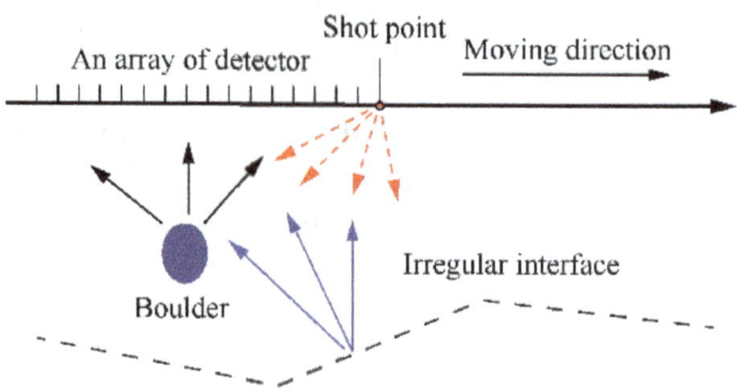

Fig. 8.5. Schematic diagram of seismic scattering technology.

(a) Seismic data sampling over water

(b) Wave detection cable over water

Fig. 8.6. Field implementation of SSP seismic scattering technology.

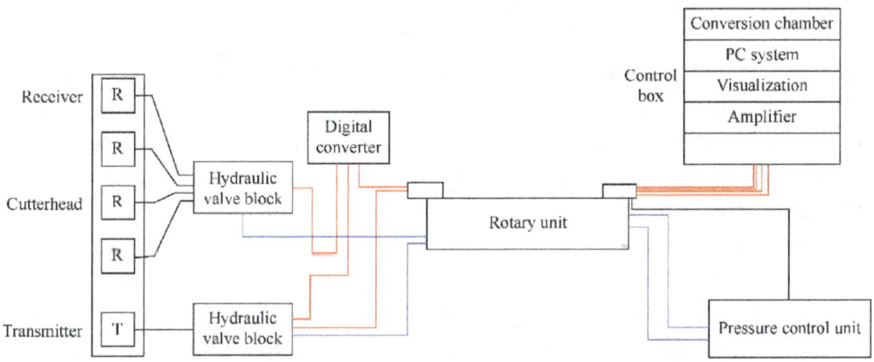

Fig. 8.7. Schematic diagram of SSP system structure.

it is unnecessary to generate elastic seismic waves by drilling and blasting in the tunnel. The structural principle of the SSP system is shown in Fig. 8.7. The whole SSP system mainly consists of the following six parts:

(i) *Seismic transmitter*: One transmitter (with hydraulic valve blocks) is installed on the cutterhead to transmit seismic wave signals to the ground ahead of the tunneling face.
(ii) *Seismic receiver*: Four receivers (with hydraulic valve blocks) are installed on the cutterhead to receive seismic reflected wave signals.
(iii) *Pressure control unit*: Installed at the back of the cutterhead, it compensates the transmitter and receives the difference between internal pressure and external pressure.

(iv) *Digital converter*: Installed at the back of the cutterhead, it digitizes the received seismic reflected wave signals.

(v) *Rotary unit*: Connected with the central rotary system of the shield machine, it transmits the digital signal to the PC system through the central rotary system.

(vi) *Control box*: Through the PC system (including visualization software), amplifier and sweeper, it receives, processes and uses all data and supplies power to all instruments after voltage transformation.

The survey is based on the seismic wave reflection principle. During the shield machine advance, the specially coded acoustic transmission signal is transmitted by the SSP transmitter from the cutterhead to the soil mass ahead through the support medium (slurry). When the seismic wave propagating in the soil mass encounters obstacles, the propagation medium changes. If this medium increases the contrast impedance of the seismic wave boundary and wavelength, the seismic wave will be reflected, and the receiver will receive the reflected wave signal. According to the reflected wave propagation time, real-time cutterhead angle, preset medium propagation speed and other parameters received by receivers at known positions, continuous and rapid analysis is conducted, and the 2D and 3D color models are established with all data within 5 m of advance to reflect the relative reflected energy. Taking a 40 m range as a unit, this system locates different media within 40 m ahead of the shield tunneling face.

The exploration principle of the SSP system is shown in Fig. 8.8. According to the geometric position of the area where the emission is

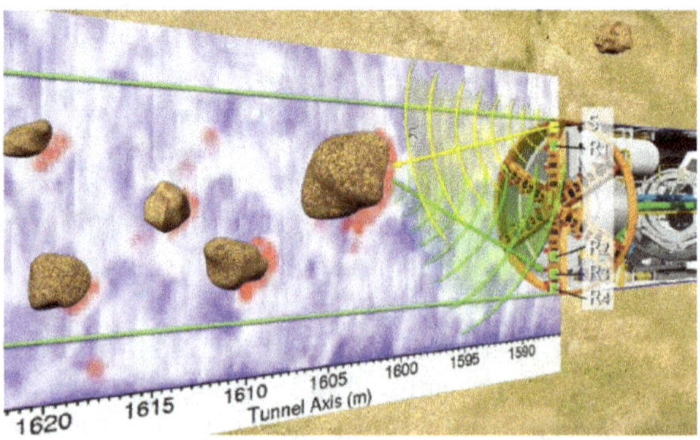

Fig. 8.8. Schematic diagram of exploration by the SSP system.

generated, the exploration data within multiple ranges of 40 m are comprehensively compared to achieve the function of predicting unfavorable formations in advance.

The SSP system works according to the following process. During shield tunneling, the PC system transmits the specially coded acoustic transmission signal to the transmitter at a signal frequency selectable between 600 and 2,000 Hz. The transmitter emits elastic waves to the soil mass ahead under the pressure stabilization of the pressure control unit. When the seismic wave propagating in the soil mass encounters obstacles, the impedance of the propagation medium (derived from the wave propagation rate in the medium) changes. If the medium increases the contrast impedance of the seismic wave boundary, the medium hardness increases, the seismic wave is reflected, and the reflected wave signal received by the receiver is enhanced in this area. The digital converter digitizes the received reflected wave signal, transmits the digital signal to the PC system through the rotary unit, and after the amplification by the amplifier, the PC system records and stores the data. When enough data are stored, the system will automatically generate the 2D and 3D data packages readable by the software for professionals to view.

During the shield construction of the Zhuhai Maliuzhou Traffic Tunnel, a set of SSP advance exploration systems was installed on the shield machine, and boulder warning information was continuously given when the shield machine advanced to cross the rock area along the western route, as shown in Fig. 8.9.

Observation conclusion	A reflector was found at TM 2506.4in migration and processing zone A; A reflector was found at TM 2497.5in migration and processing zone B; A reflector was found at TM 2517.0in migration and processing zone C;
Seismic wave velocity for migration and processing	v_s=1745m/s
Position associated with previous report	
More geological technical information	Upper sludge: grey, grayishblack, homogeneous, saturated:Middle brownish grey saturated soil, grayish black, saturated, homogeneous, course sand:Lower gravelly soil
Conclusion	Conclusion based on known data: dominated by non-homogeneous underground soil Ranked "yellow"

(a) Observation conclusion of section

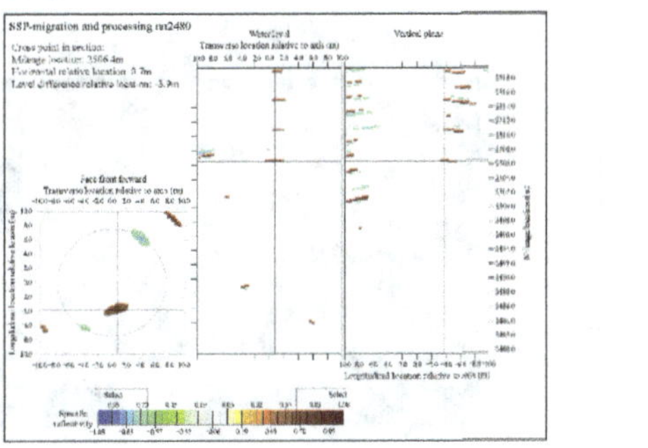

(b) Section analysis at the location of barrier

Fig. 8.9. SSP prediction of weathered granite.

During the offshore blasting pretreatment in the early stage of the actual construction, there were bedrock bulges at the cutting mileage of 2,554 m and 2,562.8 m, as shown in Fig. 8.10.

As shown in Fig. 8.11, when the shield advances to a cutting mileage of 2,554 m (ring 320) and 2,562.8 m (ring 324), the muck produced at

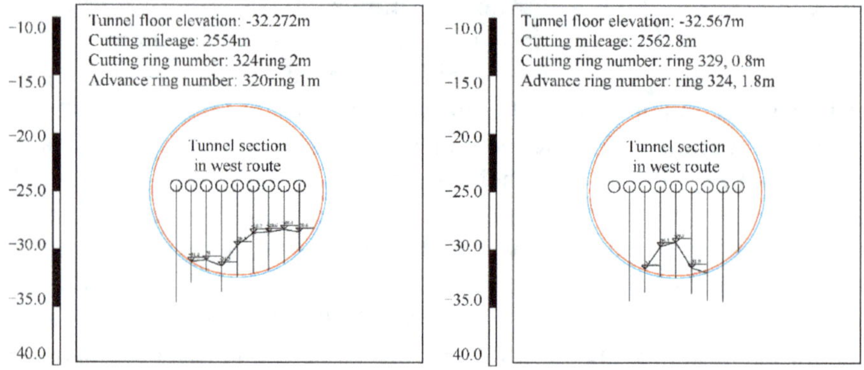

(a) Largest burial depth in rock: 3.67m (b) Largest burial depth in rock: 3.08m

Fig. 8.10. Positioning bedrock bulge by offshore drilling.

(a) Ring 308 rock (b) Ring 311 rock

(c) Ring 319 rock (d) Ring 325 rock

Fig. 8.11. Mucking of shield crossing rock area.

the slurry site is broken weathered rock, which is again judged as the bedrock bulge.

The SSP system basically predicted the mileage range and rock grade of the bedrock when the shield advanced through the bulge section of the moderately weathered granite bedrock in the Maliuzhou Waterway for the tunneling of the Zhuhai Maliuzhou Traffic Tunnel, providing a theoretical basis for the control of parameters in the actual construction process.

8.1.3.2 *Advance drilling technology*

The shield machine in the composite stratum is generally equipped with advance drills, with a drilling depth of up to 20–30 m. Generally, there are two types of drilling rigs. One is to set multiple provided holes in the shield and penetrate the drill pipe through the provided holes at an included angle of 10° (with the shield bus) to bore the soil mass ahead of the cutterhead in an umbrella shape. The other is to make the provided holes through the breast plate and the bubble chamber in a direction vertical to the cutterhead and penetrate the drill pipe through the provided holes and the cutterhead to the soil mass right ahead. The advance drilling technology can not only predict the geological conditions in advance but also inject the prepared slurry into the surrounding strata ahead of the cutterhead to improve the soil strength, keep the excavation surface stable, and avoid water inrush and quicksand on the excavation surface so as to effectively control the surface settlement. The principle of advance drilling is shown in Fig. 8.12.

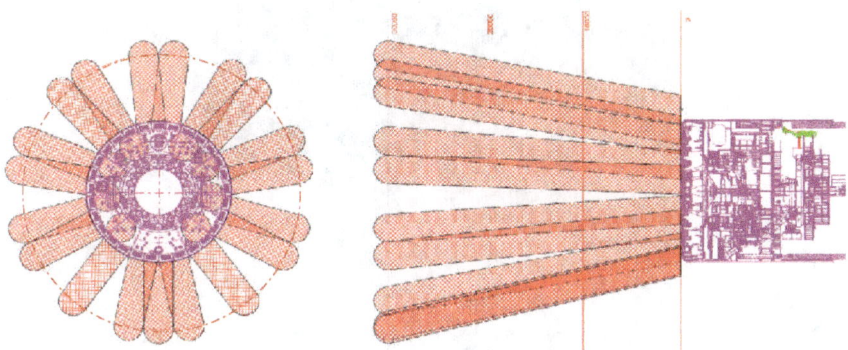

(a) Grouting section (b) Grouting profile

Fig. 8.12. Schematic diagram of advance drilling principle.

The advance drill drives drill pipes into the ground ahead of the working face, and the geological conditions ahead of the working face are predicted according to the impact pressure, torque, thrust force and the number of drill pipes sticking. The number, orientation, extrapolation angle and drilling depth of boreholes can be adjusted manually according to actual needs. The geological prediction method of advance drilling can effectively judge the interface position with the rock stratum, the degree of rock fragmentation, the porosity of the rock mass, water bearing, harmful gas bearing, etc. of the uneven rock and soil ahead. It has the characteristics of high accuracy and visual reflection of the geological conditions ahead of the working face and can effectively avoid the uncertainty and multiple interpretations of the geological prediction method in geophysical exploration.

The Shendong Bulianta Coal Mine Inclined Shaft is the first of its kind to be successfully constructed by a shield in China. The inclined shaft is 2,745 m long, the maximum burial depth is 274 m, and the gradient of the inclined shaft is 5.5°. The inclined shaft successively passes through the topsoil layer, Cretaceous Zhidan Group, Jurassic Anding Formation, Zhiluo Formation and Jurassic Yan'an Formation. During shield construction, an advance drilling rig (Fig. 8.13) was selected as the main equipment for advance geological prediction, including a data acquisition instrument, a data transmission system and other supporting equipment. The model of the advance geological drilling rig is MHP3050/550L-X, the length of each steel drill pipe is 2 m, and the diameter of the drill pipe is 64 mm.

The construction principle of advance geological prediction is "detect to clear all puzzles before tunneling". During the test, the shield machine

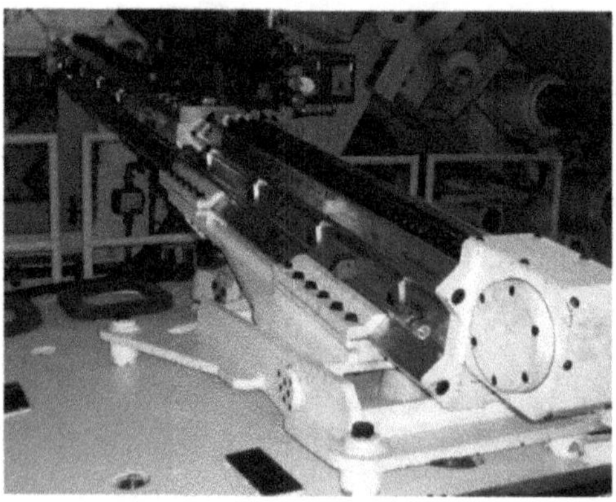

Fig. 8.13. Advance drilling rig.

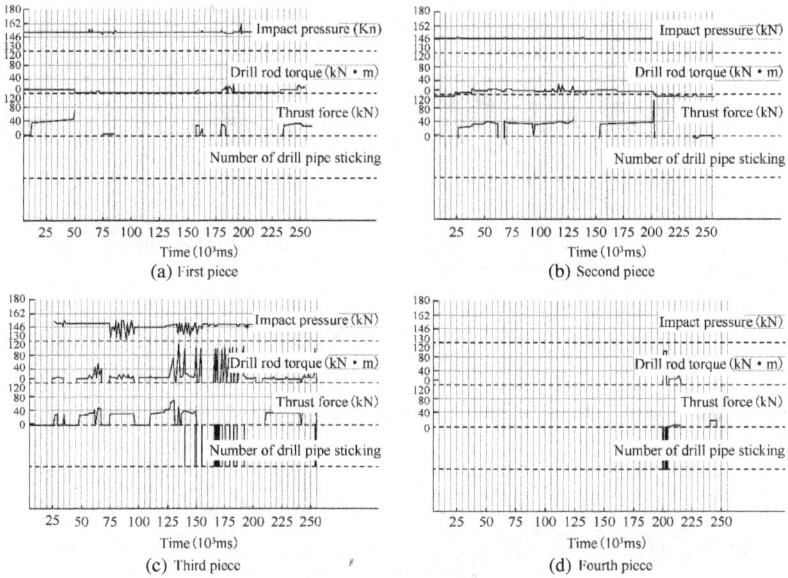

Fig. 8.14. Test data curve of advance drilling rig.

was stopped, and then the advance geological drill was installed on the segment erector to enable drilling at 360°. The site engineering application is located at a burial depth of 270 m. The hole is drilled at the 12 o'clock position, right above the working face, with an extrapolation angle of 8° for advance prediction.

All the data generated were digitally collected by the data acquisition unit, and the relevant data curves were plotted by the data processing system automatically. The test data curve of the advance drilling rig is shown in Fig. 8.14.

8.2 Cutter management

8.2.1 Arrangement of cutterhead cutters

8.2.1.1 *Function and type of cutterhead*

The cutterhead structures are divided into panel and spoke types to meet the requirements of different strata and construction conditions. See Table 8.3 for the comparison of the two cutterhead types.

Table 8.3. Comparison of two cutterhead types.

Structure type	Spoke type	Panel type
Photo		
Opening rate	Large, 70%–95%	Small, 25%–40%
Structure	Simple structure composed of several spokes	Complex structure with a panel in addition to spokes
Muck fluidity	Smooth muck flow, bonding resistance	Easy mixing and good fluidity in zones rather than center which is disadvantageous in poor fluidity and proneness to mud casing
Earth pressure in cutting chamber	Measured earth pressure equal to actual earth pressure	Differential pressure occurs due to division by the panel, with measured pressure in front of the breast plate unequal to actual pressure
Muck control	Be conveyed directly into earth chamber, almost without particle size limit	Be conveyed through cutterhead opening, with particle size limit

In order to solve the problems of severe cutter wear and frequent replacement in composite formation, cutterheads that are replaceable under atmospheric pressure have emerged. Its principle is to set a pace in the main cutter arm for the replacement of internal cutters under atmospheric pressure.

8.2.1.2 *Cutter classification and configuration*

As the excavation part of the cutterhead of the shield machine, there are different types of cutters to deal with different rock and soil conditions. As shown in Fig. 8.15, the commonly used cutters can be generally divided into hobbing cutters and cutting tools. The cutting tools include the parting cutter, scraper and advance cutter.

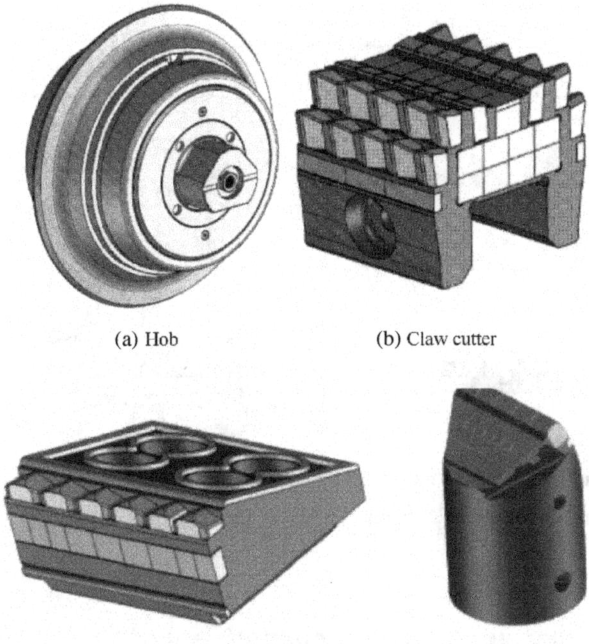

(a) Hob (b) Claw cutter

(c) Shovel blade (d) Scraper

Fig. 8.15. Cutter types.

8.2.2 Cutter management during construction

8.2.2.1 *Cutter wear monitoring*

(1) **Hydraulic wear monitoring device:** There are many kinds of cutter wear monitoring devices, among which hydraulic wear monitoring devices are the most common, as shown in Fig. 8.16. A hydraulic wear detection device consists of sensors and communication pipes installed in the cutterhead steel structure or scattered in the cutter rest. The headmost end is a wear probe, which is connected to the communication pipeline, and the tail end is a pressure sensor. If the wear probe is worn, the system liquid will leak forward. The tail end pressure sensor can recognize pressure drops in the communication pipeline and send an alarm signal.

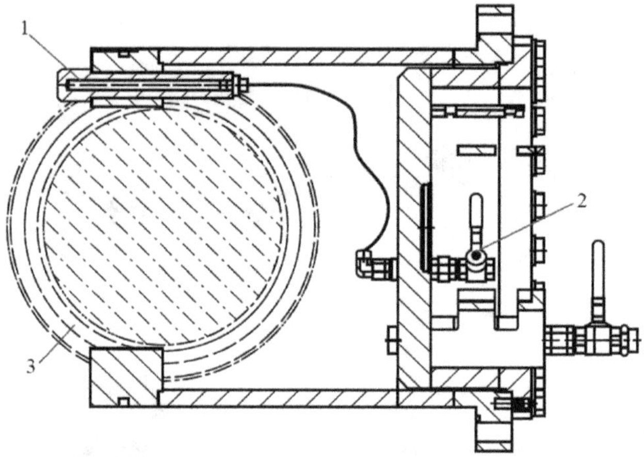

Fig. 8.16. Hydraulic wear monitoring device.

Note: 1 — Wear probe and communication pipeline; 2 — Sensor; 3 — Cutter.

(2) **Optical fiber displacement monitoring device:** The device uses a laser as the light source and an optical fiber sensor as the detection means to track and monitor the cutter state by detecting the vibration displacement of the cutter bar.

An optical fiber displacement sensor works as a non-contact measurement sensor to measure displacement, vibration and surface state. The optical fiber probe emits beams and then receives the reflected light from the measured target surface. The reflected light is converted into an electrical signal proportional to the target distance with a photosensitive element, and the output voltage can reflect the target surface position, displacement amplitude, frequency, waveform and other parameters. The principle of transmitting optical fiber displacement is illustrated in Fig. 8.17. When the optical fiber probe is very close to the target, there is little reflected light received by the receiving optical fiber; when the target moves away from the optical fiber probe, the reflected light received by the receiving optical fiber increases rapidly. Although the actual movement of the target is very small, the reflected light increment is very significant, which makes the output voltage change significantly.

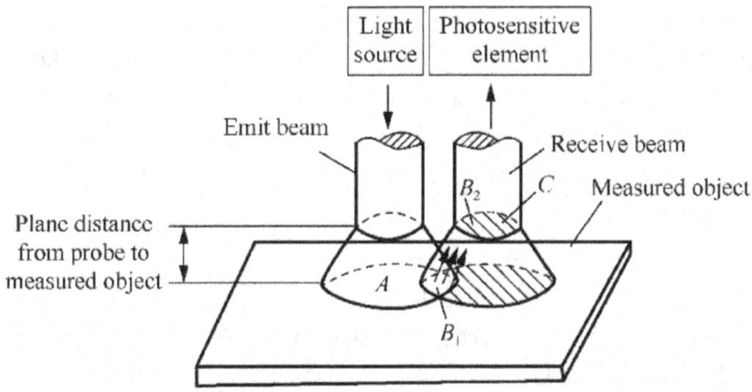

Fig. 8.17. Schematic diagram of transmitting optical fiber displacement sensor.

(3) **Hob rotation monitoring:** Once a hob in use stops rotating, the cutter wear speed will increase rapidly. Therefore, in addition to conventional wear monitoring, hob rotation should also be monitored.

The principle of hob rotation monitoring is based on the change in magnetic intensity. A receiver is arranged in the hob rest in advance, and four magnets are embedded on the hob. The magnetic intensity of the four magnets is sensed by the receiver, and whether the hob rotates is judged according to the change rate.

(4) **Cutter temperature monitoring:** Cutters may lose cutting ability under composite geological conditions. In addition to cutter wear, they may also suffer from mud lining, that is, the cutter head is completely wrapped in soil. The long-term friction between the cutter and the excavation surface causes the temperature to increase, so the cutter temperature can be monitored to determine whether it is necessary to clean and replace the cutter.

8.2.2.2 *Cutter replacement*

Shield tunneling is often plagued by cutter wear, reflected in non-ideal thrust parameters to directly affect the efficiency of the shield advance.

(1) **Cutter service life prediction:** Cutter wear is usually the dominant factor causing weakened cutting capacity of the cutterhead. When the cutter wear exceeds a certain value, the overall construction advance

parameters of the shield will deteriorate. The principle of cutter service life prediction is to calculate the theoretical cutter service life on the basis of standardizing the allowable cutter wear and combining the comprehensive wear coefficient of the actual stratum to the cutter, so as to guide cutter replacement. The main prediction methods include the following:

(i) *Advance distance prediction method*: This method is to predict the wear degree according to the advance distance of the shield. In the case of a large-diameter cutterhead, under the same conditions, the outer edge cutter will cut along a larger trajectory than the inner ring cutter. Therefore, under the corresponding stratum, the calculation principle for theoretical cutter wear is as follows:

When the cutterhead rotates, the cutter's revolution linear speed is calculated as follows:

$$v = \pi D n, \tag{8.5}$$

where

D — cutting diameter of cutter (m);
n — rotation speed of cutterhead (r/min).

The linear speed increases with the expanded cutting diameter, so in unit time, the cutter near the outer edge "passes" a longer cutting distance than the cutter near the center area, so wear worsens accordingly.

In the process of advance, the friction between the working face and the blade is the main factor in cutter wear. The sand and gravel in the stratum impact the cutter and accelerate its wear. The greater the impact force on the outer edge of the cutter, the more severe the wear.

Excessive cutter wear reduces the excavation efficiency of the shield machine and brings a risk of damage to the cutterhead. Therefore, a suitable wear limit is often specified for the cutterhead. Exceeding this limit not only means that the excavation diameter is reduced but also may cause direct friction between the cutterhead and the soil mass and damage the cutterhead. As a general rule, if the edge cutter wear is more than 10 mm, the front area cutter wear is more than 15 mm, and the center area cutter wear is more than 20 mm, the cutters must be replaced.

The cutter wear triggering early warning is usually calculated using the following empirical formula:

$$\delta = \frac{K_n \pi DnL}{v}, \tag{8.6}$$

where

δ — cutter wear (mm);
K_n — comprehensive wear coefficient of the nth cutter on the same cutting trajectory (10^{-3} mm/km);
D — cutter cutting trajectory diameter (m);
n — rotation speed of cutterhead (r/min);
L — tunneling distance (km);
v — advance rate of shield (mm/min).

The relational expression between the comprehensive wear coefficient K_n of the Nth cutter on the same cutting diameter trajectory and the wear coefficient K_1 of only one cutter on the same diameter trajectory is as follows:

$$K_n = \alpha K_1, \tag{8.7}$$

where

$\alpha = e^{-0.333\ln N}$

The wear coefficient $K_1 \approx 3 \times 10^{-3}$ mm/km of a single scraper cutting in clay and silty fine sand formation under protection by an advance cutter is $K_1 \approx 12.5 \times 10^{-2}$ mm/km in sandy formation.

It can be seen from the above calculation formula for cutter wear that the cutter wear is not only related to the geological conditions of excavation but also related to the rotating speed of the cutterhead of the shield machine, the advance rate, the cutting trajectory diameter where the cutter is located and the number of cutters with the same diameter. The larger the cutting trajectory diameter of the cutter, the more severe the wear. The overall wear of cutters can be reduced by increasing the number of cutters with the same cutting trajectory diameter.

(ii) *Prediction method of excavation volume*: For cutters with a certain width in the excavation direction, such as cutting tools, the service life of the cutter and the wear of the shield cutter δ can be calculated and analyzed according to the volume of soil excavation involved by the cutters using the following formula:

$$\delta = K\pi\,(R^2 - r^2)\,L, \tag{8.8}$$

where

δ — wear amount (mm);
K — wear coefficient (mm/m³);
R — radius of cutter cutting outer circle (m);
r — radius of cutter cutting inner circle (m);
L — advance distance (m).

In the formula, $\pi\,(R_2 - r_2)$ L is the soil cutting volume involved by the cutter. Similarly, in order to obtain the distance L (i.e. the service life of the cutter) of the shield tunneling when the cutter reaches the wear limit, the wear coefficient K needs to be calculated first. The formula is properly changed to derive $K = \frac{\delta}{\pi\left(R^2 - r^2\right)L}$, and the wear coefficient K under different working conditions can be analyzed and calculated according to this formula and in consideration of the actual cutter replacement data.

The Wuhan Sanyang Road River Crossing Tunnel was bored by a φ 15.76 m slurry balanced shield machine with an outer diameter of 15.2 m, an inner diameter of 13.9 m and a single tube length of 2,590 m. The shield was launched from the launching shaft in Wuchang. The first stage (0–400 m) was cut in the full section sandy geological conditions, with quartz content of 68.83–71.51%, which tends to cause cutter wear; the second stage (400–1,600 m) was bored in a composite stratum with sandy geological conditions in the upper layer and bedrock in the lower layer. The maximum height of the bedrock layer accounted for 80% of the tunnel section. The shield machine passed through the upper soft and lower hard composite strata. The skeleton particles in the weakly cemented conglomerate crossed are mostly hard rocks and are large-sized (the maximum particle size can reach about 20 cm). The skeleton composition is mainly composed of hard rocks such as quartzite and limestone. The quartz content of quartzitic cobbles and gravels is up to 93–98%, extremely tending to cause wear of the cutterhead

Table 8.4. Analysis on service distance of cutters.

Cutter	Allowable wear (mm)	Theoretical advance distance (mm)
Edge cutter	10	242
Front cutter	15	363
Center cutter	20	484

and cutters; the third stage (1,600–2,590 m) was bored still in the sandy geological conditions.

Referring to the common empirical formula of cutter wear for early warning, $K_1 \approx 12.5 \times 10^{-3}$ mm/km is taken in the 0–400 m full section sandy formation. It is specified that replacement is necessary when the edge cutter wear is more than 10 mm, the front area cutter wear is more than 15 mm, and the center area cutter wear is more than 20 mm. The advance rate is taken as 15 mm/min, and the cutterhead speed is taken as 1 r/min. According to the formula $L = \delta v/(K_N \pi D n)$, the theoretical advance distance can be calculated, as shown in Table 8.4.

It can be seen from the calculated value that the cutter wear for this distance in the sandy formation is not expected to exceed the allowable value. If the wear exceeds the allowable value after the cutter is extracted, the cutter wear analysis is carried out to identify the cause of abnormal wear. In this project, the silicon dioxide content in the full-section sandy formation is about 70%, which tends to cause cutter wear. The wear coefficient K_n is adjusted according to the actual cutter wear data so as to guide the subsequent cutter replacement.

After confirming the allowable value of cutter wear, the theoretical advance distance L of the cutter can be obtained, and in consideration of the cutter width, it is derived as follows:

$$L = \frac{V}{\pi\left(R^2 - r^2\right)}, \tag{8.9}$$

where

L — tunneling distance (m);
V — excavation volume (m^3);
R — radius of cutter cutting outer circle (m);
R — radius of cutter cutting inner circle (m).

Due to serious cutter wear, cutter optimization is carried out. For comparison and selection among different cutter types, this method is used to compare the excavation volume of shell cutters before and after improvement in different strata, and then appropriate cutters are selected for construction. See Fig. 8.18 for comparison of excavation volume per unit wear of cutterhead and cutter.

It can be seen from Fig. 8.18 that the improved shell cutter presents good adaptability in both sandy ground and composite formations.

(2) **Cutter replacement management platform:** Based on the Internet and big data, a platform for cutter replacement management is established and furnished with cutter management software, enabling the collection and analysis of cutter replacement detection data, the prediction of the next ring number for cutter replacement according to the multiple cutter replacement data, and the arrangement of the cutter replacement work in a planned way to avoid unnecessary cutter extraction and replacement, which is conducive to the overall planning of construction.

(a) Shell cutter before improvement (b) Shell cutter after improvement

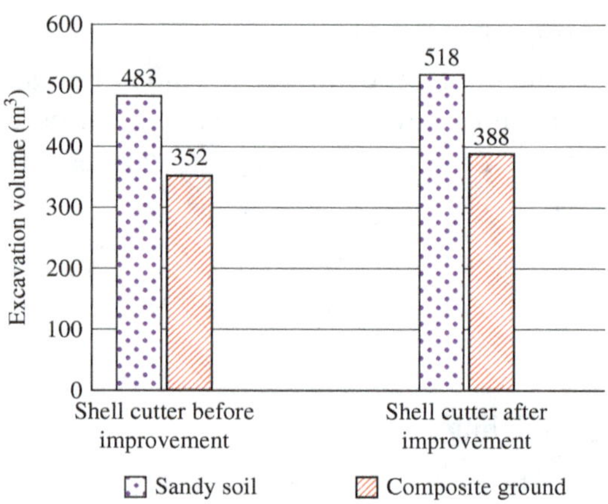

Fig. 8.18. Comparison of excavation volume per unit wear of cutterhead and cutter.

The ring number, time, cutter position, cutter type in use, cutter type after replacement, cutter wear amount and cutter usage of each cutter replacement can be entered into the information input interface of the platform in real time for the staff to query at any time. Moreover, the service life of the cutter in use can be calculated according to the wear history of the cutters replaced, and when the cutter is close to the theoretical life, a cutter replacement reminder can be given to ensure that there is no omission in cutter replacement. Through wear life analysis, the adaptability of various cutters in different strata can be generated to guide subsequent cutter improvement.

(3) **Service life extension of cutter**

(i) *Construction parameter management*

(a) Slurry quality control: High-quality slurry can reduce the cutter temperature, lubricate the cutter head and have better muck-carrying capacity, without causing muck accumulation at the shield bottom, and avoid secondary wear on the surrounding cutters. The cutterhead flushing system configured can effectively move the soil cut by the cutter into the slurry tank, preventing soil consolidation on the cutter from affecting its cutting ability. Taking the Sanyang Road Tunnel in Wuhan as an example, the opening rate within a range of 5.5 m of the cutterhead center is 0, so mucking is completely dependent on the central flushing system, at a flow rate up to 1,000 m^3/h and a flushing pressure of about 10 bar, which effectively solves the problem of mud lining on the central cutter and greatly reduces the generation of compression force for shield advance.

(b) Thrust parameter control: Cutterhead speed, advance rate, cutterhead torque and shield thrust will affect cutter wear, and the control of cutter penetration at 15 mm/rad is taken as the thrust standard. When the cutterhead speed reaches 1.7 rad/min and the advance rate reaches 25 mm/min, the alloy of the shield cutter will fly away; when the cutterhead speed reaches 1 rad/min and the advance rate reaches 15 mm/min, such a flying-away defect basically does not occur.

(c) Shield attitude control: Proper control over the shield attitude to make the cutterhead evenly stressed can prevent eccentric wear on the cutter. If the shield attitude is not properly controlled, the cutter will be subjected to uneven stress at different positions, and it is prone to eccentric wear due to frequent sticking at the position with large thrust. In particular, the edge cutter is more prone to wear due to its large linear speed, which greatly reduces the service life of the cutter.

(ii) *Cutter selection and production*

A set of scientific and reasonable manufacturing processes must be formulated for cutter production and processing. The cutter manufacturing process is the key to control cutter quality, in which the inlaying method and inlaying quality of alloy directly affect the service life of cutters. Secondly, cutters made of different materials and production processes should be selected for different geological conditions. If the same cutter is used in different geological conditions, its service life will be seriously affected. In hard rock tunneling, the cutting cutter is required to have good wear resistance and impact resistance, so the alloy strength and cutting cutter base material should be carefully selected.

8.2.3 Replacement of cutter under atmospheric pressure

8.2.3.1 *Cutter replacement tooling*

Cutter replacement tooling includes cutter support parts, cutter jacket, telescopic oil cylinder, oil cylinder connecting sleeve, base and other parts, as shown in Fig. 8.19.

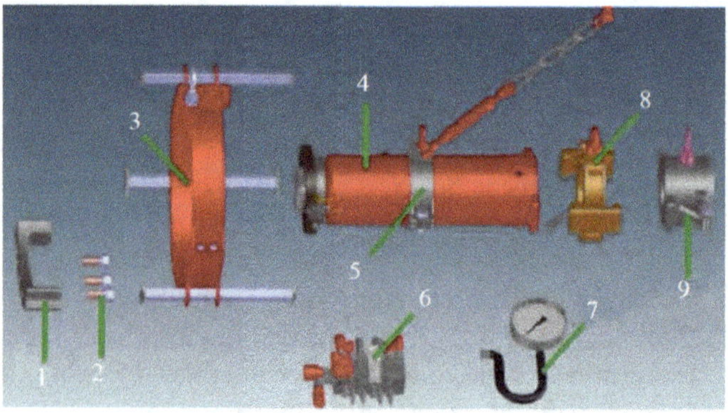

Fig. 8.19. Schematic diagram of cutter replacement tooling composition.

Notes: 1 — Cutter pot bottom support; 2 — Bottom support bolt; 3 — Cutter pot clamp; 4 — Telescopic cylinder; 5 — Telescopic cylinder clamp; 6 — Hydraulic valve block controller; 7 — Pressure gauge; 8 — Telescopic cylinder clamp; 9 — Cutterhead bearing seat.

8.2.3.2 *Preparation before cutter replacement*

(1) Stop the shield machine at the pre-selected position, properly increase the pressure of the bubble chamber, circulate slurry for 20–30 min, and clear muck from the excavation chamber.
(2) Open the flange tank sealing door, and conduct ventilation and cooling.
(3) Check the hydraulic system, flushing system and hoisting system related to cutter replacement to ensure their functions are intact.
(4) Before the cutter replacement construction, carry out detailed safety disclosure and technical disclosure to the corresponding operators.
(5) Prepare spare cutters and accessories necessary for replacement, such as bolts, oil cylinders and globe valves.
(6) Prepare lighting fixtures, hoists, boards, safety belts and other materials, as well as grinders, electric welders, ventilators, air guns and other machinery and materials.
(7) Fully estimate the possible emergencies and formulate emergency measures.

8.2.3.3 *Cutter replacement standard*

(1) The hob ring must be replaced in the case of eccentric wear, falling off, cracks, looseness, displacement, etc. Replacement is compulsory when the edge hob wear is more than 10 mm, and the front hob wear is more than 15 mm.
(2) Replacement is necessary when two of the standard alloy teeth of the scraper reach a wear amount of more than 15 mm and the edge of the toothed cutter reaches a wear amount of more than 10 mm.

The wear comparison of various cutters is shown in Fig. 8.20.

8.2.3.4 *Cutter replacement process*

The state of a hob before disassembly is shown in Fig. 8.21. By rotating the cutterhead, the position of the cutter to be replaced is adjusted to the lowest point, all equipment is debugged, and the tools and materials are prepared before cutter replacement under atmospheric pressure.

See Table 8.5 for the specific disassembly and replacement process.

(a) Inclined shell cutter (b) Hob (c) Scraper

Fig. 8.20. Comparison of cutter wear.

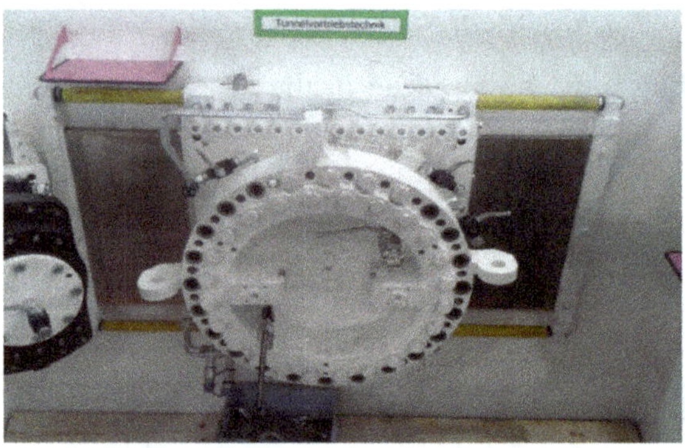

Fig. 8.21. Status of hob before disassembly.

8.2.3.5 *Special working conditions and measures*

(1) **Stuck cutter pot:** The hydraulic oil pressure is appropriately increased within the allowable range. If the cutterhead is retractable, it can be properly retracted. On this basis, tightening the bolts and removing the tooling enables the cutterhead to swing several times

Table 8.5. Process for replacement of cutter under atmospheric pressure.

S. No.	Description	Sketch
Step 1	Remove the bolt ① and washer in the rear cover of the cutter box and the cover ②	
Step 2	Load the telescopic cylinder ① into the cutter box, and fasten the bolt ②	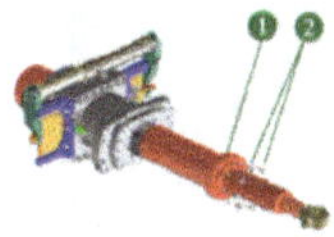
Step 3	Mount the outer cover and connect the hydraulic piping ①, mount the gate plate rod ② and hydraulic cylinder ③	
Step 4	Pull back the worn cutter by the retreating cylinder, and close the gate ②	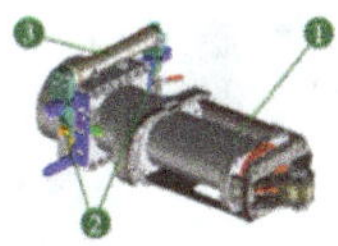
Step 5	Open the globe valve ① on the cutter box to release pressure and then close the valve	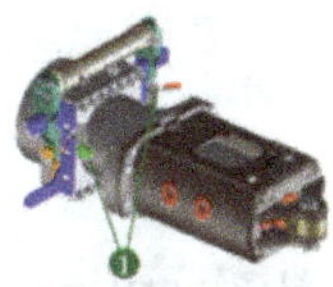
Step 6	Remove the worn cutter and outer cover of the cylinder. Replace the cutter and perform installation in an order reverse to the above	

within a 30° range on the left and right to improve desiltation at the bottom of the excavation chamber; the cutterhead on the fixed cutter box is rotated to a certain angle to shift the cutter replacement from the middle bottom position of the cutterhead to the lower left or right position to avoid the bottom siltation area.

(2) **Failure in normal gate closing:** The gate is flushed and cleaned repeatedly until full gate closing is confirmed. If cobbles and gravel cannot be cleared by simple flushing, the cutter box is pushed in again. The chamber is cleaned before proceeding with cutter extraction and replacement.

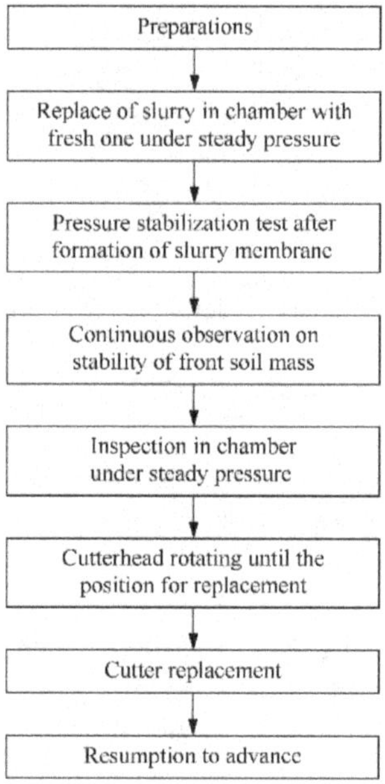

Fig. 8.22. Process of cutter replacement under pressure.

(3) **Hydraulic piping of control cylinder damaged or burst:** The globe valve at the oil cylinder is closed immediately and the hydraulic oil pipe is replaced; in the configuration with a globe valve, emergency fixing measures are taken to prevent the cutter box from retreating due to pressure relief.
(4) **Damage and roughening of cutter pot:** The damaged and roughened parts are ground smoothly. If necessary, the cutter pot may be replaced with a new one.

8.2.4 Cutter replacement under pressure

8.2.4.1 *Process of cutter replacement under pressure* (*Fig. 8.22*)

8.2.4.2 *Preparation for cutter replacement under pressure*

(1) **Main construction equipment (Table 8.6)**
(2) **Main construction materials (Table 8.7)**

Table 8.6. Main equipment for construction under pressure.

S. No.	Equipment	Number for service	Purpose	Remarks
1	Air compressor	5 (set)	Pressurized air supply to chamber	4 for emergency
2	Air tank	8 (No.)	Pressurized air supply to chamber	1 for emergency
3	Cooling system	1 (set)	Cooling of air compressor	
4	Generator	1 (set)	Emergency power supply	
5	Double-headed truck	1 (set)	Horizontal transport	
6	Traveling crane	1 (set)	Vertical transport	
7	Electric welder	2 (set)	Welding	
8	Fresh slurry mixing system	1 (set)	Muck replacement in chamber and front slurry membrane	
9	Emergency treatment equipment	1 (set)	Pressurized treatment	

Table 8.7. Main materials for construction under pressure.

S. No.	Item	Number for service	Purpose
1	Cutter	Several	Cutter replacement
2	Medical oxygen supply	2 cylinders/shift	Breathing under reduced pressure
3	Fresh slurry materials	60 m^3/d	Slurry replacement in chamber
4	Shield tail grease	1 t	Shield tail seal

(3) **Compressed air station setting:** Air is supplied from the air compressor equipped in the shield machine or another air compressor set on the ground. The air compressor of the shield machine adopts high-voltage power supply mode, and the shield machine frame is equipped with a diesel generator connected to the air compressor as an emergency power supply.

(4) **Sludge replacement in the chamber:** In order to ensure the stability of the soil mass ahead and the safety of operators during the pressurized construction process, fresh slurry is supplied to replace the slurry in the chamber and form an effective slurry membrane on the front to prevent the front soil mass from collapsing or falling due to the air pressure loss in the chamber, which will harm the operators in the chamber.

With the fresh slurry pumped to the rear chamber and subsequent liquid level rising to +1 m, slurry is discharged from the top to the rear chamber until the liquid level reaches 0, and slurry pumping and discharge are repeated in circulation until the viscosity of the slurry discharged from the top reaches 35 s, and then pressure is maintained for 12 h. After the pressure stabilization is completed, the joint pressure test of the connected front and rear chambers is conducted for 2 h, and the chambers cannot be opened until it is confirmed that there is no abnormality.

(5) **Determination of air pressure in chamber:** The chamber inlet pressure is calculated according to the burial depth of the shield and the groundwater level at each stop. When the chamber inlet pressure exceeds 5 bar, operation is required under saturated pressure and in the process, as shown in Fig. 8.23.

(6) **Evaluation standard for personnel safety for chamber entry and exit in the pressurized construction process:** After the pressure in the front cabin drops to the set value, observe the change in the liquid level. The chamber door cannot be opened until the liquid level is stable for 2 h and the pressure sensor at the top of the front chamber is kept unchanged.

When the air flow meter fluctuates in amplitude more than 200 m^3/h or the liquid level in the chamber fluctuates in amplitude more than 0.05 m/min, the occupants should be organized and evacuated quickly, while the liquid level and support pressure are restored and fresh slurry is injected into the front chamber. When the support reaches its effect, the chamber is re-opened.

The staff entering the chamber should check the slurry membrane quality. If it is found with cracks or signs of leakage and a landslide on the front, evacuation should be organized quickly, while fresh slurry should be supplied for replacement.

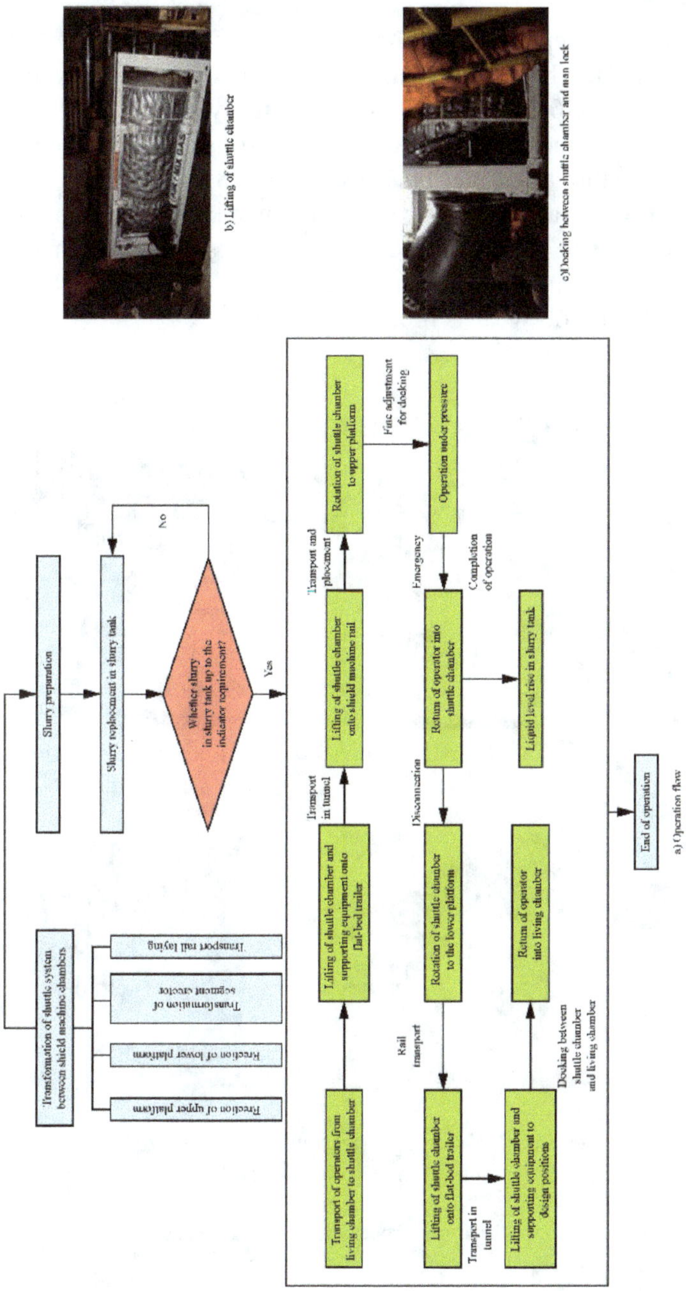

Fig. 8.23. Process of operation under saturated pressure.

8.2.4.3 *Cutter inspection and replacement*

(1) **Pressurization and depressurization process for chamber entry and exit (Fig. 8.24):** The pressurization rate in the man lock should be controlled at 0.05–0.10 MPa/10 min, and the working time for operators in the chamber and depressurization time should comply with the relevant national regulations. See Table 8.8 for details.
(2) **Cutter replacement standard:** The hob ring must be replaced in the case of eccentric wear, falling off, cracks, looseness, displacement, etc. Replacement is compulsory when the edge hob wear is more than 10 mm and the front hob wear is more than 15 mm.

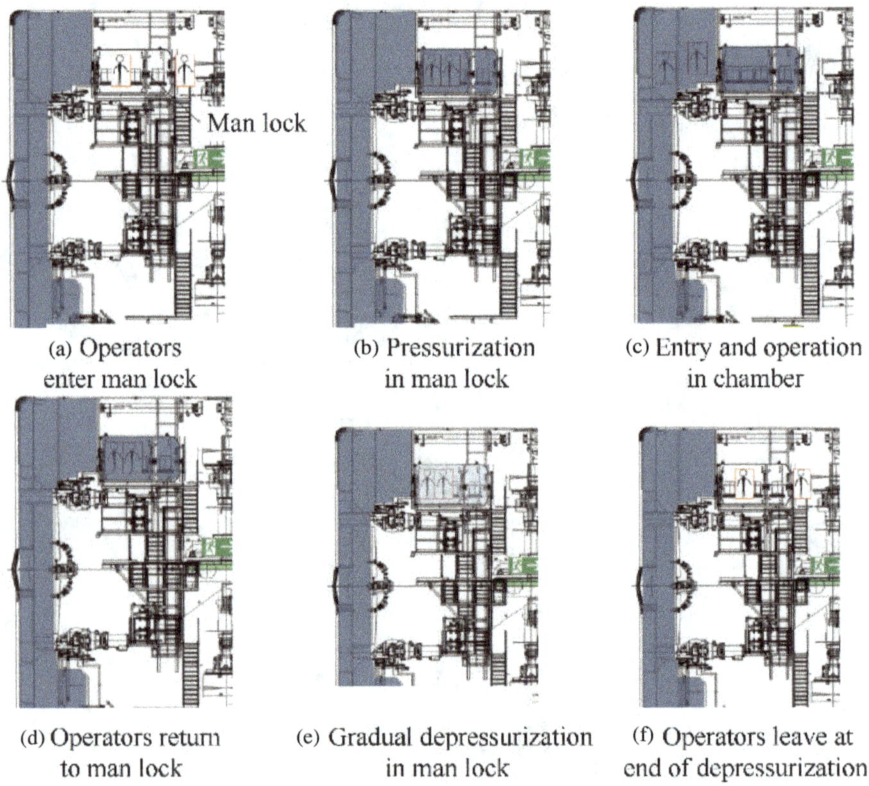

Fig. 8.24. Pressurization and depressurization process for operator transfer into and out of the chamber.

Table 8.8. Working time and pressure reduction time for pneumatic operations.

Working pressure P(MPa)	Working time (h)	First decompression time (min)	Sectional pressure reduction (MPa): 0.24 Retention period (min)	0.24 Decompression time (min)	0.21 Retention period (min)	0.21 Decompression time (min)	0.18 Retention period (min)	0.18 Decompression time (min)	0.15 Retention period (min)	0.15 Decompression time (min)	0.12 Retention period (min)	0.12 Decompression time (min)	0.09 Retention period (min)	0.09 Decompression time (min)	0.06 Retention period (min)	0.06 Decompression time (min)	0.03 Retention period (min)	Time from the last decompression stop to complete relief (min)	Total decompression time (min)
$0<P\leq0.12$	0~4		—	—	—	—	—	—	—	—	—	—	—	—		—	—	—	12
$0.12<P\leq0.15$	1		—	—	—	—	—	—	—	—	—	—	—	—		—	—	—	12~15
	2	4	—	—	—	—	—	—	—	—	—	—	—	—		—	10	3	17
	3	4	—	—	—	—	—	—	—	—	—	—	—	—		—	15	3	22
	4	3	—	—	—	—	—	—	—	—	—	—	—	—	10	3	10	3	29
$0.15<P\leq0.18$	1	5	—	—	—	—	—	—	—	—	—	—	—	—	—	—	10	3	18
	2	4	—	—	—	—	—	—	—	—	—	—	—	—	5	3	10	3	25
	3	4	—	—	—	—	—	—	—	—	—	—	—	—	10	3	15	3	35
	4	3	—	—	—	—	—	—	—	—	—	—	10	3	10	3	20	3	52
$0.18<P\leq0.21$	1	6	—	—	—	—	—	—	—	—	—	—	—	—	—	—	15	3	24
	2	4	—	—	—	—	—	—	—	—	—	—	5	3	10	3	20	3	48
	3	3	—	—	—	—	—	—	—	—	5	3	10	3	15	3	25	3	70
	4	3	—	—	—	—	—	—	—	—	10	3	15	3	20	3	30	3	90
$0.21<P\leq0.24$	1	6	—	—	—	—	—	—	—	—	—	—	—	—	10	3	15	3	37
	2	5	—	—	—	—	—	—	—	—	—	—	10	3	20	3	25	3	69
	3	3	—	—	—	—	—	—	5	3	10	3	15	3	25	3	40	3	113
	4	3	—	—	—	—	—	—	5	3	15	3	25	3	30	3	45	3	138

(Continued)

Table 8.8. (*Continued*)

Working pressure *P*(MPa)	Working time (h)	First decom-pression time (min)	Sectional pressure reduction (MPa) 0.24 Retention period (min)	0.24 Decom-pression time (min)	0.21 Retention period (min)	0.21 Decom-pression time (min)	0.18 Retention period (min)	0.18 Decom-pression time (min)	0.15 Retention period (min)	0.15 Decom-pression time (min)	0.12 Retention period (min)	0.12 Decom-pression time (min)	0.09 Retention period (min)	0.09 Decom-pression time (min)	0.06 Retention period (min)	0.06 Decom-pression time (min)	0.03 Retention period (min)	Time from the last decom-pression stop to complete relief (min)	Total decom-pression time (min)
0.24<P≤0.27	1	6	—	—	—	—	—	—	—	—	—	—	5	3	10	3	15	3	45
	2	4	—	—	—	—	—	—	5	3	10	3	15	3	25	3	30	3	104
	3	4	—	—	—	—	—	—	10	3	15	3	25	3	35	3	45	3	144
0.27<P≤0.30	1	7	—	—	—	—	—	—	—	—	—	—	5	3	15	3	25	3	61
	2	5	—	—	—	—	—	—	5	3	10	3	20	3	30	3	45	3	130
	3	4	—	—	—	—	5	3	15	3	20	3	30	3	40	3	60	3	192
0.30<P≤0.33	1	7	—	—	—	—	—	—	—	—	5	3	10	3	15	3	25	3	74
	2	5	—	—	—	—	5	3	10	3	15	3	25	3	35	3	50	3	163
	3	4	—	—	5	3	10	3	15	3	25	3	35	3	40	3	55	3	207
0.33<P≤0.36	1	8	—	—	—	—	—	—	5	3	10	3	15	3	20	3	25	3	85
	2	5	—	—	5	3	10	3	15	3	20	3	30	3	40	3	50	3	196
	3	5	5	3	10	3	15	3	20	3	30	3	40	3	45	3	55	3	249

Replacement is necessary when two of the standard alloy teeth of the scraper reach a wear amount of more than 15 mm and the edge of the toothed cutter reaches a wear amount of more than 10 mm.

8.2.4.4 *Safety protection measures for entry and operation in chamber*

(1) **Equipment support**

(i) Confirm the mixing quality of fresh slurry and that the pressure injection pipeline and equipment are in normal operation.
(ii) Calibrate the safety instruments and equipment, such as safety valves and pressure gauges, used for pneumatic equipment regularly, and perform routine maintenance and trial operations as required. Starting of the safety valve should be tried once a day to check its reliability, and any problems found should be handled in time.
(iii) Arrange specially designated personnel to regularly remove the oil and water in the tank during the use of the air tank.
(iv) Frequently detect the compressed air consumption in the pressurized work area, and properly adjust the air supply parameters according to the actual consumption. In any case, the air supply should not be less than 20 m^3/person per minute.

(2) **Personnel management**

(i) Ensure the quality of the slurry membrane after replacement with fresh slurry in the excavation chamber.
(ii) All high-pressure operators must undergo strict physical examination, training and education (including pressurized exercise and oxygen sensitivity test), and full-time safety officers and experienced foremen are designated for high-pressure operations. During pressurized operation, a guard must be assigned to monitor the change of pressure in the chamber at all times.
(iii) Pressurization, depressurization and working time for operation in the chamber should be in strict accordance with the relevant specifications and requirements related to pressurized construction.
(iv) The staff entering the chamber should first test whether there is toxic and harmful gas in the chamber and then start operation in the chamber after confirming safety. After entering the chamber, the effectiveness of the slurry membrane is confirmed first.

If necessary, reinforcement measures can be taken immediately, such as spraying cement mortar on the soil mass exposed at the cutterhead.

(v) No fire is allowed in the chamber, for which a fire extinguishing system is established and low-voltage explosion-proof lighting is used. If it is necessary to use electric welding and gas cutting, more stringent safety inspections of the equipment are performed, the ventilation capacity is enhanced, and fire-fighting equipment is added. Construction must not be started until it is approved by the relevant authority.

(3) **Emergency measures**

(i) During the pressurized operation, the water supply and electricity management departments should be kept in timely contact to avoid sudden power failure and water supply failure. Generators, sufficient fuel, sound labor protection facilities and special air pressure management and medical personnel must be available on site to ensure the health of construction personnel.

(ii) A double-power supply is adopted to ensure the construction safety of pressurized operation; all fixed or mobile electrical equipment and lighting must meet the safety requirements and should be inspected regularly.

8.3 Shield tail seal management

See Table 8.9 for the development history of the shield tail seal. The shield tail seal was originally designed as a rubber lip seal and has been gradually developed into the standard shield tail brush widely used today in engineering practice, composed of a parallel spring plate in the front plate and a staggered spring plate at the back. In 2005, in the Hallandsas Railway Tunnel in Sweden, in order to meet the water and soil pressure-bearing capacity of more than 1 MPa, the first generation of reinforced shield tail brush was produced, and the front plate was modified into a staggered spring steel plate. In 2014, shield tail leakage occurred at the deepest point of the tunnel underneath the Bosphorus Strait in Istanbul, Turkey, and the second generation of reinforced shield tail brush was produced with further improvement to the shield tail brush. The front plate was modified into two interlaced spring steel plates covered with another spring steel plate.

Table 8.9. Development history of shield tail seal.

Time	Type	Tunnel	Shield type	Center pressure	Photo
1998	Rubber lip seal	Tweede Heinenoord Tunnel, Netherlands	Slurry balanced; ϕ8.55 m	0.4 MPa	
1999	Standard shield tail brush	Cairo Metro Line 2, Egypt	Slurry balanced, ϕ9.43 m	0.35 MPa	
2005	First-generation reinforced shield tail brush	Hallandsas Railway Tunnel, Sweden	Slurry balanced, ϕ10.53 m	1 MPa	
2014	Second-generation reinforced shield tail brush	Tunnel underneath the Bosphorous Strain in Istanbul, Turkey	Slurry balanced, ϕ13.66 m	1.2 MPa	

8.3.1 Shield tail sealing principle

Super-large-diameter shield machines are widely used in the construction of large underwater tunnels crossing rivers, lakes and seas, featuring long tunneling distance, high water head pressure, and cutting in complex ground. The sealing effect of the shield tail is particularly important for engineering safety. Large-diameter shield machines are generally conical; that is, the diameter of the shield tail is smaller than that of the front shield and the middle shield to prevent the shield from jamming. However, such a design increases the passage of slurry backflow, bringing more leakage risks to the shield tail.

(a) Position of shield tail seal

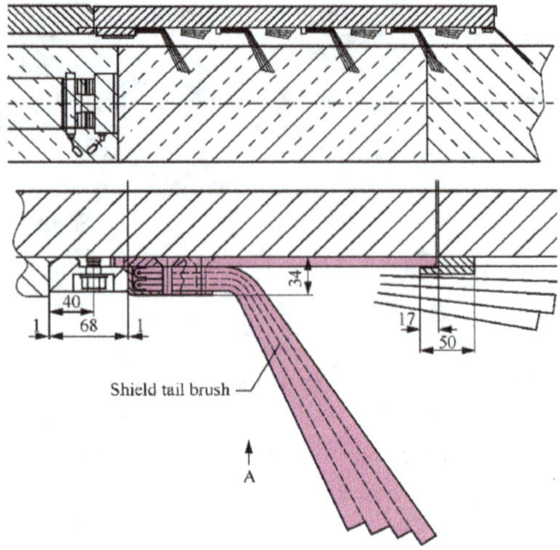

(b) Structure of shield tail seal

Fig. 8.25. Structural diagram of shield tail seal (unit: mm).

The structure of the shield tail seal is shown in Fig. 8.25. The shield tail seal system is generally composed of the steel structure of the shield tail shell, the shield tail seal, and the shield tail seal grease.

(1) The steel structure of the shield tail shell is the load-bearing component of the shield tail, which bears the external water and soil pressure and also provides space for the erection of segments into rings.
(2) The shield tail seals are generally designed as a rubber lip seal or shield tail brush composed of steel plates and bundled wires. At present, shield tail brushes are mostly used.
(3) The shield tail seal grease is mainly used to lubricate, seal and prevent the shield tail from deformation, to effectively protect the shield tail brush and to build joint isolation against the backflow of soil sediment and grouting materials.
(4) The sealed annular cavity between two adjacent shield tail brushes is pumped with shield tail sealing grease by a shield tail grease pump and shield tail grease pipe to ensure sufficient grease filling in the cavity, build a certain grease cavity pressure and resist external water and soil pressure.
(5) The grease pipe at the shield tail is generally embedded in the shield tail steel structure or laid in an exposed manner on the inner surface

of the shield tail steel structure. The number of grease pipes generally depends on the diameter of the shield tail.

8.3.2 Seal control at shield tail

8.3.2.1 *Selection of shield tail brush*

The shield tail brush is generally welded or assembled. The welded shield tail brush is directly welded to the shield tail, and the assembled shield tail brush is fixed to the shield tail by fixing bolts, nuts and pressing plates. The advantage of the assembled shield tail brush lies in its convenient replacement without hot work required at the shield tail, and the assembly accuracy is high.

The following factors are considered when selecting a shield tail brush:

(1) the clearance between shield tail steel structure and segment;
(2) the minimum turning radius of the tunnel;
(3) the maximum allowable external water and soil pressure;
(4) the length of shield tail steel structure.

8.3.2.2 *Selection of shield tail seal grease*

The shield tail seal grease plays a role in rust prevention, corrosion prevention and wear reduction on the wire brush and steel structure, as well as in improving the construction quality and working efficiency of the shield machines.

Shield tail seal grease should have the following characteristics:

(1) high water pressure resistance and good water stop performance;
(2) good pumping performance;
(3) easy operation for filling between wire brushes;
(4) good resistance to cement slurry invasion;
(5) high compliance with the uneven parts of concrete segments;
(6) non-erosion to the seal packing at joints of concrete segments;
(7) low adhesion to the surface of the concrete segment;
(8) biodegradable, harmless to water and the environment with test proof, and non-irritant, as proven by oral allergy tests with rats.

8.3.2.3 *Grease pipe setting at shield tail*

Each shield tail brush is set with shield tail grease pipes, spaced generally not less than 2.5 m.

8.3.2.4 *Control of construction parameters*

(1) **Management of grease injection for shield tail seal:** Grease injection at the shield tail seal should be controlled in both the injection amount and injection pressure. The injection amount per ring is calculated according to the grease loss. Moreover, the cavity is kept full of grease, and the pressure in the cavity is maintained sufficient to resist the external water and soil pressure.

(i) *Manual application of grease*: Before the shield launching, each shield tail brush is applied manually with shield tail grease, selected with high viscosity and strong adhesion to the steel wire to prevent other hot work. Application of shield tail grease is shown in Fig. 8.26. Grease is applied in four passes, with one layer at the front and rear outer steel sheets of the shield tail brush and one layer at both sides of the middle wire mesh, and each layer should be applied to the bottom; grease is applied at the root of the wire brush and filled in the gap between the wire brushes to prevent groundwater, slurry and synchronous grout from flowing into the wire mesh and causing rusting.

(ii) *Calculation of shield tail grease consumption per ring*: The shield tail seal grease consumption per ring is mainly composed of two parts: grease loss on the outer arc face of segments and filling in the chamfer of the circumferential and longitudinal seams on the outer arc face of segments. The calculation formula is expressed as follows:

(a) The formula for calculating the grease loss on the arc face of segments is

$$V_1 = CL\sigma, \tag{8.10}$$

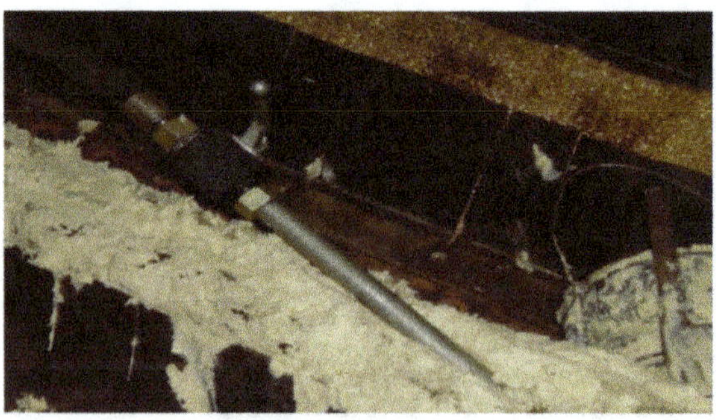

Fig. 8.26. Application of shield tail grease.

where

V_1 — volume of grease loss on the arc face of segment (m^3);
C — circumference of outer arc surface of segment (m);
L — ring width of each ring of segments (m);
σ — thickness of grease loss on the arc face of segment (mm), taken as 2 mm from domestic experience.

(b) The formula for calculating the filling amount of the chamfer of the circumferential and longitudinal seams on the outer arc surface of segments is

$$V_2 = V_c + V_d, \tag{8.11}$$

where,

V_2 — filling volume of chamfer of circumferential and longitudinal seams on the outer arc face of segments (m^3);
V_c — chamfer volume of circumferential seam on the outer arc face of segment (m^3);
V_d — chamfer volume of longitudinal seam on the outer arc face of segment (m^3).

(c) The formula for calculating the total amount of grease injected into each ring shield tail seal is expressed as follows:

$$m = \rho(V_1 + V_2), \tag{8.12}$$

where

m — injection amount of sealing grease at the tail of each ring shield (kg);
ρ — Grease density (kg/m^3), generally taken as 1,300 kg/m^3.

(iii) *Pressure control in shield tail grease cavity*
The pressure control of the grease cavity depends on the external water and soil pressure, the number of grease cavities and the quality of the shield tail brush.

(a) The difference between the pressure of the last grease cavity in the heading direction and the external water and soil pressure shall not be less than 100 kPa.
(b) According to the number of grease cavities, the pressure of each grease cavity is controlled and distributed in certain steps, as shown in Fig. 8.27.

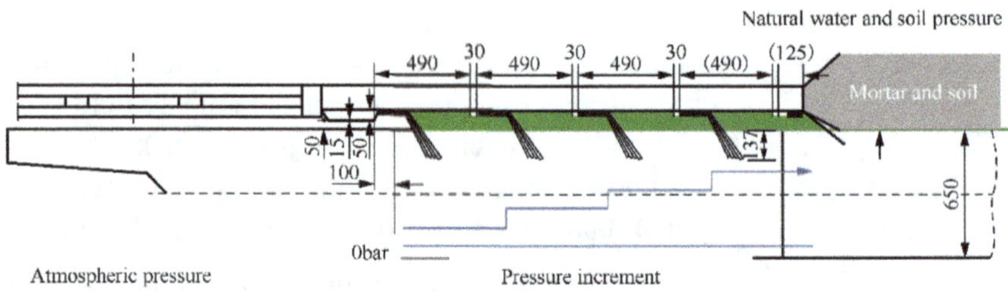

Fig. 8.27. Stepped pressure increase of shield tail seal chamber (unit: mm).

(iv) *Shield tail seal grease injection control*

(a) Under normal tunneling conditions, the automatic mode is selected to inject grease to supplement the loss in the grease cavity.
(b) In the case of shutdown, the manual mode is selected to inject makeup grease at the point with two low pressures in the grease cavity.
(c) The injection pressure should not be too high, and should be controlled according to the set pressure of each grease cavity.
(d) Grease replacement should be conducted under clean conditions to avoid sand and other granular matter being mixed into the grease bucket.

(2) **Clearance control at shield tail:** The shield tunneling is carried out along the design alignment as far as possible to avoid alignment deviation. However, in the actual tunneling process, due to geological conditions and other reasons, the attitude of the shield often deviates from the design alignment, resulting in too small shield tail clearance and segment damage.

A too small shield tail clearance will cause the shield tail brushes to be excessively squeezed, resulting in relative movement with segments, which will cause the spring plate and steel wire of the shield tail seal to fall off, undermining the sealing effect. The broken pieces of segments fall into the back of the segments and are squeezed into the grease tank through the shield tail brush, blocking the grease flow channel and leading to the failure of the shield tail seal.

A too large shield tail clearance will cause the shield tail brushes to fold instead of fully squeezing, which will damage the spring steel plate and steel wire of the shield tail brush, undermining the sealing effect.

If the shield machine deviates from the design alignment, do not rush to conduct a deviation correction. In order to ensure the good working performance of the shield tail brush seal and also to ensure the segments are free of damage, the shield tail clearance should be controlled as far as possible to maintain a more uniform shield attitude.

The shield tail clearance is deemed the main basis for segment selection. After erecting the last segment, the shield tail clearance at the top, bottom, left and right positions of the segment must be measured, and the tunneling of the shield machine should be adjusted according to the shield tail clearance and tunnel alignment to maintain a rational clearance.

8.3.3 Shield tail brush replacement

8.3.3.1 *Failure modes and reasons of shield tail seal*

A shield tail seal is deemed failed when its sealing effect cannot meet the requirements of use, which will lead to water leakage or slurry leakage at the shield tail and even water and sand gushing in serious cases.

During shield tunneling, the common shield tail seal failure modes are as follows:

(1) The shield tail brush is in elastic failure and cannot recover the elastic deformation itself.
(2) The grease cavity is under insufficient pressure or contains an insufficient amount of grease.
(3) The grease cavity is mixed with any foreign matter to form a leakage channel.
(4) The steel wire or spring plate of the shield tail brush is worn and falls off, causing damage to the shield tail seal.

The above shield tail seal failures are mainly caused by the following reasons:

(1) Due to a large change in shield tail clearance, the shield tail brush is caught in elastic failure due to plastic deformation caused by excessive squeezing from the eccentric segments.
(2) If the shield tail seal grease is poor in quality and fluidity, it cannot completely fill the grease cavity.
(3) A too small amount of grease is injected into the shield tail seal.
(4) Due to the failure to timely clear the segment fragments or sludge caused by the erection of segments at the shield tail in time, such

fragments or sludge carried into the grease cavity along with the shield tail tunneling cause abnormal wear of the shield tail brush.

(5) Under improper pressure control during the synchronous grouting, the slurry spills over the shield tail brush and is then solidified into blocks, and such slurry blocks without clearing for a long period of time not only cause abnormal wear of the shield tail brush but also lead to insufficient grease filling, undermining the sealing effect.

8.3.3.2 *Risk analysis of shield tail brush replacement*

There are certain risks in replacing the shield tail brush during shield tunneling. First of all, the shield tail is in a high-pressure external water head environment. Any damage to the shield tail brush will create a leakage channel, and if the shield tail brush is replaced under this working condition, the removed segment may cause leakage to the shield tail; second, if the shield tail brush is to be exposed for replacement, the stroke of the thrust cylinder must be greater than that under normal construction conditions, and the effective length of the overlap with segments within the scope of the shield tail decreases after the segment is removed, so the shield machine and segment attitudes may change suddenly, causing more serious shield tail leakage.

8.3.3.3 *Shield tail brush replacement process*

Generally, a large-diameter shield machine working under external high pressure is designed for the replacement of two shield tail brushes conducted under atmospheric pressure, and such a design has been applied in many projects. The specific replacement process is shown in Fig. 8.28.

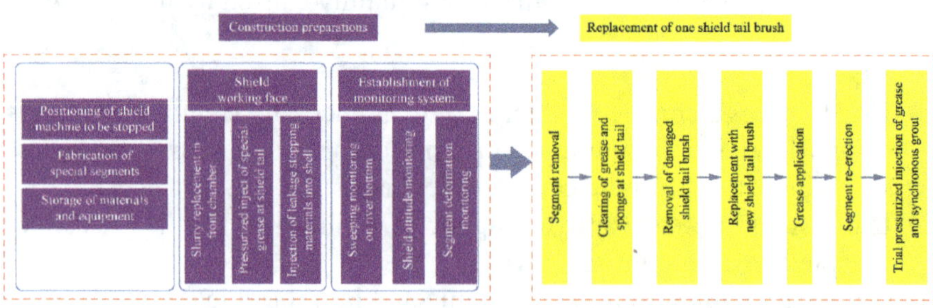

Fig. 8.28. Shield tail brush replacement process.

(1) **Preparations:** including positioning of the shield machine to be shut down, timing of shutdown and replacement, selection of segment removal method, scheme preparation and technical safety disclosure for approval, division of tasks, personnel allocation and material and equipment preparation.
(2) **Shield advance measures before replacement:** including cutting water pressure control before shutdown, synchronous grouting control, slurry quality control, pressurized injection of special waterproof grease at shield tail, segment arc face and segment erection control.
(3) **Segment removal and shield tail brush replacement:** including construction guarantee measures, segment removal sequence selection and shield tail brush replacement.
(4) **Resumption of advance:** including the preparation and implementation for resumption of advance.
(5) **Emergency response.**

8.3.3.4 *Preparations before replacing the shield tail brush*

(1) **Positioning of shield machine to be shut down:** The shield tail brush should be replaced in the section with good geological conditions and a stable stratum, and the area with shallow overburdens and the surrounding sensitive buildings (structures) should be avoided. If unstable strata and sensitive buildings (structures) cannot be bypassed, corresponding safety technical measures must be taken to ensure the safety of the shield tail brush replacement process.
(2) **Selection of segment removal method:** The segment removal method should be selected according to the actual conditions on site to ensure that a whole ring of shield tail brushes can be replaced, the replaced tail brushes are maintained in integrity at the longitudinal seam overlap, and the first ring is formed as soon as possible after an emergency occurs.

(3) **Preparation of materials**

(i) *Fabrication of special segments*: One ring of special segments is produced for erection into the third ring ahead of the current ring (separated from the shield tail). The segment is provided with grouting holes, with an additional eight grouting holes (except for the closure block) per segment for pressurized injection of polyurethane for plugging in emergencies, as shown in Figs. 8.29 and 8.30.

(ii) *Preparation of materials*: Before construction, materials should be prepared, including emergency plugging grease (CES plugging grease), high-viscosity fresh slurry, slurry for pressurized injection into the shell, polyurethane and with a certain margin for various materials.

(iii) *Equipment preparation*: See Table 8.10 for shield tail brush replacement supplies.

(4) **Special ground reinforcement:** In some areas with poor geological conditions, such as high-water-head sandy soil layers with high formation permeability coefficients, two shield tail brushes need to be

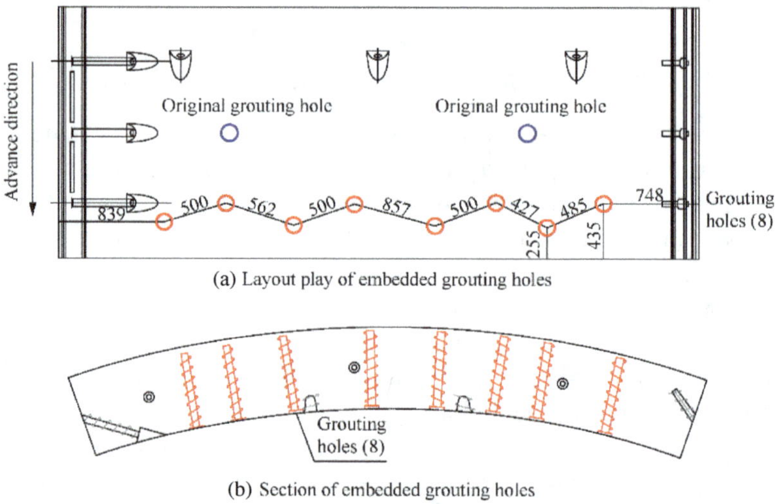

(a) Layout play of embedded grouting holes

(b) Section of embedded grouting holes

Fig. 8.29. Schematic diagram of special segments with additional grouting holes (unit: mm).

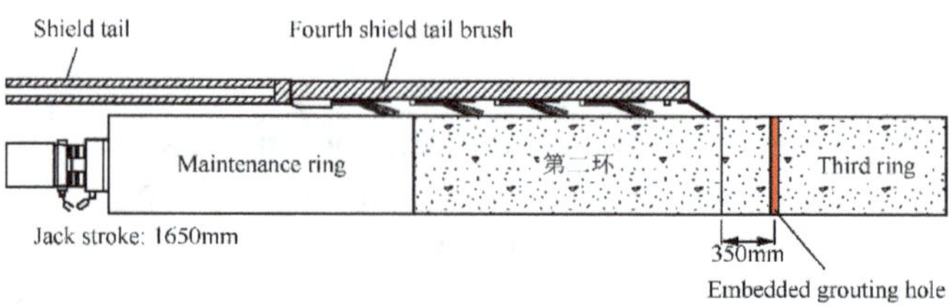

Fig. 8.30. Schematic diagram of special ring segment position.

Table 8.10. Supplies for shield tail brush replacement.

S. No.	Item	Number
1	Carbon electrode	—
2	M12 nut	—
3	Welding screw	—
4	Pressing plate	—
5	Ceramic ring	—
6	Torque wrench	2 (piece)
7	Open-end and monkey wrench	10 (piece)
8	Carbon-arc gouging welder	1 (set)
9	Electric welder	1 (set)
10	Oxygen acetylene cutter	1 (piece)
11	Polisher	1 (set)
12	Asbestos cloth	20 (coil)
13	Power extension cord	2 (piece)
14	Carbon dioxide cylinder	3 (set)
15	Oxygen cylinder	3 (set)
16	Acetylene cylinder	3 (set)
17	Fire extinguisher	10 (set)
18	Plastic woven bag and other cleaning tools	Several

replaced. If no measures are taken to directly replace the shield tail brush, there will be great risks. Therefore, the soil mass at the shield tail must be frozen and reinforced so as to replace the shield tail brush under relatively safe conditions.

(i) *Freezing scheme*: During the prefabrication of concrete segments, the freezing pipes should be embedded in the concrete segments in advance, as shown in Fig. 8.31.

After the segments embedded with freezing pipes (segments for the frozen ring) and the segments at the maintenance position of the shield tail brush (segments for the working ring) are erected, refrigerant is circulated in the freezing pipes to freeze the stratum near the shield tail and segment joint, and then the segments of the maintenance ring are removed to inspect and repair the steel wire brush. After the maintenance work is completed, the segments are re-erected in the maintenance ring, and hot water is circulated in the freezing pipes to melt the frozen soil nearby. After the frozen

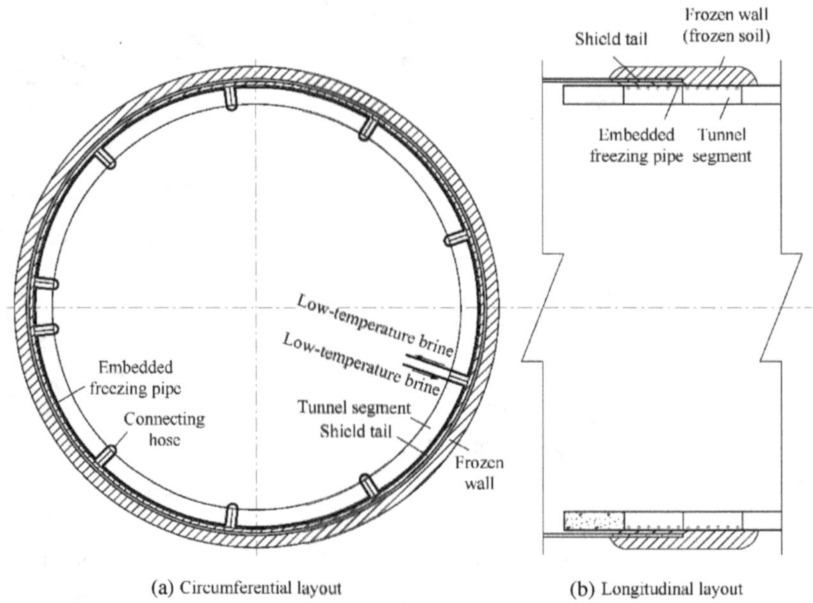

Fig. 8.31. Schematic diagram of embedded freezing pipe.

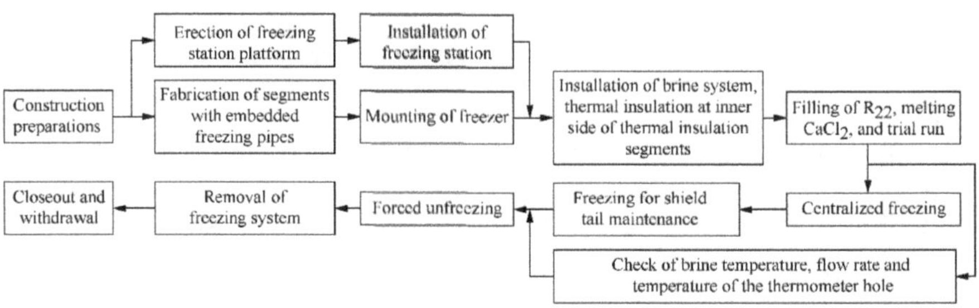

Fig. 8.32. Construction process of freezing reinforcement.

soil at the interface between the shield tail and the frozen wall thaws, the normal shield advance is resumed.

(ii) *Freezing process flow*: The construction process of freezing reinforcement is shown in Fig. 8.32.

(iii) *Fabrication of special segments necessary for freezing reinforcement*: The freezing pipes and the inner arc face connecting steel plates are embedded in the ordinary segments. Totally, three rings of segments with embedded freezing pipe (two rings for standby) and six rings of segments with embedded steel plate (including

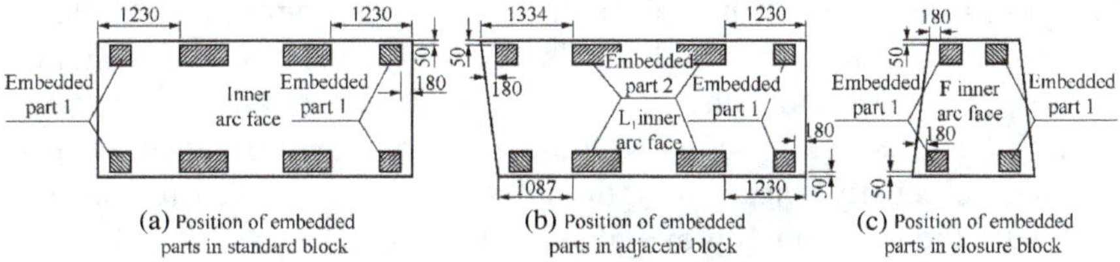

Fig. 8.33. Schematic diagram of embedded parts on inner arc face (unit: mm).

Fig. 8.34. Standard block embedded with freezing pipe.

three rings of embedded freezing pipe segments) should be fabricated. The freezing pipes are made of 38 × 3 mm seamless steel pipes, equidistant at six circles, as shown in Fig. 8.33.

As shown in Fig. 8.34, taking the standard segment block with embedded freezing pipes as an example, the length of a freezing pipe is 5,882 mm (including the elbow), and six freezing pipes are set for each block. A total of 42 freezing pipes are set for standard segment blocks in a ring and welded to the seamless steel pipes at the 90° elbow, and the seamless steel pipe is bent at a radius of 7,510 mm.

8.3.3.5 *Shield advance control before replacement*

(1) **Cutting water pressure control:** The cutting water pressure value should be set reasonably, and pressure fluctuation should be minimized to ensure stability going forward. When the shield machine advances in the maintenance ring, the top pressure is controlled steadily. When the shield machine advances to the maintenance position, the liquid level is raised to make the top pressure slightly higher than the top pressure when advancing.

(2) **Synchronous grouting control:** Synchronous grouting should be conducted timely, uniformly and sufficiently during shield tunneling. During construction, the grout with good performance is prepared, and the effective grouting volume and grout performance, such as specific density, shear strength and slump, are checked regularly to ensure the grout quality of synchronous grouting.
(3) **Slurry quality control:** The test frequency of slurry should be increased to ensure the quality of slurry and the stability of the soil mass ahead during a long-term shutdown. The slurry index should be more than 21 s.
(4) **Shield tail special plugging grease injection:** The three rings before shutdown should be strictly controlled with the grease injection amount, and special plugging grease should be injected into the grease cavity. Before the removal of the maintenance segment, grease is injected to maintain pressure in the grease cavity. During the replacement of the shield tail brush, pressure in the grease cavity of the shield tail brush rather than that to be replaced is closely monitored, and pressurization is conducted in time in cases of too low pressure.
(5) **Segment erection control:** Segments are the basic components that constitute the advance axis of the shield machine. Therefore, it is of great significance to strictly control the segment erection quality and ensure a reasonable advance of the segment rings to prevent the shield machine from sinking. In the construction process, the segment rotation angle should be reasonably selected to ensure the necessary advance of the rings. During erection, careful operation should be carried out to ensure the flatness of the rings. Moreover, the axial position of segments should be accurately positioned to prevent damage to the shield tail brush caused by a too small local clearance between segment and shield shell.

8.3.3.6 *Segment removal and shield tail brush replacement*

(1) **Construction guarantee measures**

(i) *Pressurized injection of "Clay Shock" slurry into shell*: The "Clay Shock" method is a novel method to feed under pressure two kinds of liquid, i.e. high-concentration slurry material and plastic strength regulator (i.e. sodium silicate), to the designated position through piping, mix the two kinds of liquid into a high viscosity plastic gelling agent at an appropriate proportion, and then inject the mixture through the radial hole. The mixed-flow plastic gelling agent is not prone to dilution by water, with a viscosity that remains constant.

(a) *Injection material*: The "Clay Shock" slurry is injected through the vertical grouting hole provided on the shield shell to prevent the slurry from channeling back. The water–cement ratio of the "Clay Shock" slurry is 5:1, and the ratio of liquid A to liquid B is 20:1. Before use, the mix proportion test should be carried out in advance to check the effect. The liquids A and B of the "Clay Shock" slurry are shown in Fig. 8.35. The mixed "Clay Shock" slurry is shown in Fig. 8.36.

Fig. 8.35. Liquids A and B in "Clay Shock" slurry.

Fig. 8.36. "Clay Shock" mixture.

(b) *Injection process control*: Consumption should be calculated according to the theoretical clearance outside the shield shell of the shield machine, and the injection pressure should be slightly higher than the external water and soil pressure.

(ii) *Injection equipment*: The double-liquid slurry injection equipment configured for the shield machine can be used for injection. Test runs must be carried out before normal operation, and two globe valves must be installed before unblocking the holes connected with blocked pipelines to prevent the external slurry from flowing in.

(iii) *Slurry replacement in the front chamber with fresh one*: The high-viscosity fresh slurry mixed with high polymer materials and bentonite is injected into the front chamber through advance boreholes, which are pressed from the bottom to the top until the slurry index is stable, while an appropriate amount of plugging agent may be added to the slurry tank. After the replacement with fresh slurry, the excavation tank slurry index should be more than 35 s.

(iv) *Monitoring of main construction parameters during replacement*

(a) *Excavation face stability monitoring*: During the shutdown and replacement of the shield tail brush, the liquid level of the bubble chamber, pressure, top pressure, jack oil pressure and mileage should be recorded every 1 h; the quality of slurry in the chamber should be checked every few hours; the equipment maintenance personnel should regularly check the working condition of the air compressor daily to ensure that there is always an air compressor in working condition.

When the liquid level in the bubble chamber decreases by 1 m, the slurry tank should be replenished until the set value; when the bubble chamber pressure is lower than the set value of 0.05 bar, the operation of the air compressor should be checked immediately to ensure the stability of the excavation face.

(b) *Shield machine attitude*: The automatic measurement system for the shield advance is used to monitor the status of the shield machine when the shield tail brush is replaced. A specially designated person should be assigned by the team members to track the attitude of the shield machine and record it every hour, including mileage, head and tail elevation, plane, gradient, stroke, oil pressure and other changes. Before replacement, it should be ensured that the clearance around the shield machine is filled evenly.

(c) *Grease pressure at shield tail*: During the shutdown and replacement of the shield tail brush, the pressure in the grease cavity is recorded once every 1 h. If a pressure drop is found, the waterproof grease should be replenished in time.

(d) *Relevant monitoring during replacement*: During the replacement of the shield tail brush, surface settlement, building (structure) deformation, tunnel settlement and convergence, opening of circumferential and longitudinal seams and faulting of ends should be closely monitored within the main influence area. See Table 8.11 for details.

Table 8.11. Monitoring items and frequencies during replacement.

Monitoring item	Monitoring range	Layout of monitoring points	Monitoring frequency
Surface deformation	From 50 m before cutting to 50 m after cutting, within 10 m on both sides along the axis	1 section per 5 rings, more sections in sensitive section	Properly frequent
Building (structure) deformation	From 50 m before cutting to 50 m after cutting, within 10 m on both sides along the axis	More monitoring points for sensitive buildings	Properly frequent
Tunnel settlement	20 rings ahead of maintenance ring	1 section per 5 rings, 1 monitoring point respectively for vault and invert per section	2 rounds of monitoring/d in grouting period; 3 rounds of monitoring/d in maintenance period; 1 round of monitoring/d in resumption period
Tunnel convergence	3 rings ahead of maintenance ring	1 section per 1 ring	2 rounds of monitoring/d in grouting period; 3 rounds of monitoring/d in maintenance period; 1 round of monitoring/d in resumption period

(Continued)

Table 8.11. (*Continued*)

Monitoring item	Monitoring range	Layout of monitoring points	Monitoring frequency
Opening of circumferential and longitudinal seams, faulting of ends	3 rings ahead of maintenance ring	2 points/segment	2 rounds of monitoring/d in grouting period; 3 rounds of monitoring/d in maintenance period; 1 round of monitoring/d in resumption period
Shield tail clearance	—	19 points	1 round of monitoring respectively before and after removal of each segment

(e) *Grease cavity leak detection*: Before the segment is removed, the provided grouting hole in one ring ahead of the maintenance ring is opened to check whether there is leakage in the three cavities ahead of the shield tail.

(2) **Sequence of segment removal:** The sequences of segment removal and shield tail brush replacement are shown in Fig. 8.37.

(3) **Other precautions**

(i) *Manual grease application*: After the replacement of the shield tail brush, the repaired shield tail brush should be coated with shield tail grease, with the coating requirements consistent with those of the launching construction.

(ii) After the segment is removed without leakage, any foreign matter at the shield tail should be cleared immediately. After clearing, the shield tail brush is removed and replaced. The shield tail brush fixed with welded screws is cut with a cutter and re-screwed. When replacing the shield tail brush in the bottom area, it should be ensured that the working area is dry. When replacing the shield tail brush at the waist and top, pay attention to the measures against falling and fire prevention measures.

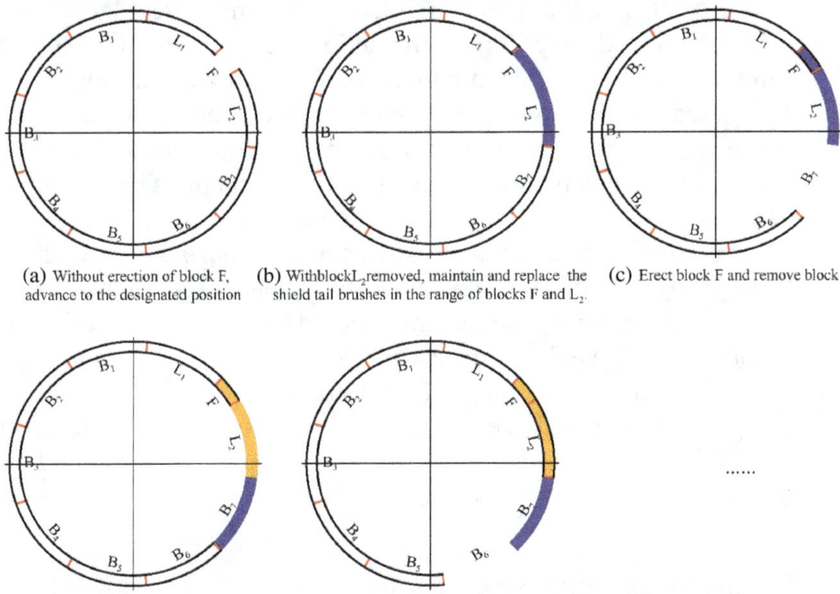

(a) Without erection of block F, advance to the designated position
(b) With block L_2 removed, maintain and replace the shield tail brushes in the range of blocks F and L_2.
(c) Erect block F and remove block B_7
(d) Maintain and replace the shield tail brushes in the range of block B_7.
(e) Erect block L_2, remove block B_6, maintain shield tail brushes in the range of B_6
(f) Maintain the other shield tail brushes, until an integral ring

Fig. 8.37. Sequence of segment removal and shield tail brush replacement.

Table 8.12. Emergency measures for replacement of shield tail brush.

Leakage	Measures
A small amount of clean water	Inject special waterproof grease into the first two cavities, inject synchronous grout at corresponding positions
Grout and sand leakage	(1) Inject special waterproof grease into the first two cavities, inject synchronous grout at corresponding positions; (2) Erect segments immediately into rings, resume advance as soon as possible when conditions are available; (3) Inject double-liquid slurry or polyurethane through secondary grouting holes in the ring just behind the shield tail for plugging, in case of continuous leakage.

(iii) When replacing the shield tail brush in the tunnel, ventilation, fire prevention and water inrush prevention measures should be taken in the tunnel. A sufficient number of fire extinguishers should be prepared and placed near the welding position and connected to a nearby water pipe. A large-flow pump is installed at the shield tail as an emergency measure. Any open fire found must be put out in time.

(iv) The protection of equipment is strengthened during construction. The key parts of the segment erector, such as the control valve group and electrical components, should be covered with asbestos cloth. During welding, protect the sensitive electrical equipment to prevent the power supply from burning.

(v) To prevent large leakage of the shield tail during the replacement of the shield tail brush, a 20 m^3 synchronous grout should be reserved in the grout bucket for emergencies.

8.3.3.7 *Emergency response measures*

A special emergency plan should be formulated. If the shield tail leaks during the replacement of the shield tail brush, the replacement should be stopped immediately, and emergency measures are taken. The replacement of the shield tail brush is continued after it is confirmed that there is no leakage. See Table 8.12 for details.

Chapter 9

Typical Engineering Case

With the development of equipment and construction technology in recent years, shield machines have been applied in a large number of ultra-long, super-large-diameter and deeply buried tunnels, with a higher and higher degree of difficulty in engineering construction. Due to different geographical environments and geological and hydrological conditions, the construction of different tunnels faces varied technical difficulties. Engineers have put forward numerous technical solutions and accumulated a lot of valuable experience, covering all kinds of difficulties. This chapter focuses on five typical tunnels in China and abroad and discusses the shield equipment and construction technology applied in each tunnel from the perspective of engineering technology in the expectation of providing some references for the construction of similar projects in the future.

9.1 Large-diameter ultra-long tunnel — Shanghai Yangtze River Tunnel

9.1.1 Project overview

The Shanghai Yangtze River Tunnel and Bridge Project starts from Wuhaogou in Pudong in the south, crosses the South Port of the Yangtze River, passes Changxing Island and the North Port of the Yangtze River, and ends at the east end of Chongming Island, connected with Chennai Highway, with a total length of 25.5 km and in two-way six lanes. The project is divided into the Yangtze River Tunnel and the Yangtze River

Bridge with Changxing Island as the boundary; that is, the South Port is connected with a tunnel and the North Port with a bridge to cross the Yangtze River (Fig. 9.1).

With a total design length of 8,955.26 m, a double-tube two-way tunnel was bored across the river with two φ 15.43 m slurry balanced shield machines by continuous shield tunneling at one shot. The up-traffic circular tube is 7,471.65 m long, and the down-traffic circular tube is 7,469.36 m long. The maximum gradient of the tunnel is 2.9%, the minimum planar radius of curvature is 4,000 m, the shallowest soil cover at the river bottom is about 14.0 m, and the thickest soil cover is about 29.0 m. There are four pump houses at the lowest point in the two tunnels underwater. Meanwhile, eight connecting passages are set between the two tubes, and three escape staircases are set between the two connecting passages, with a spacing of about 280 m.

The circular section of this tunnel is divided into upper and lower decks. The highway at the upper deck is designed with three lanes, 3.75 m wide in each lane and 12.25 m wide total, with a design speed of up to 80 km/h; the lower deck is provided with the conditions for rail transit lines in the future. The internal structure of the circular tunnel was built by the synchronous construction process; that is, the "hollow rectangular" prefabricated components in the middle of the road are assembled at the No. 2 frame synchronously with the shield tunneling. The pavement slabs on both sides are cast-in-situ structures and are placed on the corbels connected with segments. The strength grade of road concrete is C40.

The circular tunnel is lined with reinforced concrete-fabricated tapered segments of universal type with a taper of 40 mm, erected with staggered joints. The outer diameter of the lining ring is 15.0 m, the inner

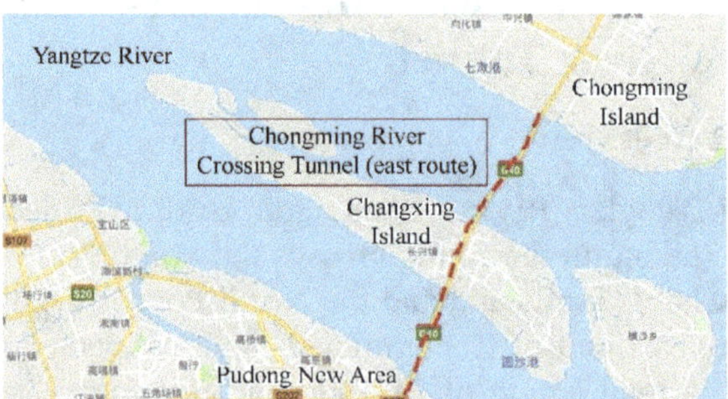

Fig. 9.1. Geographic location of Yangtze River Tunnel and Bridge in Shanghai.

diameter is 13.7 m, and the ring width is 2.0 m. The strength grade of segment concrete is C60, and the impermeability grade is S12. Each ring of lining consists of 10 segments, including seven standard blocks, two adjacent blocks and one closure block. Thirty-eight M30 longitudinal inclined bolts are used to longitudinally connect the rings of segments, and two M39 circumferential inclined bolts are used to circumferentially connect the blocks of segments. The tunnel layout section and circular structure of the tunnel lining are shown in Figs. 9.2 and 9.3.

Tunnel waterproofing is realized mainly by the self-waterproofing segments, focusing on the waterproofing of joints between segments to form a three-dimensional waterproofing system. The segments are provided with an EPDM waterstop and a hydro-expansive water strip, and the tunnel joint is attached to an elastic rubber gasket and a hydro-expansive strip to form a double waterproof system, as shown in Fig. 9.4. The EPDM used for the elastic rubber gasket is characterized by small compression permanent deformation, low stress relaxation change rate and good aging resistance.

The Shanghai Yangtze River Tunnel is mainly characterized by its large tunneling diameter, long tunneling distance at one shot and large burial depth. With an excavation diameter of up to 15.43 m, it was the largest shield tunnel in the world at that time, so there were many challenges in the stability of the shield excavation face, the erection of tapered

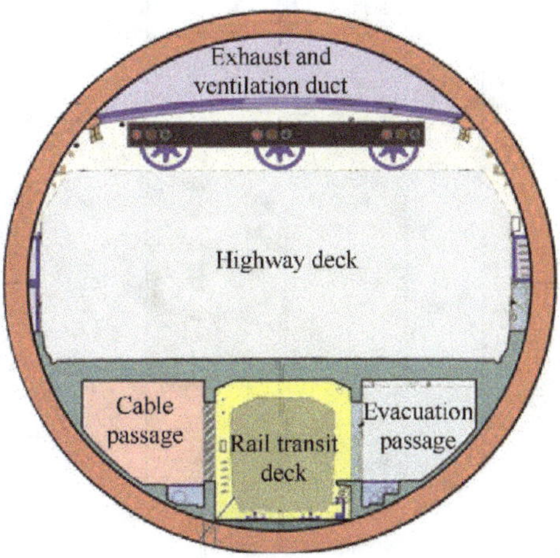

Fig. 9.2. Schematic diagram of tunnel layout section.

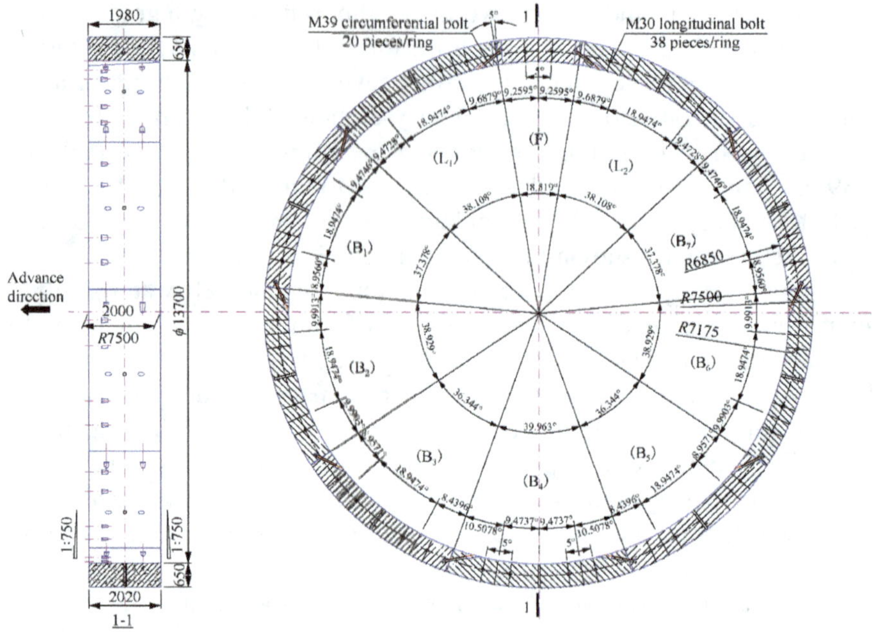

(a) Side view (b) Front view

Fig. 9.3. Schematic diagram of circular structure of tunnel lining (unit: mm).

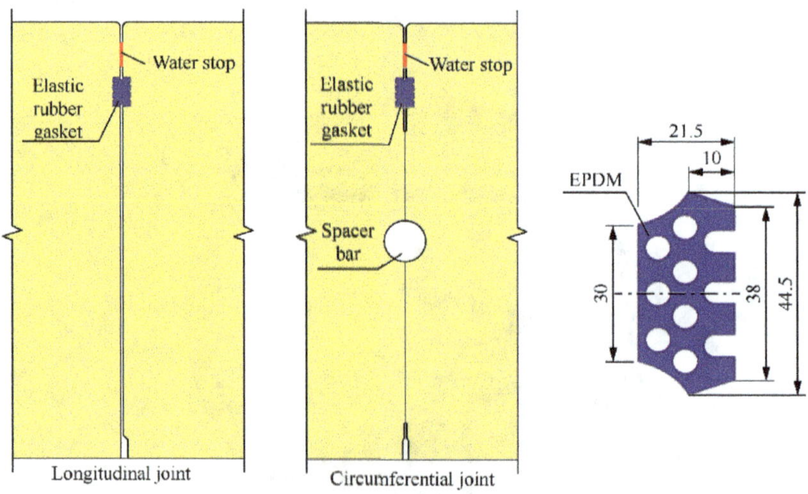

(a) Waterproofing at longitudinal joint (b) Waterproofing at circumferential joint (c) Sketch of Elastic Rubber Gasket

Fig. 9.4. Schematic diagram of waterproofing of tunnel segment joints (unit: mm).

segments with staggered joints and the stability of large-section tunnels during construction. The shield machine continuously tunneled for 7.5 km in one shot without a service shaft in the middle, which created a new world record at that time. This feature raised many puzzles on tunnel ventilation, fire prevention, accurate positioning of long-distance construction, and reasonable organization and management of transport. The maximum burial depth of the tunnel is 55 m at the river bottom. The shield construction under high water and soil pressure brought many risks to the waterproofing and seepage prevention of the tunnel and the construction of the connecting passages.

The shield was launched on September 23, 2006. The east route was successfully bored through on May 28, 2008, and the west route was successfully bored through on September 2, 2008. On October 31, 2009, the Shanghai Yangtze River Tunnel was officially open to traffic. This tunnel is the first super-large-diameter shield tunnel, taking advantage of the long-distance tunneling technology in China and creating two world records at that time, namely, the largest shield diameter and the longest advance distance at one shot. The completion of the Yangtze River Tunnel marks that China's super-large-diameter tunnel construction technology by slurry balanced shield machines has reached the international leading level.

9.1.2 Engineering geological and hydrogeological conditions

The strata crossed by the shield are mainly: ④$_1$ gray mucky clay, ⑤$_{1-1}$ gray clay, ⑤$_{1-2}$ gray mucky silty clay, ⑤$_2$ gray clayey silt with thin layer of silty clay; ③$_1$ gray muddy silty clay, ③$_2$ gray sandy silt, ⑤$_{3t}$ lens, ⑦$_{1-1}$ gray sandy silt and ⑦$_{1-2}$ gray sandy silt in some sections. There are liquefied soil, quicksand, piping, shallow gas, lenses, confined water and other unfavorable geological conditions in the soil layers along the alignment. The tunnel also passes through two submarine optical cables.

The groundwater in the shallow soil layer is phreatic water, in close hydraulic interaction with river water and is basically connected to river water. The phreatic water level is mainly affected by the tide of the Yangtze River, and the average annual water level is +2.8 m and +2.4 m, respectively. The confined water is buried in layer ⑦ and layer ⑨, with abundant water and periodic water level changes. The water head elevation is generally 0.00–8.00 m.

The main buildings (structures) along the tunnel are flood walls in the Pudong section and Changxing Island, and the rest are farmland. The water part crossed by the tunnel is mainly the Nangang waterway of the Yangtze River, which is an important channel from the Yangtze River basin area to the coastal areas of China and the oceans of the world.

9.1.3 Shield tunneling machine

Two slurry balanced shield machines with a diameter of 15.43 m manufactured by the German Herrenknecht were used for the circular tunnel (Fig. 9.5). The bubble control method was adopted for slurry pressurized control, which is more accurate and timely than the traditional method and can effectively ensure the stability of the excavation face. The total length of the shield machine is about 134 m, including the main shield and the rear supporting frame system, with a total mass of about 3,250 t.

The rear supporting system of the shield machine consists of a three-section frame, which can realize simultaneous construction of the road in the process of advance. The main power components of the shield machine are arranged on the No. 1 frame; the No. 2 frame is a connecting truss, which serves as the main transport channel for segments, slurry buckets, prefabricated "hollow rectangular" components and other materials and is used to install "hollow rectangular" prefabricated components of the road structure; the No. 3 frame is the extended working area of

Fig. 9.5. φ 15.43 m slurry balanced shield machine for Shanghai Yangtze River Tunnel.

various service pipelines. The frame travels on the auxiliary rail, and the transport vehicles of various materials run directly on the road pavement built with the installed "hollow rectangular" prefabricated components.

9.1.4 Shield construction technology for ultra-long tunnel

9.1.4.1 *Shield tunneling machine support*

(1) **Main bearing seal:** The main bearing seal consists of two sets of internal and external sealing systems. The external layer is used as a seal in the direction of the excavation chamber, and the internal layer is used to maintain the atmospheric pressure inside the shield. The external seal separates the main bearing from the excavation chamber under external pressure. In the form of a large-diameter shaft seal, it is composed of four layers of lip seals and one leading labyrinth, forming four separated areas. These four layers of seals act on a surface hardened wear-resisting ring, which provides a variable contact surface for the first layer of lip seal (Fig. 9.6).

    Four ultrasonic sensors in the sealing device can monitor the wear degree of the main seal. The wear extent found reaching the set value or continuous grease ingress monitored in the oil leakage tank during operation indicates that the sealing surface of the main seal needs to be replaced.

(2) **Shield tail seal:** The shield tail seal is composed of three rows of shield tail brushes. In order to ensure that the shield tail does not leak in case of an emergency, an emergency bladder and shield tail

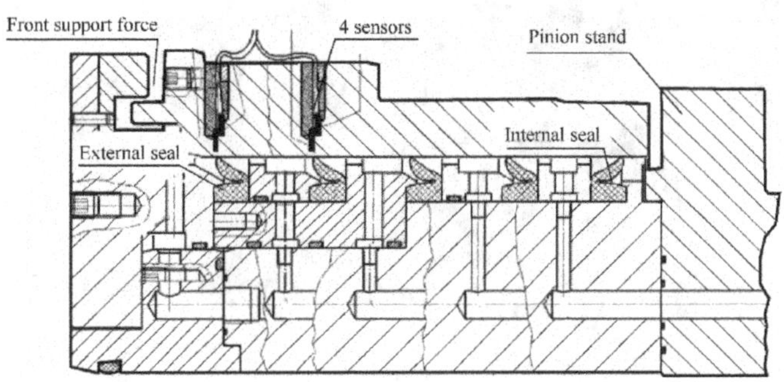

Fig. 9.6. Lip seal system.

freezing device are also set at the rear of the three rows of shield tail brushes. The shield tail brushes are arranged as shown in Fig. 9.7. As the shield tunneling route is long, the shield tail brush may lose its sealing effect due to excessive wear, so it needs to be repaired and replaced if necessary. The difficulty in replacing the shield tail brush lies in the sealing and water stop at the shield tail after the segments are removed. The frozen soil curtain formed by the freezing method has the advantages of good continuity, uniformity, water sealing, flexibility and high strength.

(3) **Cutter replacement under atmospheric pressure:** A wear detection system is set in the shield machine to detect the wear of the cutterhead, as shown in Fig. 9.8. When the cutter is worn to the coil, the coil current will be interrupted, and the detection results can be directly

Fig. 9.7. Shield tail brush layout.

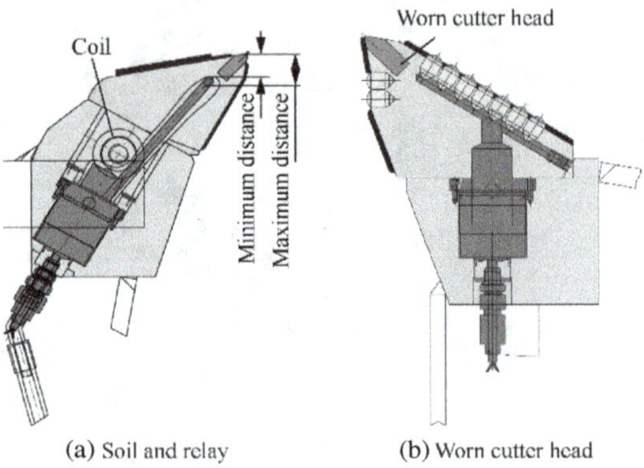

Fig. 9.8. Wear monitoring system.

viewed on a display device. The cutter wear detection device is installed on eight scrapers and two shovel blades at selected positions and connected to a plug at the rear of the cutterhead.

The cutter replacement is carried out under atmospheric pressure without entering the excavation chamber. The workers enter the cutterhead compartment from the middle of the main drive, unfasten the bolts on the fixed plate, on which the cutter arm connects the lifting frame (with bolts) to the cutter and operate the lifting frame to pull out the cutter. They then close the pressure isolation door and replace the worn cutter with a new one. Next, they lift the cutter to the installation position at the back of the pressure isolation door, then open the door and install the cutter. The schematic diagram of cutter replacement is shown in Fig. 9.9.

9.1.4.2 *Tunnel stability control*

The tunnel section up to 176.6 m^2 is not conducive to axis control. Especially in the shield launching and receiving stages, the tunnel tends to float due to shallow soil cover. The soil cover of the shield launching section is less than 8 m; the minimum soil cover in the section crossing the river is about 14 m, both smaller than the shield diameter. On the one hand, a too large hydraulic pressure fluctuation should be avoided at the cutting during the shield tunneling; otherwise, the disturbance to the soil

Fig. 9.9. Schematic diagram of cutter replacement.

mass ahead will be increased, resulting in the loss of the soil mass ahead. On the other hand, due to the shallow soil cover, the tunnel rings formed tend to float with slurry channeling backward.

In order to control tunnel floating, measures were taken to strengthen the management of synchronous grouting in an attempt to build grout resistance to liquefaction when encountering slurry with a shortened gelling time and maintain a certain fluidity so that the grout can be evenly distributed over the tunnel by synchronous grouting. According to the actual conditions of the stratum crossed, the performance of single-fluid synchronous grouting was improved and optimized to make the fluidity and yield value of grout more suitable for filling behind segments by the super-large-diameter slurry balanced shield machine and the anti-floating ability building of the tunnel; moreover, the synchronization was improved between grouting and shield tunneling so that the structure void can be filled with grout in time; in addition, all longitudinal and circumferential bolts were tightened and prefabricated components of the tunnel road were laid in time.

9.1.4.3 *Efficient transport of materials*

The structure of the approach section and buried section was completed before the shield construction, so the tunnel segment, synchronous grout, prefabricated components and other engineering materials were directly transported into the tunnel through the approach section and buried section by special vehicles. See Fig. 9.10 for the construction transport in the tunnel.

After the prefabricated road structure was transported by a special truck to the place ahead of the shield machine frame 2, it was hoisted by the lifting equipment on the connecting beam between the shield machine frames 1 and 3. The segments were transported by a truck to the connecting beam between the shield machine frames 1 and 3 through the tunnel pavement, transported to the segment transport mechanism by the lifting equipment on the connecting beam, and then transported to the erection area by the segment transport mechanism. During shield tunneling, engineering materials such as segments were transported by special trucks (Fig. 9.11) to avoid the problem of derailment in the process of transport by an electric locomotive, and double-headed trucks were used to transport at a high efficiency.

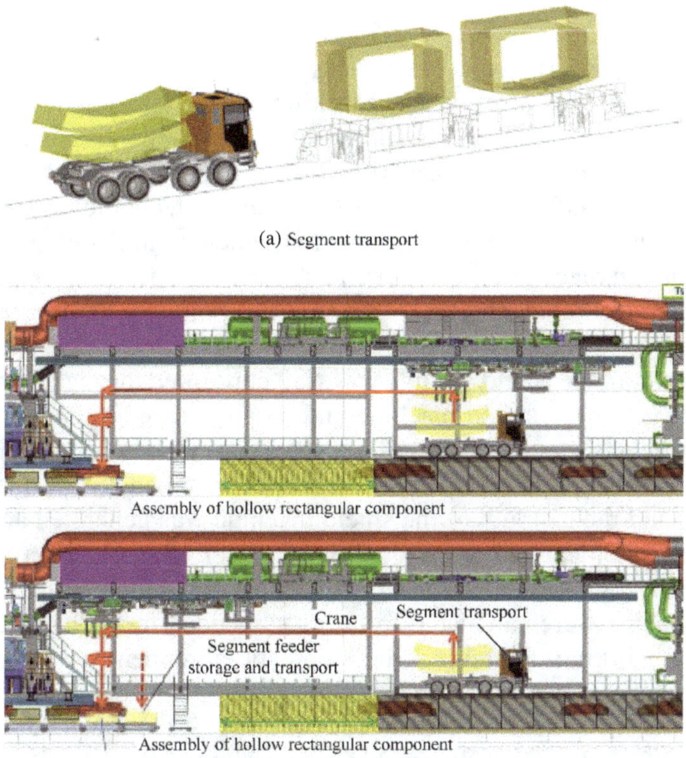

(a) Segment transport

(b) Lifting of internal components

Fig. 9.10. Schematic diagram of construction transport in tunnel.

Fig. 9.11. Schematic diagram of construction transport in tunnel.

9.1.4.4 *High precision survey alignment*

The tunnel engineering survey includes horizontal control, shaft connection, underground control, vertical jacking, shield machine attitude control and segment attitude. The purpose of survey control is to ensure accurate tunneling. The horizontal control along the axis of the Yangtze River Tunnel was conducted using GPS survey technology. Compared with the linear length of the traditional horizontal control survey, its baseline vector can reach hundreds of meters or even thousands of kilometers, contributing to the advantages of few control points, high accuracy, low cost, etc.

In addition, the shield attitude was well controlled through an accurate initial survey of the shield machine and an automatic alignment system. The rotation angle and gradient of the shield machine were automatically generated through the electronic clinometer installed inside the shield machine, and the azimuth of the shield machine was generated according to the laser incidence angle created by the laser transmitter on the laser target on the total station, and the cutting and shield tail coordinates of the shield machine were calculated from the embedded formula. Figure 9.12 shows the working diagram of the automatic alignment device used for the Shanghai Yangtze River Tunnel.

In addition, a segment selection system (Fig. 9.13) was independently developed for this tunnel. In combination with the attitude parameters of the shield tunneling process, any deviation of all the continuous three-ring segments produced by assembling with different closure block positions from the tunnel design axis was scientifically calculated by comparison to greatly improve the scientificity and correctness of segment erection.

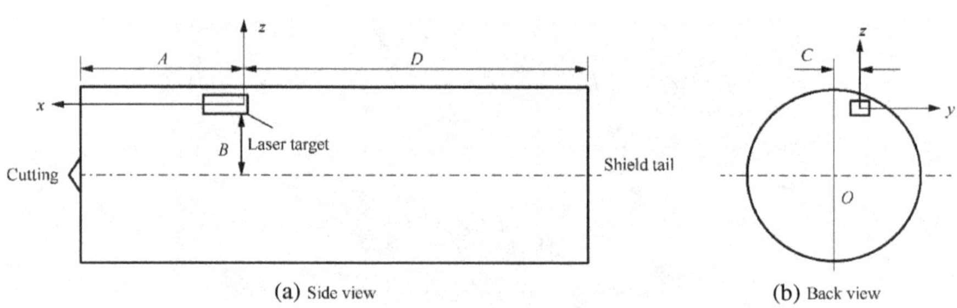

Fig. 9.12. Working diagram of automatic alignment device of Shanghai Yangtze River Tunnel.

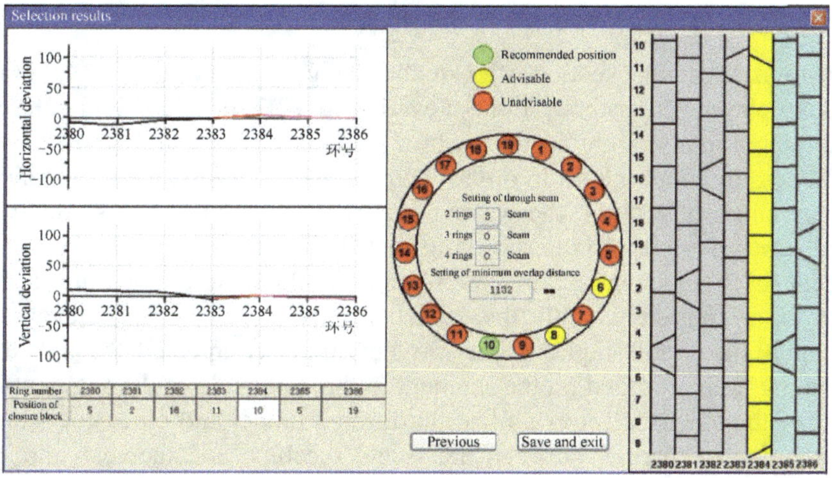

Fig. 9.13. Segment selection system.

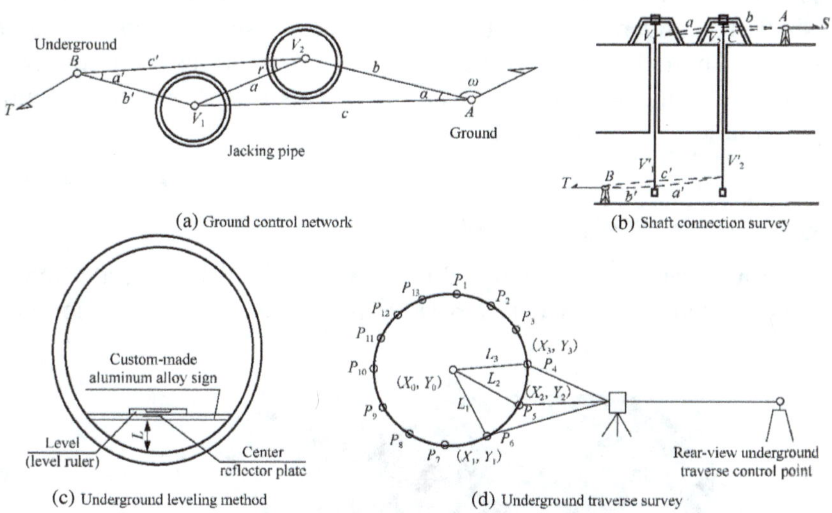

Fig. 9.14. Survey of shield vertical jacking.

In order to ensure accurate tunneling, when the shield machine advanced to 7 km, two pipes of 600 mm in diameter were jacked, 16 m apart, out of the ground through the tunnel roof, which were used to transfer the coordinates of the ground control network down to the tunnel through the shaft connection survey method, as shown in Fig. 9.14.

9.1.4.5 *Water-borne shield machine receiving*

The tunnel entrance section is located in ②$_3$ gray sandy silt and ④$_1$ gray mucky clay layer, with a ground elevation of +2.7 m. The shield entered the tunnel on a straight section with a gradient of 2.9% and a soil cover thickness of about 6.8 m. In order to ensure the safety of shield tunneling in the entrance section, the water-borne receiving scheme was adopted. Water was poured into the receiving shaft before the cutting cabin of the shield machine leaned against the tunnel portal concrete to a water level elevation consistent with the external groundwater level elevation. In addition, double-liquid grout was continuously injected through the grouting hole provided on the segment to the outer side of the segment in rear of the shield tail, to stabilize the formed tunnel and block the water and soil loss channel between the shield machine and the rear unreinforced soil. After the above work was completed, the shield machine cuts the carbon fiber reinforced concrete to enter the tunnel, then cuts the M5 cement mortar layer and sits on the mortar base to slowly enter the receiving shaft. The advance rate in the tunnel entrance section was controlled at 5–10 mm/min, the active grout was used for synchronous grouting, and the shield machine advanced in clean water. The water-borne tunnel entry is shown in Fig. 9.15.

9.1.4.6 *Synchronous construction of road structure*

The internal structure of the tunnel is shown in Fig. 9.16. In order to reasonably shorten the construction period, the simultaneous construction of

Fig. 9.15. Water-borne shield receiving in entrance section.

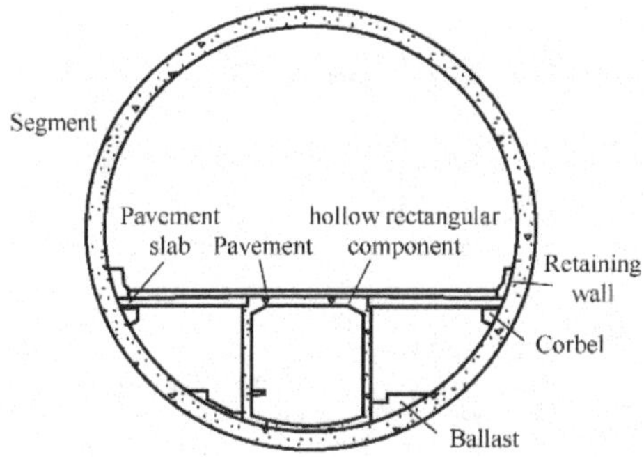

Fig. 9.16. Schematic diagram of tunnel internal structure.

shield tunneling and the internal road structure was adopted, and the large tunnel space also created favorable conditions for synchronous construction. The synchronous construction was conducted in the following steps: advance of the shield tunneling machine → hoisting of the prefabricated "hollow rectangular" structure of the road structure in place → repetition of the above steps in a cycle→ fabrication of both sides of the road and the pavement slab after a distance of 200 m → cast-in-situ anti-collision curbstone on both sides → construction of the road pavement plain concrete pavement after the shield tunneling machine enters the tunnel.

The "hollow rectangular" prefabricated components were hoisted simultaneously with the advance of the shield machine. The width of the prefabricated components is 2 m. The hoisting construction area is located in the No. 2 frame. Generally, one "hollow rectangular" prefabricated component is hoisted for each ring of advance of the shield machine. The prefabricated components hoisted in place were used as the transport channel in the tunnel. In order to ensure the rapid tunneling of the shield machine and the smooth horizontal transport road, the construction of the pavement structure lagged behind the No. 3 frame by about 200 m, and the cast-in-situ structure construction of the pavement on both sides was carried out by using the typified forms and the quick-release supports. The build structure was maintained for five days before form removal and curing, and three lanes were opened to traffic 28 days later. The coordination diagram of synchronous construction and vehicle transport is shown in Fig. 9.17.

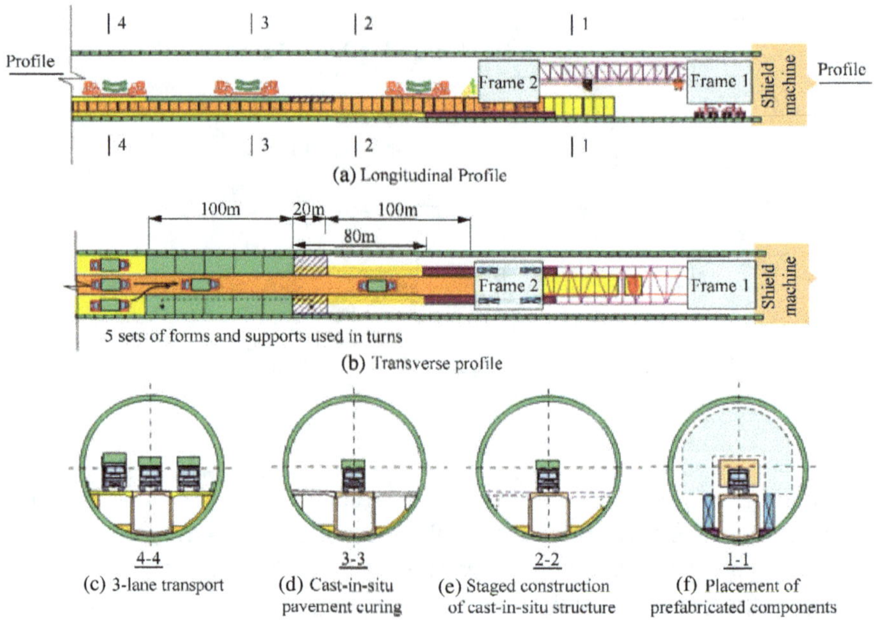

Fig. 9.17. Schematic diagram of synchronous construction and vehicle transport coordination.

9.2 Large-diameter EPB shield tunnel — Shanghai Bund Tunnel

9.2.1 Project overview

The Shanghai Bund Tunnel runs through the whole Bund area from north to south, as a part of the three north-south east lines in the three north-south and three east-west traffic trunks network in Shanghai, and is known as the "heart bypass" project to relieve the traffic jam in the city center. The tunnel, upon completion, has effectively alleviated the traffic congestion in the Bund area, improved the Bund environment, enhanced urban functions and served the traffic of the Shanghai World Expo. The cross-section of single-tube, double-deck and six-lane was adopted for the first time in China. The Bund Tunnel starts from the Laotaiping Road of the South Bund in the south, runs northward along Zhongshan 1st Road and Zhongshan 2nd Road, crosses the Suzhou River under the Waibaidu Bridge, passes through East Changzhi Road along Dongdaming Road, connects the North Bund area, connects Wusong Road at the north side of the Wusong Road Gate Bridge, and runs northward along Wusong Road to Yuhang Road, with a total length of 3,326 m. The entire Bund Tunnel is divided into the south section and the north section from Tiantong Road to Dongmen Road, with Fuzhou Road receiving shaft as the boundary. The section from Laotaiping Road to Fuzhou Road (south section) was constructed by open-cut technology; the shield method was applied for construction from Fuzhou Road to Dongdaming Road (north section). The overall location of the Bund Tunnel is shown in Fig. 9.18.

The north section of the Bund Tunnel was constructed by the shield method, for a total length of 1,098 m. The layout plan is shown in Fig. 9.19. The shield construction is mainly divided into three stages. The first stage is the construction of the tunnel exit section: from the tunnel exit of the shield machine to the crossing of Pujiang Hotel and Shanghai

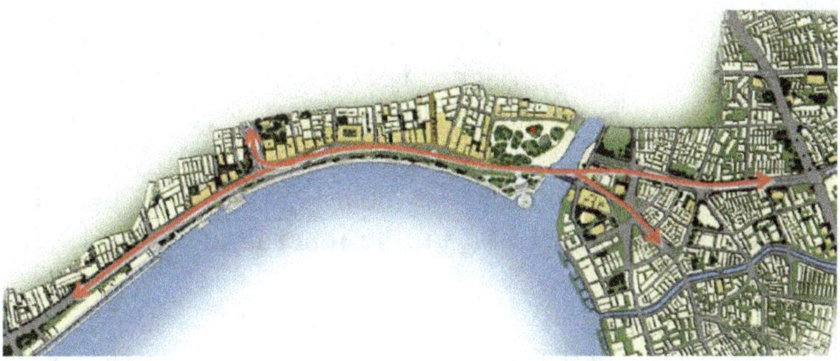

Fig. 9.18. Overall location of the Bund Tunnel.

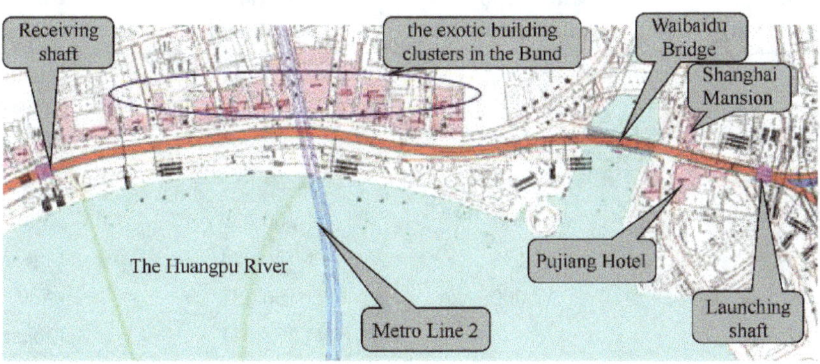

Fig. 9.19. Layout plan of circular tunnel section of Bund Tunnel.

Mansion; the second stage is conventional section construction: crossing the Waibaidu Bridge, the underpass of East Beijing Road and East Nanjing Road, crossing Metro Line 2 and crossing the exotic building clusters in the Bund horizontally to 50 m in front of the tunnel portal; the third stage is tunnel entry construction: from 50 m in front of the tunnel portal to the end of the shield receiving in the tunnel. An earth pressure balanced shield machine with a diameter of 14.27 m designed and manufactured by Mitsubishi Corporation of Japan was adopted. Two working shafts were set within the scope of the shield construction section: the Tiantong Road working shaft as the launching shaft and the Fuzhou Road working shaft as the receiving shaft. The maximum longitudinal slope of the main line of the shield tunnel is 5.0%. The plane axis of the whole tunnel can be roughly divided into eight sections, with a minimum turning radius of 550 m; the vertical axis of the tunnel is divided into six sections, with a minimum vertical curve radius of 1,500 m. See Table 9.1 for detailed data.

The whole shield tunnel section comprises 557 rings of segments (including a temporary ring). Each ring is composed of nine tapered segments of universal type, including closure block F (1: block F), adjacent blocks L (2: L_1, L_2), and standard blocks B (6: B_1, B_2, B_3, B_4, B_5, B_6), erected with staggered joints. The segment taper is 79.72 mm. The outer diameter of the segment is 13.95 m, the inner diameter is 12.75 m, the

Table 9.1. Curve radii and gradient in sections along axis of Bund Tunnel.

Vertical axis		**Plane axis**	
Section length (m)	**Curve radius (m) or gradient**	**Section length (m)**	**Curve radius (m)**
221.8	5%	3.632	Transition curve
194.4	2000	137.997	Linear
231.202	−5%	237.656	550.889
75.3	1500	200.039	500
362.854	0.3%	36.224	Linear
12.444	2,000	166.421	500
		178.611	Linear
		137.42	650

thickness is 0.6 m, and the ring width is 2 m. The segments are connected by bolts longitudinally and circumferentially. Thirty-four M30 longitudinal bolts are used to longitudinally connect the rings of segments; two M36 circumferential bolts are used to circumferentially connect the segment blocks.

The Bund Tunnel is the first project in China bored by the super-large-diameter EPB shield method. More than one third of the sections are shallowly covered by soil, and the soil cover at the top surface of the shield is only 8.52 m. The construction using the super-large-diameter EPB shield under shallow soil cover brought great challenges to surface deformation control; meanwhile, the Shanghai Bund Tunnel is located in a densely populated urban area. On one side of the tunnel is a precious historical building complex, and on the other side is the Huangpu River. The shield machine passes through the narrow zone in the middle, crossing the famous Waibaidu Bridge and Metro Line 2 is in operation. In this quite sensitive construction environment, almost all the technical difficulties in earth pressure balanced shield construction were encountered. The success of the Shanghai Bund Tunnel provides a reference for similar projects in the future and is of great significance for the application promotion of large-diameter EPB shield machines in China.

The construction of the tunnel commenced on August 18, 2007, with the shield machine launched on January 23, 2009, completed on August 17, 2009, and put into operation on March 28, 2010.

9.2.2 Engineering geological and hydrogeological conditions

The soil cover thickness of the tunnel is 8–24 m. The tunnel is mainly distributed in ②$_{3-1}$ gray clay silt mixed with silty clay, ②$_{3-1}$ gray sandy silt, ③ muddy silty clay, ④ muddy clay, ⑤$_1$ silty clay and ⑤$_3$ silty clay mixed with clay silt.

The groundwater in the shallow soil layer of the land area along the project is phreatic water. The burial depth of the water level is 0.90–2.50 m. The phreatic water level is mainly affected by atmospheric precipitation, surface runoff, etc., and varies in different ranges. The confined water is distributed in layer ⑦ (⑦$_1$, ⑦$_2$) and layer ⑨. Layer ⑦ is the first confined aquifer in the Shanghai area, and layer ⑨ is the second confined aquifer. The burial depth of the confined water level in layer ⑦ is 5.35–10.31 m; the burial depth of the confined water level in layer ⑨ is 13.80 m.

The shield machine launched from the Tiantong Road working shaft advanced through many historical protection buildings and important structures along the line, including the Pujiang Hotel, Shanghai Mansion, Waibaidu Bridge, Nanjing East Road Underpass, Beijing East Road Underpass, Metro Line 2 and the exotic building clusters in the Bund. See Table 9.2 for the location relationship between these buildings (structures) and the Bund Tunnel.

In addition, dozens of existing pipelines, such as gas, rainwater, sewage, telecommunications and electricity, were distributed over the tunnel.

Table 9.2. Location relationship between main buildings (structures) and Bund Tunnel.

S. No.	Building (structure)	Shortest distance to tunnel edge (m)
1	Pujiang Hotel	1.7
2	Shanghai Mansion	2.8
3	Pier foundation of Waibaidu Bridge	Underpass
4	China Everbright Bank	18.9
5	China Merchants Bank	17
6	Foreign Trade Bureau	25
7	Agricultural Bank of China	14
8	Industrial and Commercial Bank of China	21
9	Bank of China	22
10	North Building of Peace Hotel	21.7
11	South Building of Peace Hotel	22.8
12	China Citic Bank	30
13	Shanghai Household Textile Import & Export Company	20
14	AIA Building	15
15	Shanghai Media Group	19.4
16	China Foreign Exchange Trading Center	24
17	Shanghai the Bund Law Firm	20
18	Shanghai Federation of Trade Unions	9.9
19	Shanghai Customs	11
20	Shanghai Pudong Development Bank	13

The exact location of most of the pipelines was not very clear, which requires a lot of effort to perform a thorough investigation.

9.2.3 Shield tunneling machine

A large-diameter earth pressure balanced shield machine with a diameter of 14.27 m manufactured by Mitsubishi was used in the project, as shown in Fig. 9.20. The cutterhead is of spoke type, with a maximum torque of 63,436 kN · m, a total thrust of 176,800 kN and an advance rate of 5.4 cm/min.

9.2.3.1 *Thrust system*

The thrust system is equipped with 34 groups (68 sets) of jacks, distributed in eight oil pressure zones. The layout of jacks is shown in Fig. 9.21. The tunneling alignment of the shield machine can be corrected in these eight oil pressure zones, equipped with a set of cylinders each. The shield jack has a measurement synchronization function. In order to measure the earth pressure in the earth chamber, four earth pressure gauges were installed at the left and right symmetrical positions at the outer side of the bulkhead. The layout of the earth pressure sensors is shown in Fig. 9.22.

Fig. 9.20. "Bund Tongtai" earth pressure balanced shield machine.

Fig. 9.21. Layout of jacks.

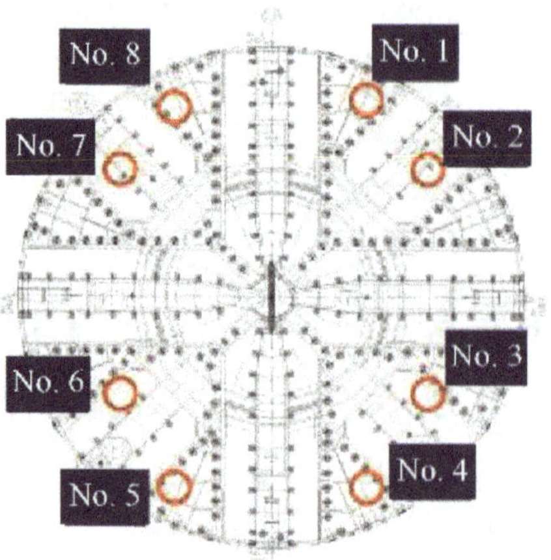

Fig. 9.22. Layout of earth pressure sensors.

9.2.3.2 *Mucking system*

The earth pressure balanced shield machine is furnished with a screw mucking system. The soil and sand conveyed into the cutterhead earth chamber after being cut by the cutterhead device are discharged to the rear through the screw conveyor set at the lower part of the bulkhead. The lower part near the rear end of the screw conveyor is equipped with a soil discharge gate, through which the cut soil is discharged. The soil discharge gate, installed at the lower part of the rear end of the screw conveyor, is opened and closed by the telescopic action of two hydraulic jacks. The gate is opened when the jacks are extended.

9.2.3.3 *Soil improvement system*

Due to the large diameter, the shield machine crossed multiple strata at the same time during excavation, with complex stratum conditions and a large pressure difference on the excavation face, which makes it difficult to maintain balance. Therefore, soil improvement is required. The soil improvement system consists of a solution tank, material tank, injection pump, foaming and control devices, air compressor, etc. on the front frame. A total of four injection pumps are configured, which are connected with eight pipelines, to inject the improvement additives into eight injection holes, including seven injection holes on the front side of the cutterhead and one injection hole on the side, or into the provided holes on the bulkhead and the screw conveyor. The layout of improvement injection holes in the cutterhead and earth chamber is shown in Fig. 9.23.

9.2.4 Tunneling technology of large-diameter EPB shield machine

9.2.4.1 *Soil improvement*

For construction using the large-diameter earth pressure balanced shield machine in the soft soil ground, special soil improvement methods are required to meet the construction requirements of the tunnel, including

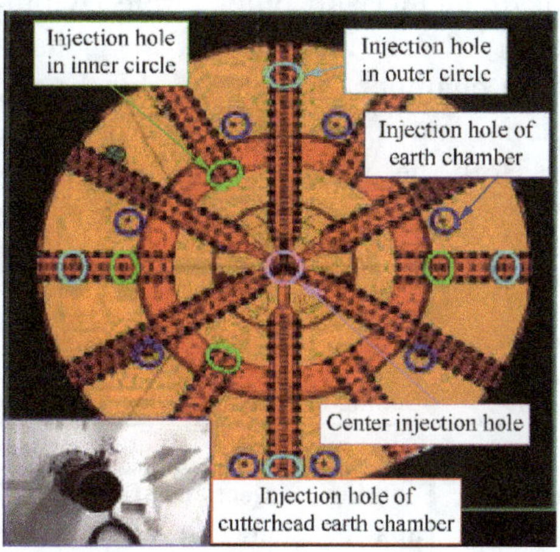

Fig. 9.23. Layout of improvement injection holes in cutterhead and earth chamber.

improving the uniformity of the soil entering the earth chamber, improving the overall plasticity, maintaining the uniform transmission of the soil pressure inside and outside the cutterhead, ensuring the stability of the excavation face and reducing the working pressure of the shield machine so as to maintain a uniform and balanced advance rate. Regulation is generally made by injecting slurry, foam and other improvement additives.

In order to ensure the safety of facilities along the Bund Tunnel in Shanghai, laboratory tests were conducted to study the adaptability of soil mass improvement in this area and select the additives used for soil mass improvement at the excavation face. Then, model tests were conducted with the selected additives to determine the construction parameters of soil mass improvement. Finally, field tests were conducted to verify the practical application effect of soil mass improvement technology. The final specific additive parameter settings are shown in Table 9.3.

In the process of shield tunneling, foam injection improvement was carried out at different excavation points ahead of the cutterhead. Foam injection holes were arranged at the center point, inner circle and outer circle of the cutterhead front. The total amount of foam injection and the distribution of injection amount at different points were set according to the actual tunneling soil conditions during tunneling and shield construction parameters, so as to ensure that the shield construction parameters are within the normal controllable range. The foam injection ratio at the center point (within 6 m in diameter), inner circle (within 6–12 m in diameter) and outer circle (within 12–14.27 m in diameter) is generally 2:3:1. In addition, a mechanical agitator was designed in the earth chamber to stir muck by the rotation of the cutterhead and the relative movement of the bulkhead on the inner wall of the earth chamber. Both the cutterhead and the bulkhead on the inner wall of the earth chamber are equipped with stirring rods to increase the mixing effect.

Table 9.3. Setting of parameters for additives.

	Foam			Bentonite slurry		
Item	**Concentration (%)**	**Foaming capacity**	**Injection rate (%)**	**Density (g/cm^3)**	**Viscosity (s)**	**Injection rate (%)**
Parameter	3–5	25	15–30	1.06	40	0–10

9.2.4.2 *Continuous mucking by belt conveyor*

The Bund Tunnel is located in the core area of the Bund with a special environment. After multiple rounds of analysis and research, it was determined to adopt a continuous mucking mode of horizontal belt conveyor + vertical lifting. The belt conveyor is used for horizontal transport, and the grab bucket is used for vertical transport, which met the schedule requirements of the Bund Tunnel. The belt conveyor system not only provides a new concept of fast, continuous and safe mucking, but also provides a good idea for the rational application of tunnel space.

The working process is as follows: the soil is unloaded from the original conveyor on the shield frame to the synchronous extension horizontal belt conveyor, and then transported along the tunnel to the sump of the working shaft by the synchronous extension belt conveyor, and then vertically transported to the overhead collecting hopper by the grab bucket and unloaded to the earth truck through the auxiliary equipment of the overhead collecting hopper for outbound transport. The mucking belt conveyor is shown in Fig. 9.24.

During the shield tunneling of the entire Bund Tunnel, the continuous mucking device has successfully transported about muck up to 2,500 m³ per day, providing strong support for the smooth progress of the project.

(a) Sump (b) Horizontal belt conveyor

Fig. 9.24. Mucking belt conveyor.

9.2.4.3 *Protection of buildings (structures) crossed*

Along the Bund Tunnel, there are dozens of historical protection buildings with different styles and a combination of Chinese and Western styles, such as the Pujiang Hotel, Peace Hotel and the Customs Building. According to the thorough investigation carried out at the early stage of construction, these buildings were found to have some problems, such as uneven settlement, inclination and aging of the superstructure. According to the location relationship between these buildings and the tunnel and in view of the foundation conditions of these buildings and their relative relationship with the tunnel, protection plans were formulated for three types of crossing. The first type is to cross the Pujiang Hotel in a very short distance, and the buildings are completely isolated by isolation piles plus pile-to-pile compensation grouting; the second type is to cross the Shanghai Mansion in a short distance, and the compensation grouting partition method is used to reduce the impact of shield machine tunneling on the buildings; the third type is to cross the exotic building clusters on the south bank of the Suzhou River for a long distance, and the parameters of the shield tunneling machine are strictly controlled so as to reduce the impact on the buildings.

The Pujiang Hotel was built in 1860. Its foundation is weak, and some sound parts of the foundations are only 3.2 m long wooden piles. The hotel is only 1.7–4.5 m horizontally away from the outer sideline of the tunnel, as shown in Fig. 9.25. Due to the complexity of shield tunneling technology, the surface settlement in the central area caused by shield tunneling cannot be completely controlled. The influence of this kind of uneven settlement on buildings with fragile structures is disastrous. In order to ensure the safety of the building when the shield machine passes the Pujiang Hotel nearby, some necessary engineering protection measures have been taken.

A special casing method — spiral mucking construction (FCEC construction) — in the outer casing was selected, as shown in Fig. 9.26. The FCEC construction method was applied by the forward and reverse synchronous double rotary full casing machine to carry out the internal and external synchronous forward and reverse drilling to the pile bottom elevation. The outer steel casing rotates in the reverse direction, and the inner spiral rotates in the forward direction for mucking through the steel pipe. Moreover, the outer steel casing acts as a retaining wall. This method can effectively reduce the disturbance to the soil mass, which is

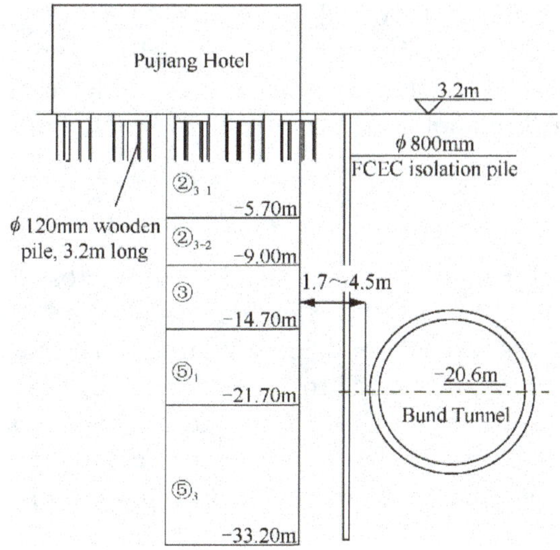

Fig. 9.25. Location relationship between Pujiang Hotel and the Bund Tunnel.

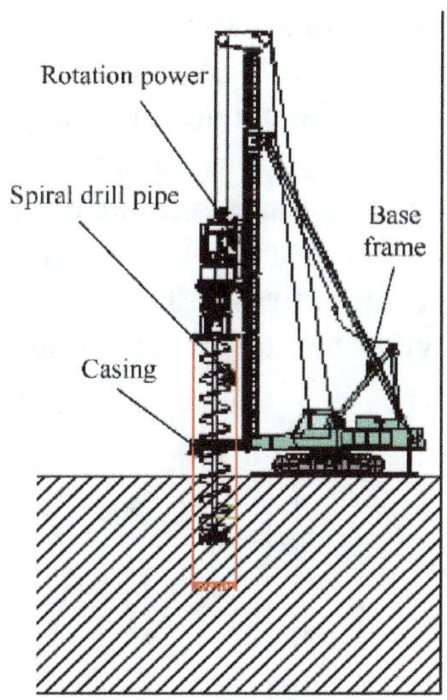

Fig. 9.26. Schematic diagram of FCEC construction method.

very useful for controlling the uneven settlement of the Pujiang Hotel. With the help of grouting between piles, the surface settlement was more effectively controlled.

The partition effect of the FCEC reinforcement method is very obvious. The settlement caused by the shield tunneling crossing the Pujiang Hotel was less than 10 mm, and the uneven settlement was less than 5 mm, and its deformation has been effectively controlled.

9.2.4.4 *Shield tunneling across operational metro line*

The shield tunnel part of the Bund Tunnel needs to cross Metro Line 2 in operation from Ring 345 to Ring 355, with a skew angle of 73°, and the top soil cover over the shield of 8.52–8.65 m, and the shield bottom only 1.46 m away from the top of Metro Line 2 tunnel, as shown in Fig. 9.27. The main soil layers in the shield machine crossing area are ②$_0$ river beach soil, ④ gray muddy clay, ⑤$_1$ gray clay and ⑤$_3$ gray silty clay.

The main construction measures taken in the process of shield tunneling are as follows:

(1) **Earth chamber pressure setting:** The value of earth chamber pressure was calculated according to the lateral static equilibrium pressure coefficient $k_0 = 0.7$ of the soil. The theoretical earth pressure value is 0.193–0.195 MPa during the shield tunneling through Metro Line 2. During the actual operation, it was adjusted ring by ring according to the analysis data of the test section and the monitoring data of the deep soil settlement points during the crossing process.
(2) **Synchronous grouting control:** The synchronous grouting was strictly controlled according to the grouting parameters of the test

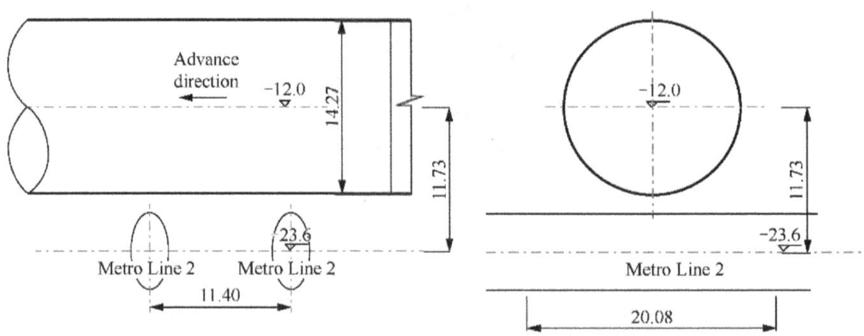

(a) Longitudinal profile of shield crossing Metro Line 2 (b) Transverse profile of shield crossing Metro Line 2

Fig. 9.27. Schematic diagram of shield crossing Metro Line 2 (unit: m).

section and the soil properties of the crossing section and axis conditions, so as to ensure the uniformity and continuity of grouting and effectively and reasonably fill the structure voids as far as possible.

(3) **Advance deviation correction:** In the process of crossing, all measures possible were taken to ensure the shield advanced at a uniform rate to reduce the amount and times of deviation correction of the shield under the condition of a well-controlled settlement ahead of the shield. Instead of sudden and sharp deviation correction, the steady slope method and gentle slope method were taken so as to reduce the impact of shield construction on Metro Line 2 and the ground surface.

(4) **Advance rate control:** The shield machine advanced in strict accordance with the principle of "constant advance rate", and the advance rate was controlled at 20 ± 3 mm/min. The constant advance rate ensured the macrodynamic mechanical balance between the shield machine and the surrounding strata.

(5) **Float control of Metro Line 2 tunnel:** The excavation volume per ring of the shield machine reached 320 m^3, and the mass of soil was nearly 285 t/m, while the mass of the shield machine itself is only 136 t/m. The mass of the shield machine is far less than that of the excavated soil. After the shield No. 1 frame is separated, the five rings between the tail of No. 1 frame and the hollow rectangular component and the inner space of the hollow rectangular component are placed with ballasts. In addition, the concrete pouring on both sides of the hollow rectangular component was rapidly implemented to increase the longitudinal stiffness of the tunnel and control the tunnel deformation.

(6) **Monitoring of Metro Line 2:** Metro Line 2 was monitored to detect settlement in the shield tunneling across Metro Line 2, in a range of monitoring with the centers in the up-traffic and down-traffic directions of the Metro Line 2 section crossed by the Bund Tunnel section as the symmetrical points, extended for 50 m to both ends of Line 2. The results of each measurement were timely summarized and submitted to the construction technology department so that the construction technicians could quickly understand the construction status and the metro deformation in the corresponding area, determine new construction parameters, grouting amounts and other information and instructions, and transmit them to the shield operation face for prompt adjustments made accordingly to ensure tunnel construction quality.

The location with the greatest change in Line 2 is at the cutting point of the shield. Due to a certain included angle between the shield machine and Line 2, the location at the left side of the shield axis suffered from the greatest deformation, instead of other locations below the shield axis. Since impact began to the down-traffic line of Line 2 when the cutting was 4–5 rings away from the tunnel of Line 2, the tunnel of Line 2 was lifted up about 1 mm when the cutting reached the tunnel. Shield tunneling reduced the soil cover load on Line 2, causing floating of the down-traffic line for about 3 mm and floating of the up-traffic line for about 4 mm, and the tunnel floating up; the advance process brought a heavier impact to cause bulge than the erection process. After the frame was separated from the crossing section, the deformation of Line 2 was gradually stable, indicating that the ballasting measures taken had an obvious effect on the deformation control of Line 2. The deformation of the tunnel of Metro Line 2 is shown in Fig. 9.28.

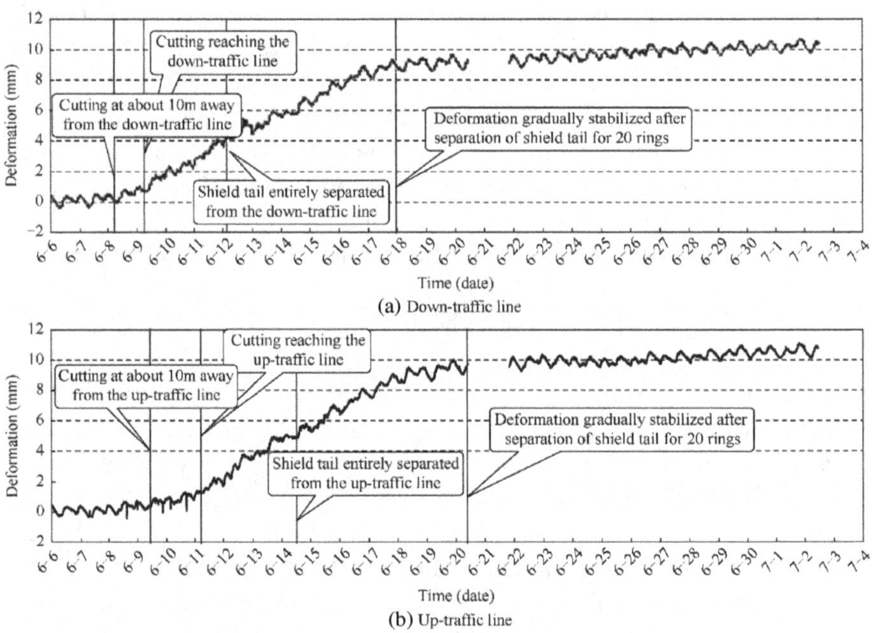

Fig. 9.28. Deformation of Metro Line 2 Tunnel.

9.3 Largest-diameter shield tunnel in China — Wuhan Sanyang Road Tunnel

9.3.1 Project overview

Wuhan Sanyang Road Yangtze River Tunnel is the Yangtze River crossing section of the Wuhan Rail Transit Line 7 Phase I Project. With a total length of the main line of the river crossing tunnel of 4,650 m, it accommodates a two-way six-lane urban trunk road with a design speed of 60 km/h and the river crossing section of Rail Transit Line 7. Four pairs of ramp tunnels are set up on Jiefang Avenue, Zhongshan Avenue, Heping Avenue and Youyi Avenue, with a total length of 2,600 m. The 2,590 m river crossing section is a circular tunnel built by the shield method for highway and railway purposes. Sanyang Road Tunnel is an important link connecting Hankou Binjiang Business District and Wuchang Binjiang Business District. Two super-large-diameter slurry balanced shield machines with a diameter of 15.76 m were used for the construction of the tunnel in the middle of the river. The shield machines were launched from the working shaft at Qinyuan Road in Wuchang, crossed the Yangtze River, and arrived at the working shaft at Sanyang Road in Hankou. The route layout of Wuhan Sanyang Road Yangtze River Tunnel is shown in Fig. 9.29.

The circular tunnel section in the middle of the river has an outer diameter of 15.2 m and an inner diameter of 13.9 m, which is divided into two layers. The upper layer is arranged with three 3.5 m wide lanes, and

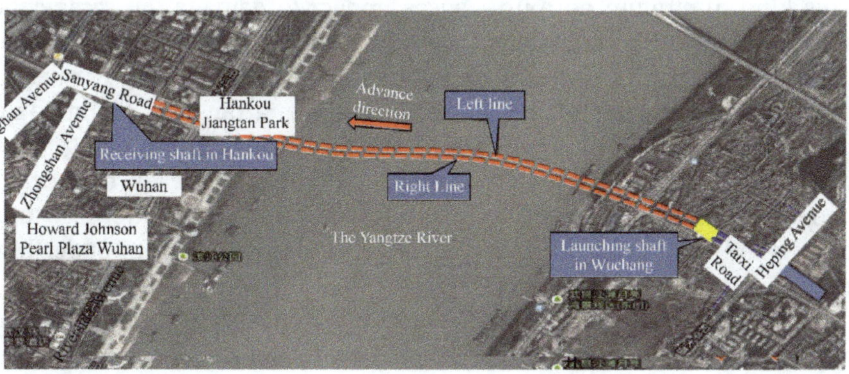

Fig. 9.29. Route layout of Wuhan Sanyang Road Yangtze River Tunnel.

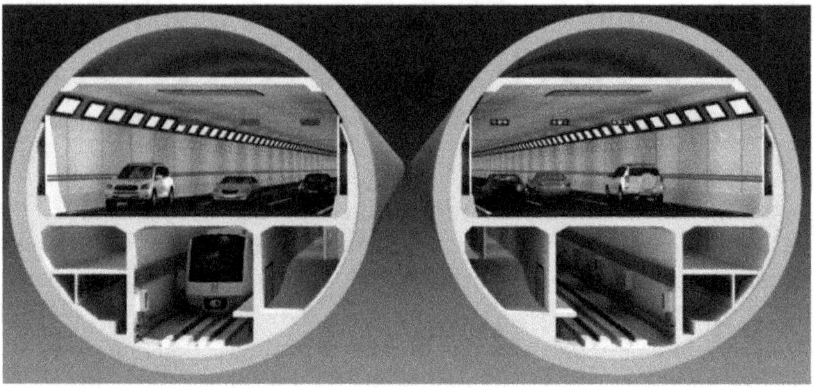

Fig. 9.30. Internal structure layout of Wuhan Sanyang Road Yangtze River Tunnel.

the lower layer is arranged with rail transit line 7, an evacuation passage and a cable gallery. The internal structure layout of the Wuhan Sanyang Road Yangtze River Tunnel is shown in Fig. 9.30.

The main challenge faced by the tunnel construction was the mudstone composite stratum crossed by the shield tunneling. The clay content in 15a and 15b silty mudstone and weakly cemented conglomerate in the tunnel section in the middle of the river is more than 15%, which is very unfavorable to the fluidity of the muck and is extremely prone to mud cake siltation in the cutterhead opening and the cutter rest. In addition, the upper part of the composite stratum is composed of soft soil, and the lower part is composed of rock with high strength and good intactness. This kind of stratum not only causes impact to damage the scraper but also produces different resistance to cause excessive mucking in the soft layer, resulting in ground sink and collapse and serious water permeability accidents. In view of the above difficulties, a range of technologies have been successfully applied in tunnel construction, including shield tunneling management, cutter management of cutterheads replaceable under atmospheric pressure, prevention and control of cutterhead mud caking and shield tail brush replacement at the river bottom under high water pressure.

The shield was launched for the right line on April 30, 2016, and was received on April 26, 2018. The shield was launched for the left line on June 25, 2016, and was received on June 12, 2018. On October 1, 2018, Wuhan Sanyang Road Yangtze River Tunnel was officially open to traffic.

9.3.2 Engineering geological and hydrological environment

The soil layers crossed by the river crossing shield mainly include: (3–2) silty clay, (4–2) silty fine sand, (4–3) medium coarse sand, (15a-1) strongly weathered silty mudstone, (15a-2) moderately weathered silty mudstone and (15b-1) weakly cemented conglomerate. The rings 1–90 are about 180 m long, built in the interlayer of <4–2> silty fine sand and <4–3> medium coarse sand; the rings 190–870 are about 1,360 m long, built in the upper silty fine sand (4–2), the lower strongly weathered silty mudstone (15a-1), moderately weathered silty mudstone (15a-2) and weakly cemented conglomerate (15b-1); the rings 1180–1295 are about 230 m long, built in the upper silty clay (3–2) and the lower silty fine sand (4–2). The rest was built in the full-section silty fine sand layer. The engineering geological profile of the Wuhan Sanyang Road Yangtze River Tunnel is shown in Fig. 9.31.

The silty, fine sand layer has a high permeability coefficient and high compactness. The mineral composition mainly comprises quartz and feldspar, partially mixed with silty clay and medium-coarse sand lens. The content of quartz is 68.83–71.51%. The strength of the rock stratum is not high, and the maximum uniaxial compressive strength is 12.2 MPa, but there are a lot of mineral components that are prone to causing mud caking in the cutterhead. In the strongly weathered silty mudstone (15a-1), the chlorite content is 15.78% and the illite content is 15.37%. In moderately weathered silty mudstone (15a-2), the montmorillonite content is 4.18%. In the weakly cemented conglomerate (15b-1), the skeleton grains are mostly hard rocks with large grain sizes (the grain size is up to about 20 cm). The skeleton is mainly composed of quartzite, limestone and other hard rocks, which are prone to cause wear of the cutterhead

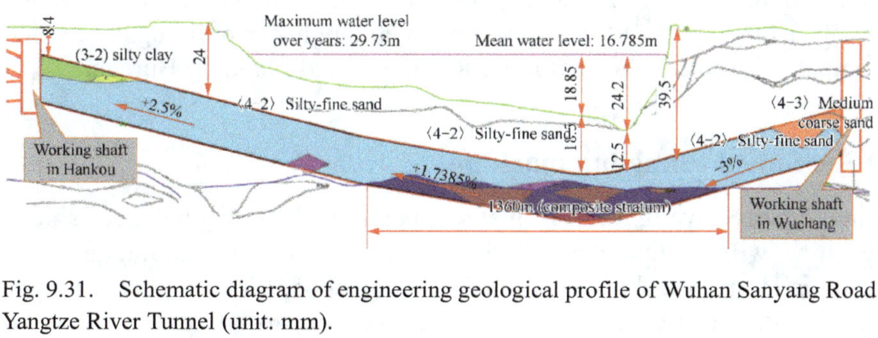

Fig. 9.31. Schematic diagram of engineering geological profile of Wuhan Sanyang Road Yangtze River Tunnel (unit: mm).

Table 9.4. Phase analysis results of soil strata of Sanyang Road Tunnel.

Sample No.	Calcite (%)	Dolomite (%)	Soda feldspar (%)	Potash feldspar (%)	Mica (%)	Montmo-rillonite (%)	Calcite (%)	Chlorite (%)	Illite (%)
⟨3–2⟩	2.73	1.42	18.77	6.06	25.49	—	—	13.81	—
⟨4–2⟩	3.89	0.35	19.21	6.94	30.67	—	—	11.34	—
⟨4–3⟩	4.84	0.22	19.83	7.78	37.26	—	—	10.00	—
⟨15a-1⟩	—	—	15.71	—	—	—	2.51	15.78	15.37
⟨15a-2⟩	32.36	0.21	40.61	—	—	4.18	—	—	—

and cutters. See Table 9.4 for the physical phase analysis and test results of the soil layer of the Sanyang Road Tunnel.

During the survey, the measured static underground water level of perched water at the site was 0.60–6.80 m in burial depth, equivalent to the elevation of 16.09–21.45 m of the Yellow Sea Datum. The pore-confined water is closely interacted with the water systems of the Yangtze River and the Han River, showing a complementary relationship. The seasonal variation of the groundwater level is obvious, and the water volume is relatively rich. Bedrock fissure water mainly occurs in the lower bedrock and is mainly replenished by groundwater infiltration and lateral seepage in the upper aquifer. The highest water level of the Yangtze River over the years is 29.73 m, the multi-year average water level is 18.97 m (Wusong elevation), and the water level fluctuates in a range of about 8 m seasonally.

The shield tunneling areas are all crowded with buildings (structures). The Wuchang section mainly includes the Wuchang–Jiujiang Railway, the Yangtze River levee, the ABC billboard foundation at Wuchang levee and many multi-story brick and concrete buildings. Hankou section mainly involves the Yangtze River embankment and high-rise buildings of timber company dormitory, Wuhan Land Resources and Planning Bureau.

9.3.3 Shield tunneling machine

The Wuhan Sanyang Road Tunnel was bored by the slurry balanced shield machine with a diameter of 15.76 m designed and manufactured by German Herrenknecht, with a maximum thrust of 193 MN, a rated torque of the cutterhead of 27 MN · m and a breakout torque of 44 MN · m.

In order to avoid the cutter replacement operation by staff in the pressurized cabin in a high-pressure environment, the cutterhead was designed to support replacement under atmospheric pressure and with interchangeability between hobs and toothed cutters, with an opening ratio of 29% and a telescopic range of 0–400 mm.

The shield machine was designed with a six-section frame with a total length of about 150 m. In order to solve the problem of material transport in shield construction and internal structure construction, the segment transfer frame was lengthened, and a synchronous slurry storage frame was added to ensure that the shield machine can achieve the excavation rate of 8–10 rings/d under the condition of meeting the synchronous construction of internal structure.

Frame 1 accommodates the shield machine operation room, power, auxiliary and control systems; frame 2 accommodates the traveling beam system, air compressor and AB hydraulic system; frame 3 accommodates the crane beam system and the hose reel for industrial water and vapor; frame 4 is the pipeline extension system; frame 5 is a synchronous grout transfer system; frame 6 accommodates a generator, secondary ventilation and a fire sprinkler system.

9.3.3.1 *Cutterhead, cutter*

The cutterhead of the shield machine is a full-section atmospheric pressure retractable cutterhead, which is equipped with 28 central claw cutters, 48 shell cutters, 3 diameter maintaining hobs, 52 soft soil scrapers replaceable under atmospheric pressure, and 160 fixed scrapers. Thanks to a full-section replaceable cutter rest installed, there is no opening at the center of 5 m. The initial cutters on the shield machine panel were configured in layers. The first layer is the advance cutter, including the claw cutter, shell cutter, and hob. The cutter is 225 mm higher than the panel and is replaceable under atmospheric pressure; the second layer includes the scraper replaceable under atmospheric pressure, a fixed scraper and a shovel blade. The cutter on this layer is 185 mm higher than the panel, and the upper and lower cutters are 40 mm apart. The advance cutter cuts the soil mass into blocks, creating good cutting conditions for the rear cutters. Schematic diagrams of the cutterhead and cutters of the shield machine are shown in Fig. 9.32.

(a) Cutterhead (b) Dual-blade hob (c) Claw cutter (d) Replaceable scraper

Fig. 9.32. Schematic diagrams of cutterhead and cutter of shield machine.

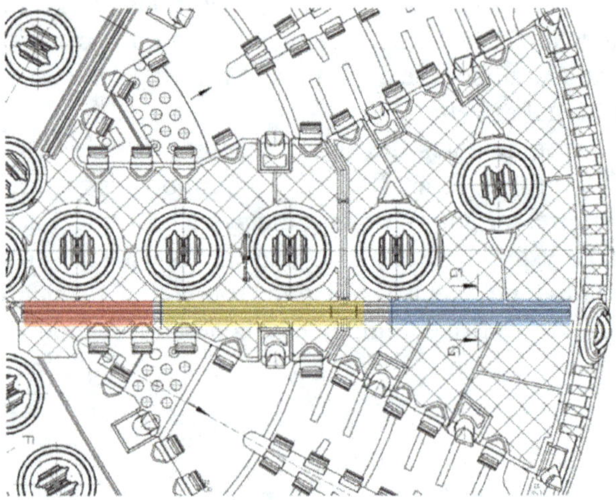

Fig. 9.33. Cutter head wear warning (colored area).

The cutterhead is equipped with various wear alarm devices. Six hydraulic warning lines are set on the front of the cutterhead, two hydraulic warning lines are set on the back of the cutterhead, and each replaceable cutter is set with an alarm line. The cutterhead wear warning device is shown in Fig. 9.33.

9.3.3.2 *Main drive*

The main drive of the shield machine has a telescoping function. The system monitors the stroke difference of the cutterhead telescoping cylinders in real time during the advance process. When the stroke difference is greater than 3 mm, the cutterhead telescoping cylinders automatically adjust and correct the deviation. Once the pressure of the automatically adjusted cylinder exceeds the rated value, the shield machine will be stopped automatically to ensure the safety of the cutterhead bearing. The main drive diameter is 7.6 m, the total installed power is 5,600 kW, the maximum speed is 3.1 r/min, and the total thrust is 79,168 kN with 18 sets of telescopic cylinders. The main drive seal can achieve the sealing effect of static 12 bar and dynamic 8 bar. The schematic diagram of the main drive device is shown in Fig. 9.34.

For the maximum design water pressure of 0.65 MPa, the shield machine main drive bearing sealing system is equipped with three lip seals and a set of pressurization systems to prevent water and soil ahead from entering the main drive bearing compartment.

9.3.3.3 *Central flushing system*

The design flow of slurry feed and discharge of the shield machine is 3,000 m^3/h, the slurry feed pipeline is DN600 mm, and the slurry discharge pipeline is DN500 mm. In order to ensure the slurry flow in the area without opening at the center of 5 m and the cutterhead flushing, a P0.1 pump is set on the shield machine, with a maximum flow of 1,000 m^3/h and a pressure of 8 bar. The pump suction port is on the slurry feed manifold, and the outlet manifold is divided into 12 pipelines for the

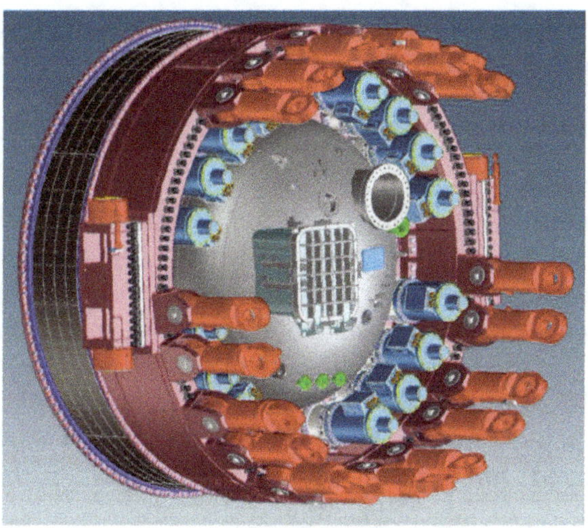

Fig. 9.34. Schematic diagram of main drive.

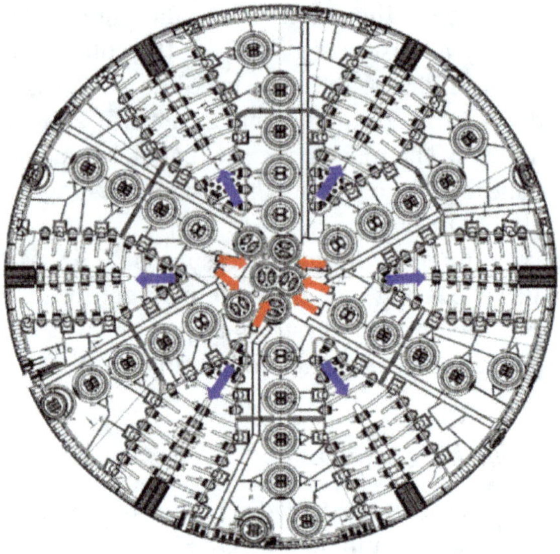

Fig. 9.35. Schematic diagram of cutterhead flushing (sequence indicated in arrows).

six flushing ports and the six flushing ports at the cutterhead opening, as shown in Fig. 9.35.

9.3.4 Large-diameter shield tunneling technology in composite stratum

9.3.4.1 *Shield tunneling management*

In the mudstone–conglomerate composite stratum, the balanced pressure of the shield cutting is set slightly higher than the active earth pressure, the cutterhead speed is generally controlled at 1.0–1.4 r/min, and the tunneling rate is controlled at about 15 mm/min. During the tunneling construction, the cutterhead torque and total extrusion stress should be strictly controlled. When the torque of the cutterhead exceeds 25 MN · m or the extrusion stress exceeds 50,000 kN, the center mud cake should be cleared or the cutter should be replaced.

An integrated slurry treatment system is adopted, including a slurry separation system, a slurry preparation system, a slurry mixing system and a PLC centralized control system. The slurry separation system adopts ZXS II — 3000/30 slurry separation equipment, which is mainly composed of a pre-screening unit, primary cyclone sand removal unit, secondary cyclone mud removal unit, vibrating screen dehydration unit, sand washing unit of the slurry storage tank, etc. The maximum slurry treatment capacity is 3,000 m^3/h. In the process of use, the slurry at the bottom outlet of the secondary cyclone has a large water content and slurry loss. The secondary cyclone underflow will be connected to the primary cyclone for retreatment and recycling.

9.3.4.2 *Prevention and control of cutterhead mud caking*

The cutter layout was improved based on the thrust test, and the options of full-disc hobs, alternate layout of hobs and advance cutters and full-disc advance cutters were respectively tried. The results show that the cutterhead eccentric wear was serious in the case of the hob layout in this stratum, and the lower fixed scrapers were seriously damaged, so the option of full-disc advance cutters was used.

The effect of the central flushing system in the initial installation state is not significant. The 1,000 m^3/h flow of the cutterhead flushing center and the surrounding openings are separated from the main mud inlet pipe, resulting in a small slurry feed flow in the excavation chamber and insufficient muck-carrying capacity. The central flushing system was changed to internal circulation mode, and the maximum flow rate of the P0.1 pump was increased from 1,000 m^3/h to 1,200 m^3/h. In addition, compressed air is added to the center flushing to increase the effectiveness of mud cake elimination and prevention.

Hydrogen peroxide is a strong oxidant which can decompose the organic matter in clay into carbon dioxide and water, thus releasing clay particles, reducing the degree of cementation and achieving the effect of removing the inherent mud cake and preventing the formation of mud cake. Based on the test results, a hydrogen peroxide solution was used to eliminate the inherent mud cake in the tunneling. The implementation process is shown in Fig. 9.36.

9.3.4.3 *Cutter management of cutterhead replaceable under atmospheric pressure*

Conventional super-large-diameter shield tunneling in composite strata is generally designed with an atmospheric pressure-replaceable cutterhead to avoid the risk of cutter replacement in high-pressure environments.

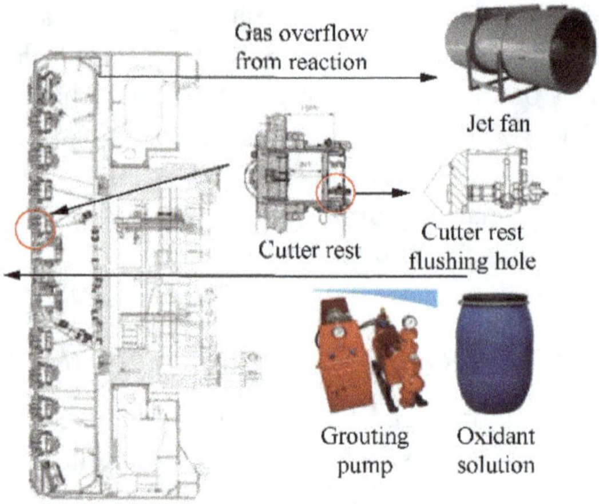

Fig. 9.36. Mud cake elimination process with hydrogen peroxide.

However, its special structure has limitations for tunneling in mineral-rich mudstone ground, which is mainly embodied by the low tunneling rate and fast cutter wear. Therefore, based on the problem that the cutterhead of the shield tunneling through the mineral-rich mudstone is prone to mud cake formation, the wear-resistant double-layer cutting advance cutter (Fig. 9.37) was innovatively designed, and the general layout form dominated by the advance cutter, with the protruding central cutters and the heightened peripheral scraper around the spokes, was proposed to achieve the integration of shield cutting and muck removal and efficiently cut the rock mass. In view of the stress characteristics of the atmospheric pressure-replaceable cutter in the cutter rest in the cutting process in composite strata, a safety device against prop-up from the cutter rest was invented to ensure reliability under the atmospheric pressure environment.

Due to frequent cutter replacements, the shield machine cutter information management system has been developed (Fig. 9.38). After each cutter replacement, relevant data is entered into the cutter information system database, including the tool form, ring number, cutter wear amount and cutter number on the cutterhead. The cutter management system scientifically judges the degree of cutter wear according to the historical cutter wear data and the current excavation volume.

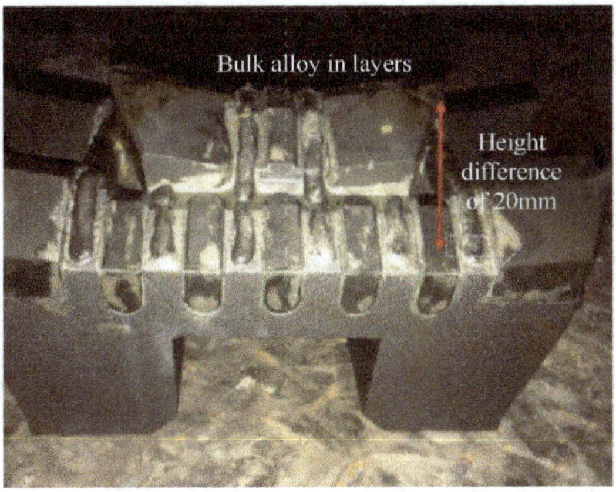

Fig. 9.37. Double-layer cutting advance cutter.

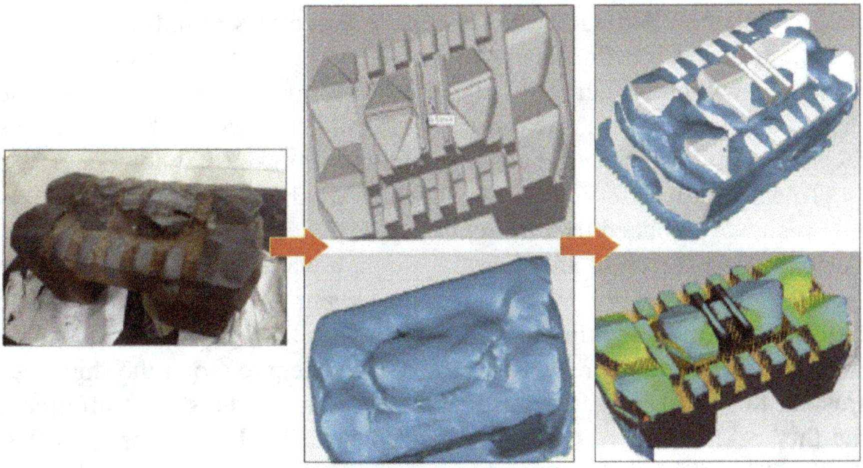

Fig. 9.38. Shield cutter information management system.

9.3.4.4 *Shield tail brush replacement at river bottom under high water pressure*

The shield tail for Wuhan Sanyang Road Tunnel is equipped with four shield tail brushes and 1 shield tail steel plate, for a total of four rows. During shield construction, synchronous grout leakage was found at the

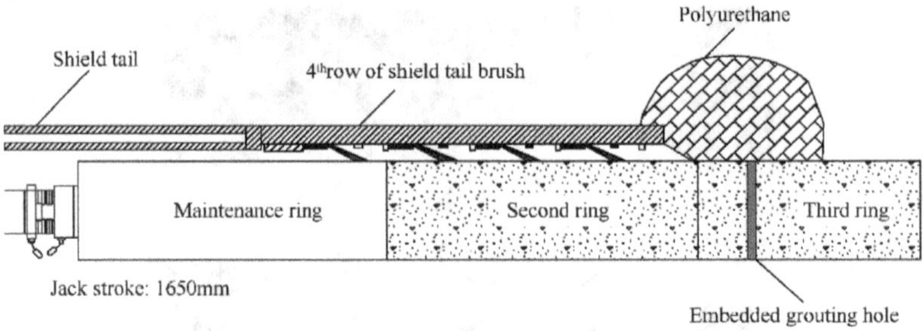

Fig. 9.39. Inspection for shield tail brush of ring 221.

shield tail many times, and measures such as supplementing grease at the shield tail, filling steel wire balls, and filling sponges were taken to basically control the leakage of the shield tail in the follow-up advance process. From ring 174, synchronous grout leakage was found at the tail of the shield in the left line many times. After the completion of ring 221 tunneling, without splicing the closure block and with the stroke of the jack reaching 1,350 mm, the fourth shield tail brush within this range was checked, and it was found that some shield tail brushes failed (Fig. 9.39).

To ensure the safety of subsequent shield tunneling, the fourth shield tail brush was replaced. Before replacing the shield tail brush, the maintenance ring was erected, and then the shield advanced for 75 cm to make the fourth shield tail brush reach the replacement position at the rear of the maintenance ring. Working principle for removing segments in the shield tail and repairing the wire brush: remove a segment to repair the shield tail brush at this position, replace the wire brush, apply the shield tail grease and then re-erect the removed segment; repeat the above steps in cycles. Before the segment is removed, the provided grouting hole in the ring ahead of the maintenance ring is opened to check whether there is leakage in the front three rows at the shield tail.

Chapter 10

Outlook

10.1 Future demand for underground space

With the advancement of urbanization, the scale of cities has expanded rapidly, and their populations have become more and more dense, which has brought many serious problems to the development of cities, such as traffic congestion, shortage of land resources, environmental pollution and ecological imbalance. Among them, the traffic problem is the acutest. Since the ground-level land resources are precious, the expansion of ground roads and urban viaducts can no longer meet the needs of new-pattern urban construction featuring intensive use, ecological livability, and harmonious development. The contradiction between people's increasing travel demands and inadequate urban traffic developments has become quite prominent, imposing serious restrictions on further economic and social development as an obstacle to the sustainable development of modern cities. Seeking space underground and building a multi-functional and multi-level three-dimensional urban traffic network have become the core tasks in the future construction of urban traffic facilities. It is the trend of urban development in the future to shift traffic completely underground, clear the urban surface land for landscaping and expand the living space of residents.

In the 1980s, the International Tunneling Association put forward the slogan "vigorously develop the underground space and start a new cave dwelling era for mankind," which was widely received by governments of all countries, promoting the rapid development of underground space.

In the 1990s, China also carried out a series of theoretical and practical studies on underground space, systematically analyzing the conditions for large-scale development of underground space in China, including its economic and technical feasibility. According to the results of such studies, tunnel construction is a strategic solution to the problems of traffic congestion, environmental pollution, energy waste, disaster prevention and mitigation, etc. It is beneficial in multiple aspects to carry out large-scale underground space construction in an orderly manner without missing opportunities. With China's future economic growth and further urbanization development, it is inevitable to develop a "high-density, efficient, energy-saving and modern" urban space. It has become an inevitable trend for China's modern urban development to seek space, land and resources underground. The 21st century is a century of "underground space" for human beings. With reference to the experience and lessons learned from the development of large cities in developed countries, major cities in China have plotted the three-dimensional urban development strategy for coordinated development of ground and underground spaces, launching intensive development of the underground space in urban dense areas to build an urban underground traffic network and meet the increasing traffic demands of cities.

Due to their geographical advantages, such as adequate water sources, a suitable climate and convenient shipping traffic, the areas of the Yangtze River basin, the Pearl River basin and the Yellow River basin in China have been densely populated since ancient times and have gradually grown into economically developed zones linked by rivers. Among them, the Yangtze River, as the longest river and with the largest drainage area in China, is known as the "Golden Waterway." After thousands of years of development and construction, it has become the region with the strongest industries, the most developed agriculture, the most prosperous commerce and the most advanced science and education in China. Since reform and development, the Yangtze River Economic Belt has gradually become the region with the strongest strategic support. In 2014, the State Council issued the *Guiding Opinions on Promoting the Development of the Yangtze River Economic Belt by Relying on the Golden Waterway* and the *Comprehensive Three-Dimensional Transportation Corridor Planning of the Yangtze River Economic Belt* (2014–2020). The "Yangtze River Economic Belt Strategy" has officially become one of China's seven development strategies with regional themes. In 2016, the population of the Yangtze River Economic Belt reached 600 million, and the aggregate economic output reached 33.3 trillion yuan. This region, occupying one-fifth of the country's land area, carries and contributes nearly half of the country's population and aggregate economic output. Among them, the Yangtze River Delta region is the most dynamic, most open and most innovative region in China. It has created more than one-fifth of the country's aggregate economic output with 1/26 of the country's drainage area. The China International Import Expo debuted in Shanghai in November 2018, announcing that the "integrated development of the Yangtze River Delta" has become a national strategy. In the spatial layout of "the Belt and Road" and "one axis and two wings," the "integrated development of the Yangtze River Delta" will become a leading power in the Yangtze River Economic Belt.

The Pearl River Delta has always been one of the core economic regions in China, playing a leading role in the country's regional development. The Guangdong–Hong Kong–Macao Greater Bay Area has a population of more than 66 million and an aggregate economic output of 1 trillion US dollars, emerging as a complex integrating the financial industry, professional services, port shipping and high-tech industries. In July 2017, the National Development and Reform Commission and the local governments of Guangdong, Hong Kong and Macao signed the *Framework Agreement on Deepening Guangdong-Hong Kong-Macao Cooperation in the Development of the Greater Bay Area* in Hong Kong, marking the all-around launch of the construction of the Greater Bay Area. At the regional strategic level, building the Guangdong–Hong Kong–Macao Greater Bay Area can drive the regional functions of high-end regions in national development. The Pearl River Economic Belt as the hinterland can drive the development of the south-central and south-west regions of China to establish an important economic support belt radiating to Southeast and South Asia and form an important fulcrum in cooperation with the international platform of the "21st Century Maritime Silk Road".

Promoting regional economic integration lays the basis for integrating the regional economy and promoting the coordinated and sustainable development of the national economy. The integration of resources, exchange of information and circulation of goods within the region have become increasingly important for the Yangtze River Economic Zone, the Guangdong–Hong Kong–Macao Greater Bay Area and the Beijing–Tianjin–Hebei Economic Circle. However, with the deepening of economic system reform, backward regional transportation infrastructure construction has become a major bottleneck, restricting regional economic development. The obstacles in rivers and straits, in turn, inhibit the exchange of goods and information between two sides. The inconvenience of river-crossing traffic has become a huge obstacle to regional economic development, so the construction of river- and bay-crossing tunnels has a wide range of spatial economic effects.

For example, in the Guangdong–Hong Kong–Macao Greater Bay Area, with the construction of the Hong Kong–Zhuhai–Macao Bridge, the Shenzhen–Zhongshan Bridge, and the Guangzhou–Shenzhen–Hong Kong Express Rail Link, the traffic durations between Hong Kong and cities in the Greater Bay Area have been shortened to 2.5 h. Relying on the Pearl River Estuary Sea-Crossing Bridge and Tunnel Project, a regional economic development axis has been formed, giving full play to the leading roles of Hong Kong–Shenzhen, Guangzhou–Foshan and Macao–Zhuhai, and improving the development systems of urban agglomerations and towns. Hainan is the largest special economic zone in China, but it is still a backward coastal province. The exchange of goods between Hainan and the mainland basically depends on waterway transportation, which is significant affected by typhoons and other climate conditions all year round, which severely restricts the economic development of both sides. Building a strategic channel of "rail, road, water and air" across the Qiongzhou Strait is of great significance for developing the port and shipping logistics industry and promoting the construction of the international shipping hub and logistics center of Hainan Island. There are 157 cities in the Bohai Rim region, which is the main trade route between China and more than 160 countries and regions in the world and is praised by economists as the "third growth pole" of China's economy after the Pearl River Delta and the Yangtze River Delta. However, it is difficult for the Bohai Economic Circle to form a continuous and unified economic circle like the Pearl River Delta and the Yangtze River Delta, and the main reason lies in the geographical separation into the north and south sides of the Bohai Bay. The straight-line distance from Yantai to Dalian is 170 km, and the shipping time is 6–8 h; the railway transportation meanders for 1,500 km and the expressways even for 1,600 km. If the all-weather Bohai Strait sea-crossing channel is completed and open to traffic, it will only take 2 h at most. Therefore, the construction of this channel across the Bohai Strait is of great significance for promoting the overall coordinated and integrated development of the regional economy in the Bohai Rim region, revitalizing the old industrial bases in northeast China, optimizing the transport structure, consolidating national defense and developing marine resources and energy.

At present, leveraging the more and more mature shield construction technology of soft soil stratum in the Yangtze River Delta region, the 15 m super-large-diameter shield tunnels completed, such as the Yangtze River Tunnel and Hangzhou Qianjiang Tunnel, have promoted the connection between Shanghai and other cities in the Yangtze River Delta, greatly served the development of the Hangzhou Bay and Yangtze River Delta industrial zones, and promoted the construction of super-large-diameter shield tunnels across the whole country. The super-large-diameter river-crossing tunnels of Wuhan Sanyang Road Tunnel, Shenzhen Chunfeng Tunnel and Jinan Yellow River Tunnel aim to clear river barriers and bring vitality to the development of more cities. However, different from the soft soil ground, the geological conditions faced by the projects are becoming exceedingly complex, which brings many new problems and challenges to the construction of super-large-diameter shield tunnels.

10.2 Development trends of shield tunneling

10.2.1 Super-large-diameter tunnel

In the recent 20 years, the shield tunnels abroad have gradually developed toward large cross-sections, and a large number of 15 m diameter subsea tunnels and urban road tunnels have been built. The super-large-diameter shield tunnels can meet larger traffic flow demands and provide more functions. For example, the SR99 Tunnel in Seattle, USA, has a shield diameter of 17.48 m, which can meet the traffic demand of buses and cars, as well as trucks on both the north and south routes; the proposed Bering Strait Tunnel is designed to directly connect Siberia, Russia, to Alaska, USA. The tunnel will be constructed with a 19.2 m shield to accommodate a high-speed railway, a highway, and several oil and gas pipelines, making full use of the underground space resources created by tunneling in one shot. The super-large-diameter shield tunnels are generally advantageous in terms of low construction costs but are accompanied by greatly increased construction difficulty. The development of shield machine manufacturing technology provides a foundation for the construction of super-large-diameter shield tunnels.

With the increasing demand for traffic flow and municipal infrastructure capacity, the contradiction between the narrow tunnel section and the growing traffic flow has become increasingly prominent. The existing 4-lane and 6-lane highway tunnels can no longer meet the needs of urban traffic development with growing urban traffic volume; with the advancement of urbanization, the municipal utility tunnels have become an important measure for innovative municipal infrastructure construction in China. The utility tunnels integrate various municipal pipelines such as communication, power, lighting, water supply and drainage, and the pipeline layout needs a large section size of the utility tunnels. In order to meet the needs of sustainable development of urban construction, the tunnels with large sections also represent an inevitable trend in the future construction of utility tunnels.

10.2.2 Deeply buried tunnel

At present, the development of underground space mainly focuses on the medium and shallow overburden areas within 40 m. The development and utilization of shallow space have become saturated since the existing pile foundations of buildings (structures) and underground pipeline networks have a great impact on the construction of new underground projects. However, most of the deep underground space is yet to be developed, with few cross-conflicts in engineering construction, which is conducive to macro control. The development of deep underground space has become an inevitable choice for future urbanization. For example, in order to solve the problem of urban waterlogging and improve disaster control capacity, Chicago, London, Paris, Singapore, Tokyo and other international metropolises have successively established "deep tunnel" drainage pipeline systems. In order to save shallow underground space resources, these tunnels were built 50 m below the surface; in order to bypass private property space at 30–40 m underground, the Diet of Japan adopted the *Law on Special Measures for Public Use of Deep Underground Space* in 2001, which defines deep underground space for public use by law. The burial depth of the new Tokyo Outer Ring Road is controlled below 40 m. Moreover, due to the fact that the burial depth for shield tunneling for tunnels crossing straits and mountains is also increasing, for example, the road tunnel underneath the Bosphorus Strait and crossing Europe and Asia has a water depth of 61 m and a maximum burial depth of 106 m, among the deepest tunnels bored by slurry balanced shield machines in the world.

In 2017, the Ministry of Land and Resources of the People's Republic of China issued the *Guiding Opinions on Strengthening Urban Geological Work*, pointing out that the rational development and utilization of deep underground space resources are encouraged for industry, warehousing and commerce. In general, the maximum burial depth of tunnels constructed using the shield method in China in the past was generally 30–40 m. With the development and utilization of shallow underground space, the construction of river-crossing tunnels has gradually developed with increasing burial depth. The maximum burial depth of the newly built Hongmei South Road Tunnel in Shanghai reaches 59 m; the Pearl River Estuary Tunnel of Shenzhen-Maoming Railway is 7.225 km long with a maximum burial depth of more than 100 m. Therefore, it is a strategic demand and inevitable trend for urban land resource expansion, functional expansion, transformation and regeneration to develop and utilize deep underground space.

10.2.3 Ultra-long tunnel

With the continuous technological improvement of shield machines and tunneling construction, the tunneling distance has been constantly broken through, and the application scope has been expanded. The developed countries have been studying the construction of undersea and mountain shield tunnels since the 1930s. In 1994, Britain and France adopted the shield tunneling method and hard rock tunneling method to jointly build a 50.5 km undersea tunnel beneath the English Channel; in 2016, the St. Gotha Tunnel was completed in Switzerland, with a total length of 57 km, among the longest tunnels so far in the world. The construction of long-distance tunnels connects islands and the mainland, breaks the barriers of mountains, and is of great significance from the perspective of political, economic and social development. In China, with the accelerated construction on the coastal areas, it is imperative to improve the traffic capacity across both sides of bays and straits. In Hangzhou Bay, Pearl River Estuary, Qiongzhou Strait, Bohai Bay and Taiwan Strait, the construction of long-distance tunnels is an ideal solution to achieve all-weather, large-capacity and fast traffic, ensure smooth traffic channels, and reduce the impact of natural disasters such as typhoons in the future. In addition, in order to improve the congestion of ring roads in big cities during peak hours, the surface three-dimensional traffic has gradually shifted to the underground space, and the construction of long-distance and rapid transit tunnels has sprung up in big cities. Since 1995, Boston, USA, has begun to dismantle the urban viaducts built in the 1950s, and turned to the development of underground roads to eliminate noise, pollution and other adverse effects of surface traffic and release ground space for landscaping. Long-distance tunnels provide support for the construction of ecological cities and green cities.

With the recognition and active participation of more and more countries and regions in building the "Belt and Road" in the new era, the construction of the Yangtze River Economic Belt, the Bohai Rim Economic Circle and the Pearl River Economic Belt has been in full swing, giving rise to a climax in infrastructure construction. The super-large-diameter tunnels have broad prospects for future development, and "super-large, deeply-buried and ultra-long" will become the mainstream trends of tunnel development.

10.3 Outlook on intelligent shield construction

10.3.1 Practice of intelligent management and control technology

The shield tunneling industry is undergoing prosperous development with huge potential. As of December 2018, the operating mileage of rail transit in 35 cities in China had reached 5,767 km. Although the shield tunneling industry has made great progress and accumulated a lot of experience, conventional management methods are still applied in shield construction, which are far from effective control of construction risks and quality hazards bringing effects to the smooth construction progress of tunnels. The main difficulties lie in the following: difficulties sharing experience and technology among tunnels in wide geographical distribution; asymmetric information during construction and low efficiency of construction management; large construction management span and low risk control ability. With the development of new technologies such as cloud computing and big data, various industries in China have been gradually transitioning from the era of traditional manual operation to the era of intelligent management and control. As a traditional industry, the pattern of the construction industry is bound to be affected by the development of the internet.

The development of intelligent management and control technology is divided into three stages: the knowledge-driven stage, the data-driven stage and the management-driven stage. In the 1990s, shield tunneling was just emerging in China. At that time, the knowledge-driven traditional construction management mode was applied. Due to the lack of construction data, the knowledge base and model base were established only based on the experience and knowledge of experts in the field. The knowledge-driven construction management system is shown in Fig. 10.1.

In the 21st century, shield tunneling has become more and more mechanized, and with increasing more data acquirable, shield tunneling management and control have entered the data-driven stage. Through the collection of construction data, a sound data warehouse for shield tunneling is constructed. Through the study of classical parameter calculation methods and genetic algorithms, the collected data are analyzed and processed, and the best parameter matching scheme for shield construction is generated. However, the management and control mode in this stage is only driven by original data without deep data mining, and the experience of engineering cases cannot be widely applied. The data-driven construction management system is shown in Fig. 10.2.

With the continuous development of the market and technology, data-driven construction management and control have been upgraded to a standardized, refined and intelligent management concept to better facilitate construction management. Based on mass data such as quality, progress, risk and equipment during tunnel construction, big data technology, BIM technology and Internet of Things technology are applied to realize

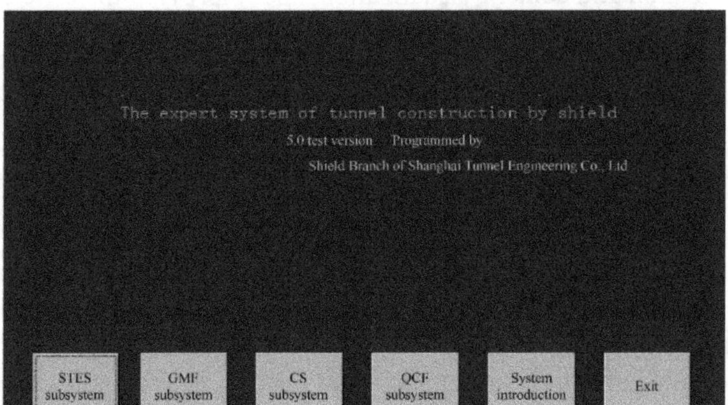

Fig. 10.1. Knowledge-driven construction management system.

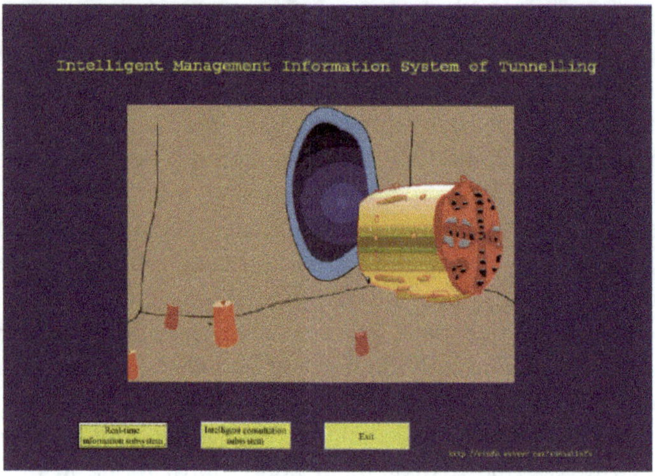

Fig. 10.2. Intelligent management information system of tunnelling.

dynamic monitoring, risk early warning and intelligent decision-making in shield tunneling. The current shield tunnel construction management and control platform, characterized by "mobility, centralization and intelligence", supported by the most advanced Internet, Internet of Things, cloud computing and big data technologies, integrated mobile app, a big-screen monitoring terminal, a smart website and a big data background processing system in a huge and efficient system, realizing parameter standardization and content visualization, and improving the efficiency of construction management to ensure the intelligent management of tunnel construction.

The intelligent management and control platform integrates data collection, storage, analysis and graphical display. Based on big data processing, it deeply mines shield tunnel construction data to guide construction. Based on more than half a century of shield construction experience, the system platform enables the management and control expert team to carry out systematic analysis of data and visual monitoring of real-time data on shield construction based on engineering interpretation of various parameters and by using visual charts and other methods.

The shield machine management and control platform integrates massive construction data in an all-round way, forming a big data system with the intelligent learning function, and is developing as a "super brain" of shield machines. Through loading, analysis and calculation, the system

(a) Construction management and control center (b) Control and management system interface

Fig. 10.3. The management-driven construction management system.

can deeply dig into the construction rules of shield tunneling, conduct dynamic analysis of tunnel construction in real time and use the intelligent analysis function to guide the optimization of construction parameters, laying a foundation for future intelligent shield research and development. The management-driven construction management system is shown in Fig. 10.3.

10.3.2 Research and development of intelligent shield machines

Since the first generation of manual shield machines, three industrial revolutions have taken place. From the emergence of mechanized shield machines to the emergence of electric shield machines, until the electronic shield machines that are centrally controlled by the system, shield machine development has experienced four generations of upgrades in a relatively slow development process. Since the beginning of the 21st century, artificial intelligence technology has ushered in a period of integrated development. The introduction and rapid development of machine learning, artificial neural networks and deep learning theories have accelerated the realization and application of artificial intelligence. Moreover, the development of the Internet of Things and big data has set off a new wave of product intelligence, promoted a new generation of artificial intelligence technology based on the new information environment and new development goals, and traditional shield machines will also be developed in the trend of intelligence.

The intelligent shield machine will integrate the four major functions of intelligent perception, intelligent execution, intelligent analysis and intelligent decision-making. The four functions of the intelligent shield machine are shown in Fig. 10.4. The new-generation remote sensing technology is used to realize multi-dimensional information perception and real-time prediction of stratum and ground environment along tunnels; improve the control accuracy and response speed of the core actuator of the shield machines, and improve the intelligent level; deeply mine the construction data of historical cases, build the model bases of different shield machine types, different geological conditions and different crossing conditions, and deepen the big data processing and analysis platform; evaluate the construction risks in advance to achieve the optimal matching of shield construction strategies based on the deep learning strategy and model of neural network. The visual identity model based on deep neural network is shown in Fig. 10.5.

In order to realize the fusion analysis and mining of real-time information and historical data of TBMs, a multi-level standard data warehouse is built. With big data technology and deep machine learning at their core, the construction parameters are analyzed to identify trends based on clustering and evaluation of project status in terms of progress, quality and risk.

As a core function of intelligent shield machines, cruise control mainly realizes intelligent control in two aspects: direction control and disturbance control. The former controls the shield attitude according to

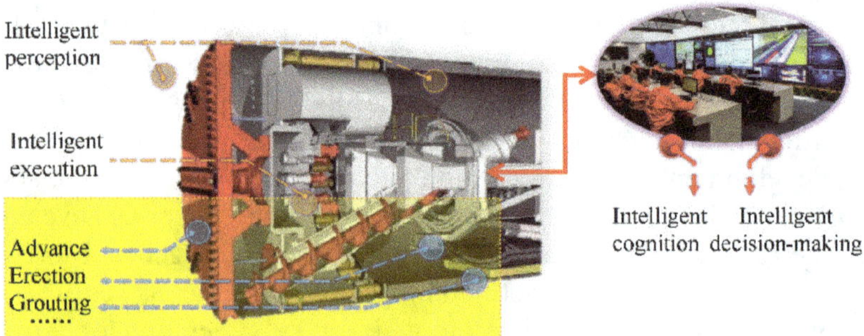

Fig. 10.4. The four functions of the intelligent shield machine.

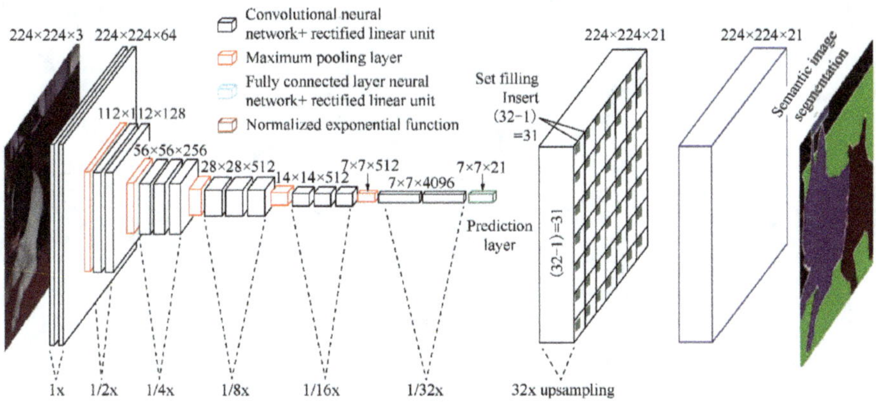

Fig. 10.5. Visual identity model based on deep neural network.

Note: Numbers in the figure represent pixels.

the deviation value from the tunnel axis to ensure that the tunnel axis is consistent with the design axis, and the latter controls the construction parameters according to the monitoring value of stratum disturbance to ensure that the shield tunneling has the minimum impact on the surrounding environment. Integrating massive data deep mining, human wisdom and experience, and artificial intelligence strategy, the trinity paradigm of shield tunneling is established, representing the dominant trend of shield intelligent cruise development.

With the increasingly mature mechanical automation technology, how to effectively apply intelligent interconnection technology to shield tunneling machine equipment for intelligent control, and how to effectively use engineering construction big data and dig into the knowledge for unmanned shield tunnel construction will become the next major technical challenges in the field of tunnel engineering and hot spots of future industry competition.

Bibliography

André Borrmanna, Matthias Flurlb, Javier Ramos Jubierrea, *et al.* Synchronous collaborative tunnel design based on consistency-preserving multi-scale models. *Advanced Engineering Informatics*, 2014, 28(4): 499–517.

Avni N, Fishbain B, Shami RU. Water consumption patterns as a basis for water demand modeling. *Water Resources Research*, 2015, 51: 1–17.

Bai Tinghui. New technology for design and construction of Shanghai metro project. *Tunnel Construction*, 2014, 34(1): 53–59.

Basarir H, Genis M, Ozarslan A. The analysis of radial displacements occurring near the face of a circular opening in weak rock masses. *International Journal of Rock Mechanics and Mining Sciences*, 2010, 40: 771–783.

Bonacina C, Comini G, Fasano A, *et al.* Numerical solution of phase change problems. *International Journal of Heat Mass Transfer*, 1973, 16(10): 1825–1832.

Cao Chengyong. *Study on Stability of Slurry Shield Excavation Face and Tunneling Parameters in Shallow Permeable Composite Strata*. Changsha: Central South University, 2014.

Chen Gang, Xie Kefan, Wu Qian. Emergency events analysis in metro construction project. *Systems Engineering Procedia*, 2012, 4: 402–408.

Chen Hui, Yang Yandong. New technology and development trend of shield machine manufacturing in China. *Tunnel Construction*, 2017, 37(3): 276–284.

Chen Kui, Hong Kairong, Jiao Shengjun. *Shield Construction Technology*. Beijing: China Communications Press Co. Ltd., 2016.

Colella F, Rein G, Borc Hiellini R, *et al.* Calculation and design of tunnel ventilation systems using a two-scale modeling approach. *Building and Environment*, 2009, 44.

Copur H, Cinar M, Okten G, *et al.* A case study on the methane explosion in the excavation chamber of an EPB-TBM and lessons learnt including some recent accidents. *Tunnelling and Underground Space Technology*, 2012, 27(1): 159–167.

Cundall PA. Computer model for simulating progressive large scale movements in blocky rock systems ISRM. In *Proceedings of the Symposium of the International Society of Rock Mechanics*. Nancy: ISRM, 1971, pp. 2–8.

Dang Xifeng. *Research on Key Technologies of Chengdu Metro Construction Based on Shield Construction and Slag Improvement*. Shijiazhuang: Shijiazhuang Tiedao University, 2017.

Ding Guangying, Yang Guoxiang, Wan Bo. Application of first domestic large slurry balance shield in second tube of Dapu Road Tunnel. *Journal of Municipal Technology*, 2009, 27(Supplement 2): 217–221.

Ding Wq, Ouyang WB, Ge SP, *et al.* Calculation method for ground settlement due to shield tunnel undercrossing existing building and its application. In *The 4th International Conference on Computational Methods*. Gold Coast, 2012.

Du Chuangdong. Shiziyang tunnel shield underground connection technology and implementation. *Tunnel Construction*, 2014, 34(8): 771–777.

Du Mei. *Risk Assessment of Underwater Tunnel Construction Schedule under the Coupling of Multiple Influencing Factors*. Chongqing: Chongqing Jiaotong University, 2017.

Duan Yagang. Research on key technologies of small diameter shield in the construction of comprehensive pipe gallery. *Journal of Railway Engineering Society*, 2017, 34(4): 65–69.

Feng Kun, He Chuan, Xiao Mingqing. Bending resistance test of segment joint on complex joint surface of shield tunnel under high axial pressure. *China Civil Engineering Journal*, 2016, 49(8): 99–110+132.

Gallardol LA, Meju MA. Joint two-dimensional DC resistivity and seismic travel time inversion with cross-gradients constraints. *Journal of Geophysical Research*, 2004, 109(B3): 1–11.

Gao Ruihua. Domestic metro shield — "Pioneer" breakthrough documentary. *Manufacturing Information Engineering of China*, 2005, 28–31.

Gokce A, Koyama F, Tsuc Hiya M, *et al.* The challenges involved in concrete works of Marmaray immersed tunnel with service life beyond 100 years. *Tunneling and Underground Space Technology*, 2009, 24(5): 592–601.

Gröger G, Plümer L. Topology of surfaces modelling bridges and tunnels in 3D-GIS. *Computers, Environment and Urban Systems*, 2011, 35(3): 208–216.

Hasanpour R, Rostami J, Ünver B. 3D finite difference model for simulation of double shield TBM tunneling in squeezing grounds. *Tunnelling and Underground Space Technology*, 2014, 40: 109–126.

He Chuan, Feng Kun, Fang Yong. Current situation and prospect of shield tunneling technology. *Journal of Southwest Jiaotong University*, 2015, 50(1): 97–109.

He Chuan. *Study on Seismic Behavior of Shield Tunnel Longitudinal Direction*. Tokyo: Waseda University, 1999.

Hegemann F, Lehner K, König M. IFC-based product modeling for tunnel boring machines. In *Proceedings of the 9th European Conference on Product and Process Modeling 2012*. London: CRC Press, 2012. pp. 289–296.

Hong Kairong, Chen Kui, Feng Huanhuan. Innovation and breakthrough of shield tunneling technology in China. *Tunnel Construction*, 2013, 33(10): 801–808.

Hu Jianqin. *Research on Crack Resistance and Mechanism of High-Performance Concrete*. Wuhan: Wuhan University of Technology, 2002.

Hu Zhiping, Cai Zhiyong, Luo Lijuan. Three dimensional stress analysis of the connection between shield tunnel and connecting passage. *Journal of Highway and Transportation Research and Development*, 2005, 22(9): 133–136.

Hudson JA, Brown ET, Fair Hurst C, *et al. Comprehensive Rock Engineering*. Oxford: Pergamon Press, 1993.

Ingerslev LCF. Considerations and strategies behind the design and construction requirements of the Istanbul strait immersed tunnel. *Tunneling and Underground Space Technology*, 2005, 20(6): 604–608.

Japan Road Association. *Road Tunnel Maintenance Manual*. Tokyo: Japan Road Association, 1993.

Jiang Zhiyi. *Research on the Structural Design Method of Double Shield TBM Segment in High Stress Compressional Formation*. Chengdu: Southwest Jiaotong University, 2017.

Jin Shihe. *Research on Risk Analysis and Control Measures for Shield Tunnel Construction of Lanzhou Metro*. Lanzhou: Lanzhou Jiaotong University, 2017.

Katayama I, Adachi M, Shimada M, *et al.* Proposition of practical Quasi-dynamic analysis method "Response Seismic Intensity Method" of underground burying structure JSCE. In *Proceedings of the 40th Annual Scientific Lecture Meeting of JSCE*. Tokyo: Society of Civil Engineering, 1985, pp. 1–9.

Lee S, Park SI, Park J. Development of an IFC based data schema for the design information representation of the NATM tunnel. *KSCE Journal of Civil Engineering*, 2016, 20(6): 2112–2123.

Li Chen, Zhang Kaiyin, Huang Jun. Stress analysis of double-layer lane slab structure inside shield tunnel. *Modern Tunnelling Technology*, 2016, 53(1): 103–110.

Li Chenghui, He Shaohui, Liu Xiabing. Study on parameter control of slurry balance shield tunneling under the Yellow River in coarse grained Sandy Pebble Stratum. *China Civil Engineering Journal*, 2017, 50(S2): 147–152.

Li Hongbo, Jia Feng, Li Jing, *et al. Key Technologies for cross Sea tunnel design of Dalian metro line 5. Rock and Soil Mechanics*, 2017, 38(S1): 395–401.

Li Xiaojun, Zhu Hehua. Development of a web-based information system for shield tunnel construction projects. *Tunnelling and Underground Space Technology*, 2013, 37(6): 146–156.

Liang Rongzhu, Xia Tangdai, Lin Cungang, *et al.* Analysis of surface deformation and horizontal displacement of deep soil mass caused by shield tunneling. *Chinese Journal of Rock Mechanics and Engineering*, 2015, 34(3): 583–393.

Liang Xiao, Guan Linxing, Wen Zhuyin, *et al.* Field test study on structural performance of rectangular shield tunnel segment lining during construction. *Tunnel Construction*, 2016, 36(12): 1456–1464.

Liu Dong. *Research on Key Technologies of TBM & EPB Dual Mode Shield Construction in Composite Stratum.* Chengdu: Southwest Jiaotong University, 2017.

Liu Dongyang. *Study on the Stability of the Excavation Face and the Law of Surface Settlement of Nanning Sandy Soil Tunnel Shield Construction.* Xi'an: Xi'an University of Science and Technology, 2017.

Liu Jianhang, Hou Xueyuan. *Shield Tunnel.* Beijing: China Railway Press, 1991.

Liu Wen, Zhao Tingsheng, Zhang Yajing, *et al.* Analysis and countermeasures on safety risk law of metro shield construction. *China Safety Science Journal*, 2017, 27(10): 130–136.

Ma Yongchi, Ma R Tin De Jong, Koppenjan Joop. Identifying organizational and contractual drivers behind subway construction accidents in China: A tale of three cities. *Journal of Urban Technology*, 2013, 20(3): 39–64.

Min Fanlu, Xu Jingbo, Song Hangbiao, *et al.* Mud film density evaluation test on slurry shield excavation face under back pressure. *China Journal of Highway and Transport*, 2017, 30(8): 216–221+246.

Qi Chun, He Chuan, Feng Kun, *et al.* Development and application of slurry balanced shield simulation test system. *Chinese Journal of Geotechnical Engineering*, 2016, 38(11): 1999–2008.

Qian Qihu. Main progress and development direction of geological prediction and information technology for tunnel engineering construction. *Tunnel Construction*, 2017, 37(3): 251–263.

Qiao Yafei, Ding Wenqi, Xie Dongwu, *et al.* A 3D finite element model for shield tunnel undercrossing the historical building in soft ground. In *3rd International Conference on Computational Methods in Tunnelling and Subsurface Engineering.* Bochum, 2013, pp. 855–862.

Raviv G, Fishbain B, Aviad Shapira. Analyzing risk factors in crane-related near-miss and accident reports. *Safety Science*, 2017, 91: 192–205.

Standing JR, Selemetas D. Greenfield ground response to EPBM tunnelling in London Clay. *Geotechnique*, 2013, 63(12): 989–1007.

Tang Zhicheng, He Chuan, Lin Gang. Model test study on mechanical behavior of segment structure of subway shield tunnel. *Chinese Journal of Geotechnical Engineering*, 2005, (1): 85–89.

Turner MJ, Cloug HRW, Martin HC, *et al.* Stiffness and deflection analysis of complex structure. *Journal of the Aeronautical Sciences*, 1956, 23(9): 805–823.

Unlu T, Gercek H. Effect of possions ratio on the normalized radial displacements occurring around the face of a circular tunnel. *Tunneling and Underground Space Technology*, 2003, 18: 547–553.

Wang Jiyun. China's super diameter shield construction experience in the past decade. *Tunnel Construction*, 2017, 37(3): 330–335.

Wang Jun, He Chuan, Hu Ruiqing, *et al.* Study on disturbance of EPB shield tunneling to upper soft and lower Hard Strata. *Chinese Journal of Rock Mechanics and Engineering*, 2017, 36(4): 953–962.

Wang Rong. *Fatigue Life Prediction of Shield Tunnel Segments with Different Reinforcement Measures Under the Dynamic Load of High-Speed Railway Trains*. Southwest Jiaotong University, 2017.

Wang Shangao, Shi Shibo, Shu Heng, *et al.* Prefabrication construction technology for internal structure of single pipe double-layer extra long shield tunnel. *Tunnel Construction*, 2016, 36(4): 451–457.

Wang Shimin, Shen Xingzhu, He Xiangfan, *et al.* Model test research on stress and failure mode of shield tunnel segment lining under different assembly methods. *China Civil Engineering Journal*, 2017, 50(6): 114–124.

Wang Xingang. *Research and Application of Functionally Graded Concrete Segments for River Sea Tunnel*. Wuhan: Wuhan University of Technology, 2007.

Wang Ying. *Research on Surface Settlement Caused by Large Diameter Slurry Shield Construction in Water Rich Sandy Pebble Stratum*. Beijing: Tsinghua University, 2014.

Wang Yuan, Liu Songyu, Tan Yuehu, *et al.* Informatization construction and analysis of shield launching working shaft of Nanjing Yangtze River Tunnel. *Chinese Journal of Geotechnical Engineering*, 2008, 30(S1): 554–558.

Wang Zhenfei, Zhang Chengping. Particle flow simulation study on instability and failure of slurry shield excavation face. *China Railway Science*, 2017, 38(3): 55–62.

Wang Zhenfei. *Study on Adaptability of Large-diameter Slurry Pressurized Balanced Shield in Beijing Sandy Pebble Stratum*. Beijing: Beijing Jiaotong University, 2014.

Wei Gang, Pang Siyuan, Zhang Shimin. Prediction of ground deformation induced by double parallel shield tunnelling. *Disaster Advances*, 2013, 6(13): 91–98.

Wei Gang, Pang Siyuan. Research on three-dimensional soil deformation caused by double track parallel shield tunnel construction. *Rock and Soil Mechanics*, 2014, 35(90): 2562–2568.

Wei Gang. Prediction of soil settlement caused by double-line parallel shield tunnel construction. *Disaster Advances*, 2013, 6(6): 23–27.

Wei Junjie, Luo Hanzhong, Huang Xingchun. Shield construction information and disturbance displacement control in complex strata. *Chinese Journal of Underground Space and Engineering*, 2010, 6(5): 1044–1052.

Wu Shiming, Lin Cungang, Zhang Zhongmiao, *et al.* Research on risk analysis and control of slurry shield tunneling under embankment. *Chinese Journal of Rock Mechanics and Engineering*, 2011, 30(5): 1034–1042.

Xia Yimin, Wang Yang, Wu Dun, *et al.* Pipeline transmission characteristics of slurry shield circulation system. *Journal of Central South University (Natural Science Edition)*, 2017, 48(11): 2889–2896.

Xu QW, Zhu HH, Ding WQ, *et al.* Laboratory model tests and field investigations of EPB shield machine tunneling in soft ground in Shanghai. *Tunnelling and Underground Space Technology*, 2011, 26(1): 1–14.

Yang Linde, Huang Kang. Study on durability failure risk of Subsea tunnel segment members. *Chinese Journal of Underground Space and Engineering*, 2004, (1): 1–4+10.

Yoshiyuki Shigeta, Toshiyuki Tobita, Katsumi Kamemura, *et al.* Propose of Tunnel Crack Index (TCI) as an evaluation method for lining concrete. *JSCE Proceedings F*, 2006, 62(4): 628–632.

Yu Haitao, Li Longjin, Cao Chunyan, *et al.* Seismic performance analysis of shield tunnel considering internal prefabricated structure. *Chinese Journal of Underground Space and Engineering*, 2016, 12(S2): 834–840.

Yuan Dajun, Shen Xiang, Liu Xueyan, *et al.* Study on stability of slurry shield excavation face. *China Journal of Highway and Transport*, 2017, 30(8): 24–37.

Zhang Haonan. Analysis on the technical configuration and characteristics of automatic assembly line for shield segment precasting construction. *Construction & Design for Engineering*, 2017, (12): 97–98.

Zheng Gang, Lu Ping, Cao Jianran. Risk analysis based on sensitivity of shield tunneling parameters to surface settlement. *Chinese Journal of Rock Mechanics and Engineering*, 2015, 34(S1): 3604–3612.

Zhong Yu, Chen Jian, Chen Guoliang, *et al.* Method of shield tunnel structure information modeling based on building information model technology. *Rock and Soil Mechanics*, 2018, 39(5): 1867–1876.

Zhou Wenbo. Discussion on application technology of super large diameter EPB shield in central urban highway tunnel. *Modern Tunnelling Technology*, 2013, 50(3): 1–7.

Zhou Wenbo. *Shield Tunneling Technology and Application*. Beijing: China Architecture & Building Press, 2004.

Zhou Wenbo. *Smart City: Municipal Engineering Construction and Management*. Beijing: China Architecture & Building Press, 2016.

Zhou Zhipeng, Li Qiming, Wu Weiwei. Developing a versatile subway construction incident database for safety management. *Journal of Construction Engineering and Management*, 2012, 138(10): 1169–1180.

Zhu Hehua, Cui Maoyu, Yang Jinsong. Research on design model and load distribution of shield lining segment. *Chinese Journal of Geotechnical Engineering*, 2000, (2): 190–194.

Zhu Hehua, Ding Wenqi, Qiao Yafei, *et al.* Shield tunnel micro disturbance construction control technology system and its application. *Rock and Soil Mechanics*, 2014, 36(11): 1983–1993.

Afterword

Founded in 1965, Shanghai Tunnel Engineering Co. Ltd. is the first professional construction enterprise engaging in shield tunneling technology research and development and construction application in China. In the past more than a decade, with the development of shield tunneling in the trend of large diameters, Shanghai Tunnel Engineering Co. Ltd. has successively undertaken the construction of 18 large-diameter shield tunnels with a diameter of more than 14 m, including Shanghai Shangzhong Road Tunnel, Shanghai Yangtze River Tunnel, Shanghai Bund Tunnel, Hangzhou Qianjiang Tunnel, Zhuhai Hengqin Maliuzhou Tunnel, and Wuhan Sanyang Road Yangtze River Tunnel. It has accumulated plenty of practice experience, achieved fruitful scientific and technological outcomes and developed a series of key technologies.

This book is compiled based on the construction experience of large-diameter tunnels and the experience and lessons learned from large-diameter shield tunneling practice in China and abroad. It further refines the construction processes and key technologies, thanks to the years of efforts of the compilation team, aiming to form a complete system of key technologies that can represent the current development of large-diameter shield tunneling in China.

Looking back on the compilation process, a compilation team of more than 40 members contributed to the creation of this book, including Zhou Wenbo and Wu Huiming, respectively the president and deputy chief engineer of Shanghai Tunnel Engineering Co. Ltd., who joined us at the planning stage of this book. The other members and their specific chapters are listed as follows: Li Lei, Bao Zhen, and Li Hongxia in charge of

Chapter 1 Introduction; Wang Jiyun, Yang Jiangang, Wang Shengyong, Chen Yu and Ma Zhigang in charge of Chapter 2 Shield Machine Selection; He Guojun, Wei Linchun and Wang Zhihua in charge of Chapter 3 Slurry System; Zhu Hailiang, Ying Zhuoqing, Gong Haiyun, Wan Yang, Fu Qingqing and Chen Yihao in charge of Chapter 4 Segment Fabrication; Huang Dezhong, Fan Jie, Huang Kai, Lu Kangmin, Zhou Chang and Liu Gao in charge of Chapter 5; Wang Jiyun, Yang Jiangang, Zhang Jiajun, Yao Jiajie and Zheng Rui in charge of Chapter 6 Internal Structure and Connecting Passage; Yang Hongyan, Chen Gang and Liu Jie in charge of Chapter 7 Information Monitoring; Fan Jie, Huang Kai, Sun Jie, Sun Han, Li Yong and Ma Zhigang in charge of Chapter 8 Risk Management; Zheng Yifeng, Xu Zhiling, Luo Sainan, Shi Wenqi and Huang Jialun in charge of Chapter 9 Typical Project Cases; Bao Zhen and Li Gang in charge of Chapter 10 Outlook.

In the compilation process of this book, some experts and technicians from relevant tunnel engineering construction companies, design institutes, research institutions and universities have also provided strong support and assistance. Among them, Academician Sun Jun from the Chinese Academy of Sciences specially wrote a foreword to this book, and Professor Zhang Mengxi from Shanghai University provided many valuable suggestions for the compilation of this book.

We extend our sincere thanks to them for their efforts and support in the successful compilation of the book!

It is expected that the publication of *Large-Diameter Shield Tunneling* could promote the technical progress of large-diameter shield tunnel construction and raise the overall development level of China's transportation infrastructure.

Index